Methods in
Molecular Biology

Volume 3

New Protein Techniques

Biological Methods

Methods in Molecular Biology
edited by **John M. Walker**

Volume I: ***Proteins,*** *1984*
Volume II: ***Nucleic Acids,*** *1984*
Volume III: ***New Protein Techniques,*** *1988*
Volume IV: ***New Nucleic Acid Techniques,*** *1988*

Liquid Chromatography in Clinical Analysis
edited by **Pokar M. Kabra** and **Laurence J. Marton,** *1981*

Metal Carcinogenesis Testing
Principles and In Vitro Methods
by **Max Costa,** *1980*

Methods in
Molecular Biology

Volume 3

New Protein
Techniques

Edited by

John M. Walker

The Hatfield Polytechnic, Hatfield, Hertfordshire, UK

Humana Press • Clifton, New Jersey

© 1988 The Humana Press Inc.
Crescent Manor
PO Box 2148
Clifton, New Jersey 07015

Printed in the United States of America

Library of Congress Cataloging in Publication Data
Main entry under title:

Methods in molecular biology.

    (Biological methods)
    Includes bibliographies and indexes.
    Contents: v. 1. Proteins—v. 2. Nucleic acids—v. 3. New protein techniques.

1. Molecular biology—Technique. I. Walker, John M., 1948–   II. Series.

QH506.M45   1984   574.8'8'078   84-15696

ISBN 0-89603-062-8 (v. 1)
ISBN 0-89603-064-4 (v. 2)
ISBN 0-89603-126-8 (v. 3)
ISBN 0-89603-127-6 (v. 4)
ISBN 0-89603-150-0 (v. 5)

Preface

In recent years there has been a tremendous increase in our understanding of the functioning of the cell at the molecular level. This has been achieved in the main by the invention and development of new methodology, particularly in that area generally referred to as "genetic engineering." Although this revolution has been taking place in the field of nucleic acids research, the protein chemist has at the same time developed fresh methodology to keep pace with the requirements of present-day molecular biology. Today's molecular biologists can no longer be content with being experts in one particular area alone. They need to be equally competent in the laboratory at handling DNA, RNA, and proteins moving from one area to another as required by the problem that is being solved. Although many of the new techniques in molecular biology are relatively easy to master, it is often difficult for a researcher to obtain all the relevant information necessary for setting up and successfully applying a new technique. Information is of course available in the research literature, but this often lacks the depth of description that the new user requires. This requirement for in-depth practical details has become apparent by the considerable demand for places on our Molecular Biology Workshops held at Hatfield each summer.

Volume 1 of this series described practical procedures for a range of protein techniques frequently used by research workers in the field of molecular biology. Because of the limitations on length necessarily inherent in producing any

v

book, one obviously had to be selective in the choice of titles for Volume 1. The production of Volume 3, therefore, allows the development of the theme initiated in Volume 1. This volume contains a further selection of detailed protocols for a range of analytical and preparative protein techniques, and should be seen as a continuation of Volume 1. Companion Volumes 2 and 4 provide protocols for nucleic acid methodology.

Each method is described by an author who has regularly used the technique in his or her own laboratory. Not all the techniques described necessarily represent the state-of-the-art. They are, however, dependable methods that achieve the desired result.

Each chapter starts with a description of the basic theory behind the method being described. The main aim of this book, however, is to describe the practical steps necessary for carrying out the method successfully. The Methods section, therefore, contains a detailed step-by-step description of a protocol that will result in the successful execution of the method. The Notes section complements the Methods section by indicating any major problems or faults that can occur with the technique and any possible modifications or alterations.

This book should be particularly useful to those with no previous experience of a technique and, as such, should appeal to undergraduates (especially project students), postgraduates, and research workers who wish to try a technique for the first time.

John M. Walker

Contents

Contributors

S. A. AMERO • Department of Biology, Washington University, St. Louis, Missouri

SAROJANI ANGAL • Celltech Ltd., Slough, Berkshire, UK

BRIAN AUSTEN • Peptide Unit, Department of Surgery, St. George's Medical School, London, UK

H. M. BAILEY • School of Life Sciences, Leicester Polytechnic, Leicester, UK

R. D. J. BARKER • School of Life Sciences, Leicester Polytechnic, Leicester, UK

ROBERT J. BEYNON • Department of Biochemistry, University of Liverpool, UK

ALEX F. CARNE • Celltech Ltd., Slough, Berkshire, UK

MICHAEL J. DUNN • Jerry Lewis Muscle Research Centre, Royal Postgraduate Medical School, UK

DAVID J. EASTY • Department of Histopathology, Royal Postgraduate Medical School, London, UK

S. C. R. ELGIN • Department of Biology, Washington University, St. Louis, Missouri

W. GAASTRA • Rijksuniversiteit Utrecht, Faculteit der Diergeneeskunde, Vakgroep Bacteriologie, Utrecht, The Netherlands

PAULO GALLO • Department of Neurology, University of Padova, Clinica Delle Malattie Nervose É Mentali, Padova, Italy

CECILIA GELFI • Chair of Biochemistry, Faculty of Pharmacy and Department of Biomedical Sciences and Technologies, University of Milano, Milano, Italy

LYNNE GOULDING • Downstream Processing Department, Celltech Ltd., Slough, Berkshire, UK

W. J. GULLICK • Protein Chemistry Laboratory, Institute of Cancer Research, Chester Beatty Laboratories, London, UK

JOHN B. W. HAMMOND • Biochemistry Department, Rothamsted Experimental Station, Hertfordshire, UK

E. L. V. HARRIS • Celltech Ltd., Slough, Berkshire, UK

LUCY F. HENLEY • Department of Zoology, University of Edinburgh, Edinburgh, Scotland

CHRISTOPHER HILL • Downstream Processing Department, Celltech Ltd., Slough, Berkshire, UK

H. H-S. IP • Research Computer Unit, Imperial Cancer Research Fund Laboratories, London, UK

ROSEMARY JAGUS • Department of Microbiology, Biochemistry, and Molecular Biology, University of Pittsburgh School of Medicine, Pittsburgh, Pennsylvania

T. C. JAMES • Department of Biology, Washington University, St. Louis, Missouri

ALAN JONES • Janssen Pharmaceutical Ltd., Wantage, UK

ANDREW KENNEY • Downstream Processing Department, Celltech Ltd., Slough, Berkshire, UK

NICHOLAS J. KRUGER • Biochemistry Department, Rothamsted Experimental Station, Hertfordshire, UK and Agricultural Genetics Company, Cambridge Science Park, Cambridge, UK

MARC MOEREMANS • Department of Life Sciences, Laboratory of Biochemical Cytology, Division of Cellular Biology and Chemotherapy, Janssen Pharmaceutica NV, Beerse, Belgium

KETAN PATEL • Jerry Lewis Muscle Research Centre, Royal Postgraduate Medical School, London, UK

MARNIX PEFEROEN • University Hospital, St. Raphael, Laboratory of Hematology, Leuven, Belgium

JEFFREY W. POLLARD • MRC Research Group in Human Genetic Disease, Department of Biochemistry, Queen

Elizabeth's College, University of London, London,
UK

PIER GIORGIO RIGHETTI • Chair of Biochemistry, Faculty of
Pharmacy and Department of Biomedical Sciences
and Technologies, University of Milano, Milano, Italy

N. W. SCOTT • School of Life Sciences, Leicester
Polytechnic, Leicester, UK

AKE SIDEN • Department of Neurology, Karolinska
Institute, Huddinge University Hospital, Huddinge,
Sweden

BRYAN JOHN SMITH • Celltech Ltd., Slough, Berkshire, UK

CHRISTOPHER F. THURSTON • Department of Microbiology,
King's College, University of London, London, UK

M. D. TREVAN • Biological Sciences, The Hatfield
Polytechnic, Hatfield, Hertfordshire, UK

JOHN M. WALKER • Biological Sciences, The Hatfield
Polytechnic, Hatfield, Hertfordshire, UK

G. B. WISDOM • Department of Biochemistry, The Queen's
University of Belfast, Medical Biology Centre, Belfast,
Ireland, UK

Chapter 1

Prevention of Unwanted Proteolysis

Robert J. Beynon

1. Introduction

Inescapably, all cells contain proteases, introducing the possibility that disruption of the tissue can bring together a protease and a protein, with the result that the latter suffers hydrolytic damage. To quote Pringle (1, 2), "Proteolytic artifacts are pervasive, perplexing, persistent and pernicious but with proper precautions, preventable." Autolysis has long been recognized as a problem during protein purification, but methods for its control are still far from perfect. Moreover, there are many circumstances other than during protein purification in which endo- or exopeptidase attack upon a pro-

1

tein can be at best a frustrating nuisance and at worst an undetected artifact that leads to erroneous conclusions.

The purpose of this chapter is to build upon the excellent papers by Pringle (*1,2*) and to provide updated information on methods for prevention of unwanted proteolysis. (Few of my colleagues have been impressed by my suggestion that an effective general purpose protease inhibitor is 2*M* sulfuric acid!) Unfortunately, no global solution to the problem exists, and to a great extent, an *ad hoc* solution depends upon elucidation of some of the properties of the protease(s) that is (are) suspected to be responsible. This chapter may differ from many others in the volume because I cannot present a "method" as much as a philosophy based upon the advice "know thine enemy." Hence, the methods include a sensitive protease assay in addition to a discussion of the handling of protease inhibitors. Largely, I shall restrict the subject matter to proteolytic artifacts that occur in vitro. Control of proteolysis of proteins in vivo is still difficult, although of increasing importance in studies that aim to express a normal or mutated gene in a foreign cell type.

Critically important but sometimes overlooked is the need to establish that the artifact is truly attributable to proteolysis. Dramatic losses of activity of a protein may be caused by proteases, but may also be caused by, among others, thermal denaturation, dissociation of a cofactor, adsorption onto surfaces, dephosphorylation, or inadvertent modification of the redox status of sulfhydryl/disulfide groups. In a crude homogenate, it may be difficult to assign changes in the properties of a protein to the action of proteases, and often, the only successful approach may require addition of potentially protective protease inhibitors. Limited exoproteolytic attack can combine dramatic changes in the biological properties of a protein with minimal effects upon physicochemical properties; such modifications are virtually undetectable by analytical methods and are best identified by judicious use of inhibitors in a diagnostic fashion. It is difficult to offer any hard and fast guidelines for circumstances in which proteolytic artifacts are most likely or preventable, but the following should be kept in mind.

a. Cells differ in the intracellular concentrations of proteases, and unwanted attack upon a protein of interest may be diminished in a cell/tissue in which protease levels are low. In many single-cell systems, mutant strains are available that are defective in the expression of protease-coding genes; this should be considered as an option.

b. Homogenization of a tissue often allows for complexation of a protease with a pool of (previously isolated) inhibitor. Such enzyme–inhibitor complexes may be dissociated later by inactivation of the inhibitor, however, or the two components may be resolved by a purification step. Thus, proteolysis may manifest itself in later stages of preparation of a protein.

c. Proteins compete for the active site of proteases, and, therefore, a purification may separate the target protein from contaminants that are protective, particularly if the protease is copurifying with the target protein. Such behavior will manifest itself as proteolytic attack that occurs as the protein become more highly purified.

d. Many proteins are made more resistant to a variety of denaturing/destabilizing assaults by complexation with their ligands. Substrate or cofactor-mediated protection of a protein from hydrolytic attack is a common observation, but care should be taken to avoid the alternative of ligand-induced labilization of the target protein.

e. Proteases are often more stable than their substrates. Thus a denaturing treatment that has the goal of inactivating the protease may have the opposite result of labilizing the target protein to the more resistant protease. This behavior can manifest itself during sample preparation for sodium dodecyl sulfate polyacrylamide gel electrophoresis; the lack of a detectable band may imply that the protease in the preparation was more tolerant than the target protein to the detergent in sample buffer. In these circumstances, the target protein, rendered vulnerable by the detergent, is exposed to a short but effective proteolytic attack.

f. Proteolytic inactivation of a protein is relatively easy to detect and, thus, to control. Far more difficult to identify and regulate is limited digestion that leads to relatively minor changes in biological properties, but that may introduce microheterogeneity in the final product.

2. Classes of Protease

Proteases can be divided conveniently into endopeptidases and exopeptidases (3). Endopeptidases are further subdivided into five classes based upon the mechanisms that they employ to achieve hydrolysis of the peptide bond. Exopeptidases are classified primarily in terms of the terminal amino acids, dipeptides, or tripeptides that they remove from the carboxyl or amino termini of a protein (4). I stress that successful control of adventitious proteolysis can best be attained if the reader has some appreciation of the mechanistic class to which the offending protease belongs. Pertinent features are given below.

2.1. Serine Endopeptidases

Serine endopeptidases achieve hydrolysis of the peptide bond by attack upon the carbonyl carbon by a nucleophilic serine residue. The active-site serine residue is a much stronger nucleophile than other serine residues in proteins, and the special properties of this residue are a consequence of electron flow to the serine side chain oxygen atom via a histidine residue. Thus, both the serine and histidine residues are effective targets for many serine endopeptidase inhibitors.

2.2. Cysteine Endopeptidases

Cysteine endopeptidases (previously referred to as thiol proteases) employ a nucleophilic cysteine residue in an analogous fashion to the serine residue above, and again, a histidine residue is implicated in the catalytic mechanism. The special properties of the cysteine residue in particular make it a valuable target for mechanism-based inhibitors.

2.3. Aspartic Endopeptidases

Aspartic endopeptidases (previously known as acid proteases) employ a pair of aspartic residues to labilize and hydrolyze the peptide bond. There are few effective inhibitors that are directed to aspartic residues, and strategies for inhibition of aspartic endopeptidases usually rely upon tight-binding transition state analogs rather than modification of the active-site residues.

2.4. Metalloendopeptidases

Metalloendopeptidases capitalize upon the electron-withdrawing properties of a metal ion (thus far, always zinc) to weaken the peptide bond. It follows that the target for effective inhibition will be the active site metal ion. It is important to discriminate between true metalloendopeptidases and metal-activated proteases that employ another mechanism, such as the calcium activated cysteine endopeptidase, calpain.

Finally, it is worth noting that a number of the newly discovered endopeptidases do not fall naturally into any one of these classes and may employ totally new hydrolytic mechanisms. These proteases do not respond in a predictable fashion to archetypical class-specific inhibitors.

In most cases, classification of exopeptidases has yet to be formalized in terms of catalytic mechanisms, but it is increasingly apparent that many of them have evolved mechanisms that are similar to the endopeptidases (4). Thus, inhibitor strategies will often be similar for prevention of endopeptidase or exopeptidase attack.

3. Measurement of Endoproteolytic Activity

Protease assays vary from the highly specific, using a substrate deliberately optimized for a single enzyme, to the most general, based on a substrate that is hydrolyzed to a greater or lesser extent by all proteases (5). Assays in the former category are of limited use for determination of (usually unknown) contaminating proteases. General protease substrates are usually proteins that are intrinsically vulnerable to proteolytic attack, such as casein, denatured proteins, or smaller peptides that do not possess the higher-order structure that confers proteolytic resistance. It is feasible to monitor the hydrolysis of unmodified proteins as the appearance of acid-soluble peptides or amino acids, determined as ultraviolet absorbing material, or capitalizing upon the properties of specific amino acid residues. These assays are relatively insensitive, however. Greater sensitivity and convenience can be attained by labeling the substrate, usually with chromogenic, fluorogenic, or radioactive moieties. Representative labeled substrates are given in Table 1, together with references to the labeling techniques.

Given here is a method for the preparation of a very sensitive radiolabeled substrate; the B-chain of insulin, radioiodinated at its two tyrosine residues (6). Digestion of the substrate releases smaller peptides that are soluble in a trichloroacetic acid concentration that precipitates the undigested substrate. The method is presented in two sections. First, the labeled substrate is not available commercially and, thus, an iodination reaction must be performed. Second, a typical assay based upon this substrate is described, although the conditions (buffer, pH, ionic strength, temperature, substrate concentration, assay volume) can be altered at will provided that certain basic conditions are met.

4. Prevention of Undesired Proteolysis

It is likely that the reader will become aware of endoproteolytic attack much more readily, since the consequences are more tangible. I shall therefore concen-

Table 1

Some Representative General-Purpose Endopeptidase Assays[a]

Substrate	Product	Detected by	Reference
Azocasein (sulfanilamide-dyed protein)	Dye-peptides	Acid-soluble 340 nm	12, 13
Fluorescamine-casein	Fluorescamine-peptides	Acid-soluble fluorescence 405/475 nm	14, 15
Fluorescein isothiocyanate-casein	FITC-peptides	Acid-soluble fluorescence 490/525 nm	16
Succinyl casein + leucine amionoopeptidase and L-amino acid oxidase	Amino acids > keto acids also generates hydrogen peroxide	Absorbance 550 nm	17
[^{14}C]-Collagen	[^{14}C]-Peptides	Soluble radioactivity	18
[^{3}H]-Elastin	[^{3}H]-Peptides	Acid-soluble radioactivity	19
[^{125}I]-Casein Sepharose	[^{125}I]-Peptides	Released radioactivity	20
Glucosidase-casein-Sepharose	Glucosidase-peptides	Released enzyme activity	21

[a]This table is far from exhaustive, but illustrates the variety of assay methods that may be employed. It should not prove difficult to identify an assay that has the appropriate degree of sensitivity and is compatible with the specific conditions of the study. Additional assays may be found in ref. 22, a valuable reference to mammalian proteases.

trate upon the control of artifactual endopeptidase action. There are in fact two strategies. The first relies upon separation of the substrate from proteases. This may be dependent upon a high resolution and exhaustive purification scheme and may require that the protease activity in addition to the protein of interest is monitored. A more sophisticated approach relies upon a step that is specifically designed a an affinity purification stage for the *protease* in which the unbound material is the fraction of interest. Affinity ligands for proteases abound, but the choice is facilitated if the catalytic mechanism of the protease is known. Such considerations justify a series of experiments to assess the effects of a series of protease inhibitors (such as those in Table 2) upon artifactual proteolysis or upon general proteolytic activity. Good affinity ligands for proteases are provided by proteinaceous inhibitors (Table 3) that are reasonable stable and that can be coupled to insoluble matrices with the minimum of chemistry. An oxirane-derivatized bead, Eupergit C, offers remarkable ease of coupling, reasonable capacities, and good mechanical and flow properties (7), and a method for preparation of an Eupergit C–protein complex is given below. Alternatively, proteinase inhibitors can be successfully coupled to cyanogen-bromide-activated Sepharose (8); some of these are commercially available. Such immobilized inhibitors are also valuable for the removal of proteases after intentional proteolytic attack upon a protein (9,10).

The second strategy permits the protease to remain in the biological sample in an inactive form, attained by judicious addition of inhibitors. The choice of inhibitors is not simple. There is no general-purpose inhibitor that can inhibit all proteases (alpha-2 macroglobulin is closest to this ideal), and thus, a mixture of inhibitors will usually be added. The number of additions is mini-

Table 2
Low Molecular Weight Protease Inhibitors[a]

Inhibitor	Specificity	Stock solution/ solvent	Stability	Effective concentration	Note
Irreversible Inhibitors					
Diisopropylphosphofluoridate (DipF), 184.2 mol. wt.	Serine proteases	200 mM in dry propan-1-ol	Long term at –20°C	0.1–1.0 mM	*b*
Phenylmethane sulfonylfluoride (PMSF), 174.2 mol. wt.	Serine proteases	200 mM in dry propan-1-ol or methanol	Long term at –20°C	1.0–10 mM	*c*
Tosylphenylalanyl-chloro-methyl ketone (TPCK), 351.9 mol. wt.	Chymotrypsin-like serine proteases	10 mM in methanol	Stable below pH 7.5	0.1 mM	
Tosyllysylchlomethyl ketone (TLCK), 369.3 mol. wt.	Trypsin-like serine proteases	10 mM in aqueous solution	Prepared fresh as needed	0.1 mM	
Iodoacetic acid, 208.0 mol. wt.	Cysteine proteases	100 mM in aqueous solution	Decomposes slowly	0.1–1.0 mM	
E-64, 357.4 mol. wt.	Most cysteine proteases	1 mM in aqueous solution	At least 1 mo. at –20°C	10 µM	
Reversible Inhibitors					
Leupeptin, 426.6 mol. wt.	Trypsin-like serine proteases, cysteine proteases	1mg/mL in aqueous solution	At least 1 wk at –20°C	25 µg/mL	*d*

Inhibitor	Protease class	Stock solution	Storage	Working concentration
Chymostatin, 582.7 mol. wt.	Chymotrypsin-like serine proteases, cysteine proteases	1 mg/mL in DMSO	1 mo at –20°C	25 µg/mL
Elastatinal, 512.6 mol. wt.	Elastase-like serine proteases	1 mg/mL in aqueous solution	At least 1 mo at –20°C	1 µg/mL
Pepstatin, 685.9 mol. wt.	Aspartic proteases	1 mg/mL in DMSO	Long term at –20°C	0.1 ng/mL
Phosphoramidon, 543.6 mol wt	Metalloendopeptidases	1 mg/mL in water	At least 1 mo at –20°C	25 µg/mL
1,10-Phenanthroline, 198.2 mol wt	Metalloendopeptidases	200 mM in methanol	Long term at 4°C	1–10 mM

[a]Listed here are the protease inhibitors that are most commonly used to supress protease action. All of them are commercially available. Suggested working concentrations are provided, together with concentrations that have proved effective toward true members of each class of protease. Further information on many of these inhibitors can be found in refs. 22–25.

[b]DipF is highly toxic and volatile, and, in my experience, vials have a tendency to develop a positive pressure. Stringent precautions should be taken with this reagent. Once dissolved in propan-1-ol, it can be stored at –20°C in tightly capped vials. Preparations of DipF contain a contaminant that is able to inhibit cysteine proteases and, if necessary, this should be inactivated before use (5).

[c]PMSF appears to be a less potent inhibitor of serine proteases than DipF, often giving partial inhibition under conditions in which the latter would inhibit completely. Higher concentrations of PMSF are therefore routinely used. Note that both DipF and PMSF are inactivated in aqueous solutions and are only active for a limited time after addition to the sample.

[d]Leupeptin, chymostatin, and, indeed, many protease inhibitors are themselves peptides and are susceptible to peptidase action. Mammalian tissue preparations contain peptidases that inactivate leupeptin and chymostatin (26–28). The reader should be alert to this possibility.

[e]1,10-Phenanthroline has a strong ultraviolet absorbance that can interfere with spectrophotometric measurements.

Table 3

Proteinaceous Protease Suitable for Coupling to an Insoluble Matrix[a]

Inhibitor	Target proteases	Reference (to inhibitor)
Soybean trypsin inhibitor	Trypsin and chymotrypsin-like serine proteases	29
Lima bean trypsin inhibitor	Trypsin and chymotrypsin-like serine proteases	29
Egg white ovomucoid	Trypsin and chymotrypsin-like serine proteases (specificity depends on species)	29
Egg white ovostatin	Most metallo-proteinases, including collagenases and gelatinases	30
Egg white cystatin	Cysteine proteases, but not calpain	31
Alpha-2macroglobulin	Virtually all endopeptidases except those with a very restricted substrate specificity	32
Alpha-1 proteinase inhibitor	Trypsin-like serine proteases	32
Pancreatic trypsin inhibitor (aprotinin)	Mammalian serine proteases	33

[a]All of these inhibitors can be coupled to insoluble matrices and used as chromatographic media to separate proteases from samples. It is advisable to determine the binding capacity of the matrix by titration with a pure protease. Capacities of 1–5 mg protease/mL of matrix are readily attainable.

mized by knowledge of the inhibitor-sensitivity profile of the sample. An additional complication may arise if the protease inhibitor also affects the biological activity of the protein of interest, a problem that is particularly prevalent when chelators or thiols/thiol-reactive agents are involved.

Irreversible inhibitors have the advantage that they need only be added at the early stages of the experiment. Reversible inhibitors must be maintained at inhibitory concentrations throughout the experiment. The best criterion for effective inhibition of problematical proteases is the disappearance of the problem, but it may be advisable to check for the presence of active proteases as well. The inhibitors listed in Table 2 (and to some extent, those in Table 3) can be added to samples in appropriate combinations at the recommended concentrations. I stress, however, that the susceptibility of proteases to each of the inhibitors extends over several orders of magnitude, and optimal protection may be attained at considerably higher concentrations.

5. Materials

5.1. Method 1. Preparation of Radioiodinated Insulin B-Chain

The following materials are used for Method 1.

1.   0.5 M Sodium phosphate buffer, pH 7.5.
2.   Insulin B-chain (performic acid oxidized or reduced/carboxymethylated).
3.   Na$^{[125]}$I (100 mCi/mL; 15 mCi/µg).
4.   0.15M KI (25 mg/mL in water).
5.   0.44M Chloramine T (100 mg/mL) prepared just before use.

6.   0.32M Sodium metabisulfite (60 mg/mL).
7.   Sephadex G-25™ (medium) column, V_t = 10–15 mL, equilibrated with:
8.   6.0 mM KI (1 mg/mL) adjusted with NaOH to pH 9–10.
9.   2% (w/v) Casein.
10. 25% (w/v) Trichloroacetic acid.

5.2. Method 2. Protease Assay Using ^{125}I-Insulin B-Chain

The following materials are used for Method 2.

1.   0.5 mg/mL [125] I-Insulin B-chain.
2.   0.04M Hepes/0.28M NaCl, pH 7.5.
3.   2% (w/v) Casein.
4.   25% (w/v) Trichloroacetic acid.
5.   Siliconized reaction tubes.
6.   1.5-mL Microcentrifuge tubes.

5.3. Method 3. Preparation of Immobilized Protease Inhibitors

The following materials are used for Method 3.

1.   Eupergit C™ oxirane acrylic beads (Rohm Pharma GmbH, Weiterstadt, FDR).
2.   1.0M Potassium phosphate buffer, pH 7.5.
3.   0.1M Potassium phosphate buffer, pH 7.5.
4.   2-Mercaptoethanol.
5.   Protease inhibitor.

6. Methods

6.1. *Preparation of Radioiodinated Insulin B-Chain*

The following applies to Method 1.

1. Dissolve 2–3 mg insulin B-chain in 0.25 mL of phosphate buffer. The insulin B-chain is not particularly soluble, and it may be preferable to make the solution slightly alkaline with NaOH to dissolve the peptide fully and then readjust the pH to 7.5 (narrow range pH papers are adequate for measuring the pH of samples).
2. Add 5 µL of 0.15M KI (25 mg/mL) to lower the specific radioactivity of the isotope.
3. Add 2 µL of Na[125]I (approximately 200 µCi/7.4 MBq).
4. Add 50 µL of chloramine T and mix completely. Allow reaction to proceed at room temperature for 15 min.
5. Add 20 µL of sodium metabisulfite to terminate the reaction.
6. Add 20 µL of 0.15M KI to act as carrier during chromatography.
7. Apply whole sample to the Sephadex G-25 column and elute with 6.0 mM KI, pH 9–10. Manually collect fractions in disposable tubes marked at approximately 1 mL.
8. For each fraction, transfer 5 µL to 1.0 mL of water. Determine the radioactivity in 100 µL of each dilution by gamma counting. This gives the total radioactivity in 0.5 µL of each original fraction.
9. Remove 100 µL of each dilution and add to 100 µL of 2% (w/v) casein as a coprecipitant. Precipitate

the sample with 200 µL of 25% (w/v) trichloroacetic acid and sediment the precipitated material by centrifugation at 10,000g for 2 min. Determine the radioactivity in 100 µL of the supernatant. This gives the acid-precipitable radioactivity in 0.125 µL of the original fraction.

10. Determine the total radioactivity and acid-precipitable radioactivity of each fraction, and pool
those that are more than 95% precipitable (*see* Fig.
1).

11. Determine the concentration of insulin B-chain
by measurement of the A_{280} nm ($E1\% = 9.92$). Use
an acrylic cuvet; it is sufficiently transparent at
280 nm and is disposable.

12. Dilute the solution to a final concentration of 0.5
mg/mL, adjust the pH to near-neutrality, and
store aliquots of the solution at –20°C in a lead-
lined box.

6.1.1. Notes on Method 1

1. Unlike many iodinations (e.g., ref. *11*), the goal here
is not maximal specific activity, but the generation
of the maximal amount of substrate labeled to a
workable specific activity. Thus, the iodinations
are performed in relatively high volumes at high
protein concentrations.

2. As supplied by the manufacturer, Na$^{[125]}$I is at a very
high specific activity and potentially hazardous.
Small microaerosols or droplets caused by careless
transfers are very radioactive and can potentially
be inhaled. Therefore, conduct all iodinations in a
fume hood that has been certified for radioactivity
work. Isotopes of iodine have been shown to pass

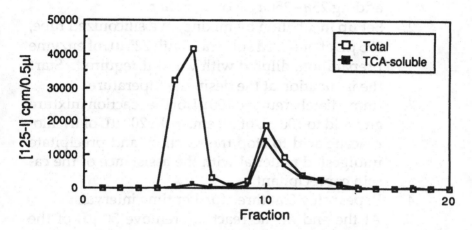

Fig. 1. Separation of radioiodinated insulin B-chain from unincorporated sodium iodide. Insulin B-chain (2–3 mg) was iodinated with 200 µCi of sodium iodide and subsequently applied to a 10 mL column of Sephadex G-25. The total and acid soluble radioactivity was determined for an aliquot of each fraction. Fractions 5 and 6 were pooled and used as substrate.

across disposable gloves, and we routinely wear two pairs of gloves when dispensing the iodine; the outer pair is discarded immediately after the radioiodine is added.

3. When the carrier iodide is added (step 6) the solution may turn pale yellow/brown, indicating elemental iodine. If this happens, add a further 20 µL of sodium metabisulfite to retain the isotope in solution as the less hazardous iodide ion.

6.2. Protease Assay Using ^{125}I-Insulin B-Chain

The following applies to Method 2.

1. For n assays, prepare buffered substrate solution by adding $25n$– $250n$ µL of 2x buffer.

2. Set up incubation by mixing, in a siliconized tube, 275 µL of buffered substrate with 225 µL of enzyme preparation, diluted with water if required. Start the incubation at the desired temperature.

3. Immediately remove 100 µL of the reaction mixture and add to 100 µL of casein. Add 200 µL of trichloroacetic acid to stop the reaction, and precipitate undigested material with the assistance of the casein coprecipitant.

4. Repeat step 3 at three further time intervals.

5. At the end of the reaction, remove 50 µL of the reaction mixture, add to 100 µL of water, and count directly, without precipitation (100% sample). This figure gives the specific radioactivity of the sample on that day and permits the calculation of extent of solubilization of insulin B-chain = (cpm in supernatant from steps 3 and 4)/(cpm in 100% sample).

6. Calculate the slope of the line (preferable by linear regression) and ensure that the time course is linear. Calculate the rate of solubilization of the insulin B-chain. Remember that the specific radioactivity of the substrate will fall over a few weeks and, thus, measurement of the rate of release of radioactivity is less acceptable.

6.2.1. Notes on Method 2

1. No strict guidelines can be given as to the choice of reaction buffer, time course of assay, need for cofac-

tors, or amount of biological material to be assayed. This will depend upon the particular samples and circumstances. I recommend that some "ranging" assays be performed first to establish reasonable working conditions.

2. The time course of the assay is linear provided that the extent of the solubilization remains below approximately 10%. More extensive digestion will usually manifest itself as a downward curvature in the process curve. For this reason, I recommend a time course assay as superior to a single sampled assay that gives no indication of the linearity of the digestion.

3. Exopeptidases are not very efficient at solubilizing radioactivity in this assay, but they can potentiate activity by attacking the new terminii that are generated by endopeptidases.

4. The substrate concentration will undoubtedly be well below the Michaelis constant for most proteases. This too may contribute to nonlinearity if the substrate is depleted significantly. The low concentration probably contributes to the tendency of the substrate to adhere to glass and plastic surfaces. It is advisable to perform the digestion reaction in siliconized, well-washed glass tubes.

5. Always include a protease-free blank to determine the intrinsic precipitability of that particular batch of the substrate (zero time) and the background rate of release of material into the trichloroacetic acid-soluble fraction.

6. With some proteases we have noticed that extensive digestion leads ultimately to a decrease in acid-soluble radioactivity. This may be caused by the formation of short peptides that recover the property of insolubility in acid solution and is a further

justification for minimizing the extent of hydrolysis of the substrate.

7.   Most of the radioactivity will be in the precipitated material—this can be disposed of as solid waste.

6.3. Preparation of Immobilized Protease Inhibitors

The following applies to Method 3.

1.   Dissolve 10–50 mg of the protease inhibitor in 4.0 mL of 1.0M phosphate buffer.
2.   Add the inhibitor solution to 1 g of Eupergit C dry beads and react for 16–72 h at 4°C.
3.   Wash the inhibitor-Eupergit beads with five washes of 100 mL of 0.1M phosphate buffer.
4.   Add 4.0 mL of 5% (v/v) aqueous 2-mercaptoethanol to block unreacted groups. Leave for 16 h at 4°C.
5.   Wash the beads as in step 3.
6.   Determine the protease binding capacity of a sample of the beads by titration with a pure protease.
7.   Store the inhibitor-Eupergit at 4°C in buffer containing a suitable bacteriostatic agent.

6.3.1. Notes on Method 3

1.   The manufacturers recommend 20–25°C for coupling, but this may compromise the stability of the inhibitor. Make sure that the inhibitor preparation does not contain any ammonium salts, since these will react rapidly with the oxirane groups of the matrix. The coupling reaction is best carried out in a wide-necked flask. The manufacturer's literature states that mixing should not be necessary, but we routinely place the mixture on a slow bottle roller.

There is very little free buffer at this ratio of beads-to-protein solution.

2. Blocking of unreacted groups is not always necessary, but is a relatively straightforward step to perform. Mercaptoethanol has the advantage that it does not disturb the hydrophilicity and electroneutrality of the matrix.

3. The water regain value of the beads is about 2.5 mL/g dry beads.

Acknowledgments

I am grateful to my colleagues whose common interest in proteases and their control has led to many interesting discussions along the lines of this contribution. I thank John Kenny for permission to reproduce a modified version of his technique for preparation of the radioiodinated insulin B-chain and to Maggie Cusack for supplying details of the modifications.

References

1. Pringle, J.R. (1974) Methods for Avoiding Proteolytic Artefacts in Studies of Enzymes and Other Proteins from Yeasts, in *Methods in Cell Biology* vol. 12 (Prescott, D.M., ed.) Academic, New York.

2. Pringle, J.R. (1978) Proteolytic Artefacts in Biochemistry, in *Limited Proteolysis in Microorganisms* (Cohen, G.N. and Holzer, H., eds.) US Department of Health, Washington, DC.

3. Barrett, A.J. (1986) The Classes of Proteolytic Enzymes, in *Plant Proteolytic Enzymes* (Dalling, M.J., ed.) CRC Press, Boca Raton, Florida.

4. McDonald, J.K. (1985) An overview of protease specificity and catalytic mechanisms: Aspects related to nomenclature and classification. *Histochem.J.* **17**, 773–785.

5. Wagner, F.W. (1986) Assessement of Methodology for the Purification, Characterisation and Measurement of Proteases, in *Plant Proteolytic Enzymes* (Dalling, M.J., ed.) CRC Press, Boca Raton, Florida.

6. Fulcher, I.S. and Kenny, A.J. (1983) Proteins of the kidney microvillar membrane. *Biochem. J.* **211**, 743–753.

7. Hannibal-Friedrich, O., Chun, M., and Sernetz, M. (1980) Immobilisation of β-galactosidase, albumin and globulin on epoxy-activated acrylic beads. *Biotechnol. Bioeng.* **22**, 157–175.

8. Otsuka, A.S. and Price, P.A. (1974) Removal of proteases from DNase1 by chromatography over agarose with covalently attached lima bean protease inhibitor. *Anal. Biochem.* **62**, 180–187.

9. Place, G.A. and Beynon, R.J. (1982) The chymotrypsin-catalysed activation of glutamate dehydrogenase. *Biochem. J.* **205**, 75–80.

10. Place, G.A. and Beynon, R.J. (1983) Chymotryptic activation of glutamate dehydrogenase. *Biochim. Biophys. Acta* **747**, 26–31.

11. Bailey, G.S. (1984) Radioiodination of Proteins, *Methods in Molecular Biology* vol. 1 (Walker, J.M., ed.) Humana, Clifton, New Jersey.

12. Barrett, A.J. (1972) A new assay for cathepsin B1 and other thiol proteinases. *Anal. Biochem.* **47**, 280.

13. Beynon, R.J., Shannon, J.D., and Bond, J.S. (1981) Purification and characterisation of a metalloproteinase from mouse kidney. *Biochem. J.* **199**, 591–598.

14. Sogawa, K. and Takahashi, K. (1978) Use of fluorescamine-labeled casein as a substrate for assay of proteinases. *J. Biochem.* **83**, 1783–1787.

15. Evans, C.H. and Ridella, J.D. (1984) An evaluation of fluorometric proteinase assays which employ fluorescamine. *Anal. Biochem.* **142**, 411–420.

16. Twining, S.S. (1984) Fluorescein isothiocyanate-labelled casein assay for proteolytic enzymes. *Anal. Biochem.* **143**, 30–34.

17. Sugiura, M., Ishikawa, M., Sasaki, M., Hirano, K., Ito, Y., and Awazu, S. (1979) A new method for protease activity measurement. *Anal. Biochem.* **97**, 11–16.

18. Johnson-Wint, B. (1980) A quantitative collagen film collagenase assay for large numbers of samples. *Anal. Biochem.* **104**, 175–181.

19. Bieger, W. and Scheele, G. (1980) A sensitive and specific enzyme assay for elstase activity using [3-H] elastase as substrate. *Anal. Biochem.* **104**, 239–246.
20. Sevier, E.D. (1976) Sensitive, solid-phase assay for proteolytic activity. *Anal. Biochem.* **74**, 592–596.
21. Andrews, A.T. (1982) A new approach to the general detection and measurement of proteinase and proteinase inhibitor activities. *Biochim. Biophys. Acta* **708**, 194–202.
22. Barrett, A.J. (1977) *Proteinases in Mammalian Cells and Tissues* Elsevier-North Holland Biomedical, Amsterdam.
23. Katunuma, N., Umezawa, H, and Holzer, H. (1983) *Proteinase Inhibitors-Medical and Biological Aspects* Springer Verlag, Tokyo.
24. Umezawa, H. (1982) Low molecular weight inhibitors of microbial origin. *Ann. Rev. Microbiol.* **36**, 75–99.
25. Dawson, R.M.C., Elliott, D.C., Elliott, W.H., and Jones, K.M. (1986) *Data for Biochemical Research* Claredon, Oxford.
26. Beynon, R.J., Brown, C.P., and Butler, P.E. (1981b) The inactivation of Streptomyces-derived proteinase inhibitors by mammalian tissue preparations. *Acta Biol. Med. Germ.* **40**, 1539–1546.
27. Brown, C.P. and Beynon, R.J. (1983) Tissue and subcellular distribution of enzymes inactivating leupeptin. *Biosc. Rep.* **3**, 179–184.
28. Place, G.A., Ning, M.C.Y.J., and Beynon, R.J. (1985) A Leupeptin-inactivating enzyme from rat liver in *Intracellular Proteolysis* (Khairallah, E.A., Bond, J.S., and Bird, J.W.C., eds.) Alan R. Liss, New York.
29. Laskowski, M. and Kato, I. (1980) Protein inhibitors of proteinases. *Ann. Rev. Biochem.* **49**, 593–626.
30. Nagase, H. and Harris, E.D. (1983) Ovostatin, a novel proteinase inhibitor from chicken egg white. *J. Biol. Chem.* **258**, 7490–7498.
31. Anastasi, A., Brown, M.A., Kembhavi, A.A., Nicklin, M.J.H., Sayers, C.A., Sunter, D.C., and Barrett, A.J. (1983) Cystatin, a protein inhibitor of cysteine proteinases. Improved purification from egg-white, characterisation and detection in chicken serum. *Biochem. J.* **211**, 129–138.
32. Travis, J. and Salvesen, G.S. (1983) Human plasma proteinase inhibitors. *Ann. Rev. Biochem.* **52**, 655–709.
33. Fritz, H. and Wunderer, g. (1983) Biochemistry and applications of aprotinin, the kallikrein inhibitor from bovine organs. *Arzneimittelforsch Drug. Res.* **33**, 479.

Chapter 2

The Bradford Method
for Protein Quantitation

John B. W. Hammond
and Nicholas J. Kruger

1. Introduction

A rapid and accurate method for the estimation of protein concentration is essential in many fields of protein study. The Lowry method (Chapter 1 in vol. 1 of this series) has been widely used, but is susceptible to interference from a wide range of compounds commonly present in biological extracts. Although interference can be avoided by trichloracetic acid precipitation of the protein prior to assay, this lengthens the procedure.

The Bradford method (1), by contrast, is not subject to interference by most common reagents or nonprotein substances likely to be present in biological samples (*see* Note 3 in section 4). It is simpler, faster, and more sensitive than the Lowry method. The assay relies on the binding of the dye Coomassie Blue G250 to the protein molecule. The cationic form of the dye, which predominates in the acidic assay reagent solution, has a λmax of 470 nm. The dye binds to protein as the anionc form, which has a λmax of 595 nm (2). Thus the amount of dye bound to the protein can be quantitated by measuring the absorbance of the solution at 595 nm.

The dye appears to bind most readily to arginine residues (but not to the free amino acid) (2). This can lead to variation in response to different proteins, which is the main drawback of the method. The original Bradford assay shows large variation in response between different proteins, with the common protein standard bovine serum albumin at the higher end of the range (3,7–9). For this reason, the standards should be carefully chosen if absolute protein concentrations are required or pure proteins are being assayed (*see* Note 1 in section 4). Several modifications to the method have been proposed to overcome this problem (*see* Note 2 of section 4). The original method still appears to be the most widely used, however, and is described here. A microassay protocol is also described.

2. Materials

1. Reagent: The assay reagent is made by dissolving 100 mg of Coomassie Blue G250 in 50 mL of 95% ethanol. The solution is then mixed with 100 mL of 85% phosphoric acid and made up to 1 L with distilled water.

The reagent should be filtered through Whatman No. 1 filter paper before storage in an amber bottle at room temperature. It is stable for several weeks, but slow precipitation of the dye will occur, so filtration of the stored reagent is necessary before use.

2. Protein standards (*see* Note 1 in section 4). Ovalbumin at a concentration of 1 mg/mL (100 µg/mL for the microassay) in distilled water is used as a stock solution. This should be stored frozen.

3. Plastic and glassware used in the assay should be absolutely clean and detergent-free. Quartz (silica) spectrophotometer cuvets should not be used, since the dye binds to this material. Traces of dye bound to glassware or plastic can be removed with ethanol or detergent solution.

3. Method

3.1. Standard Method

1. Pipet between 10 and 100 µg of protein in 100 µL total volume into a test tube. If the approximate sample concentration is unknown, assay a range of dilutions (1, 1/10, 1/100, 1/1000). Duplicate each sample.

2. For the calibration curve, pipet duplicate volumes of 10, 20, 40, 70, and 100 µL of 1 mg/mL ovalbumin stock solution into test tubes, and make each up to 100 µL with distilled water. Pipet 100 µL of distilled water into a further tube for the reagent blank.

3. Add 5 mL of protein reagent to each tube and mix well by inversion or gentle vortexing. Avoid foaming, which will lead to poor reproducibility.

4.   Measure the A_{595} of the samples and standards against the reagent blank after at least 2 min and within 1 h of mixing. The 100 µg standard should give an A_{595} value of about 0.2. Since the calibration curve is not linear, it should be determined and plotted for each set of assays.

3.2. Microassay

This form of the assay is more sensitive to protein (and interfering substances) and is thus useful when the amount of the unknown protein is limited.

1.   Pipet duplicate samples containing between 1 and 10 µg in a total volume of 100 µL into test tubes or Eppendorf tubes. If the approximate sample concentration is unknown, assay a range of dilutions (1, 1/10, 1/100, 1/1000).
2.   For the calibration curve, pipet duplicate volumes of 10, 20, 40, 70, and 100 µL of 100 µg/mL ovalbumin stock into test tubes, and make up to 100 µL. Pipet 100 µL of distilled water into a tube for the reagent blank.
3.   Add 1 mL of protein reagent, mix, and measure the A_{595} as in step 4 of the standard method. The 10 µg standard should give an A_{595} value of about 0.15.

4. Notes

1.   The assay technique described here is subject to variation in sensitivity between individual proteins (*see* Table 1). Valid comparisons can be made

between protein content of solutions of similar composition. For critical determination of the concentration of solutions containing one or a few proteins, the initial use of a second method, e.g., Lowry, to check the relative dye binding capacities of the standard protein and the unknown in the Bradford assay, is recommended. Ovalbumin is a more suitable general standard than the commonly used bovine serum albumin since its dye binding capacity is closer to the mean of those proteins that have been compared (Table 1).

2. A number of modifications have been suggested that reduce the variability between proteins (3,4). These rely either on increasing the dye content of the solution or reducing the acid content. They appear to be most effective when a relatively pure form of the dye is used (3). Suppliers' figures for dye purity may not relate well to Coomassie Blue G250 content (5), so caution is necessary in choosing a suitable dye source for these methods.

3. Although the Bradford assay is relatively free from interference by most commonly used biochemical reagents, a few may cause significant alteration in absorbance (2,6). The materials that are most likely to cause problems in biological extracts are detergents and ampholytes (Table 2). These should be removed from the sample solution, for example, by gel filtration or dialysis. Alternatively, they should be included in the reagent blank and calibration standards at the same concentration as in the sample, since such interference may cause a change in the shape of the calibration curve. The presence of a base in the assay increases absorbance by shifting the equilibrium of the free dye toward the anionic form. This may also occur with concentrated basic

buffers (2). Guanidine hydrochloride and sodium ascorbate compete with the dye for protein, leading to underestimation of the protein content (2).

Table 1

Response of a Range of Proteins in the Bradford Assay[a]

Protein	Relative A_{595}
Cytochrome c	128
Bovine serum albumin	100
Histone H1	10
H2B	102
H4	89
Carbonic anhydrase	83
Ovalbumin	64
Chymotrypsinogen A	50
Lysozyme	40
Trypsin	24
Pepsin	13
RNAse	12
Immunoglobulin G	10
Gelatin	1

[a]For each protein, the A_{595} is expressed relative to that of the same concentration of bovine serum albumin (data from refs. 3 and 7–9).

Table 2

Effects of Common Reagents and Interfering Compounds on the A_{595} in the Bradford Assay[a]

Compound	A_{595}
1M KCl	0.000
2M Tris	0.026
0.1M HEPES	0.010
0.1M EDTA	0.004
1M $(NH_4)_2SO_4$	0.000
99% Glycerol	0.012
1M 2-Mercaptoethanol	0.004
1M Dithiothreitol	0.002
1M Sucrose	0.013
5% Phenol	0.046
1% Triton X-100	0.590
1% Sodium dodecyl sulfate	0.495
10 mM Quercetin	0.676
10 mM Apigenin	0.968
10 mM Flavone	0.052
1% Ampholytes (pH, 3.5–10.0)	0.089

[a]Data from refs. *1*, *2*, and *6*.

References

1. Bradford, M.M. (1976) A rapid and sensitive method for the quantitation of microgram quantities of protein utilizing the principle of protein-dye binding. *Anal. Biochem.* **72**, 248–254.
2. Compton, S.J. and Jones, C.G. (1985) Mechanism of dye response and interference in the Bradford protein assay. *Anal. Biochem.* **151**, 369–374.

3. Reade, S.M. and Northcote, D.H. (1981) Minimization of variation in the response to different proteins of the Coomassie Blue G dye-binding assay for protein. *Anal. Biochem.* **116**, 53–64.

4. Peterson, G.L. (1983) Coomassie Blue Dye Binding Protein Quantitation Method, in *Methods in Enzymology* vol. 91 (Hirs, C.H.W. and Timasheff, S.N., eds.) Academic, New York.

5. Wilson, C.M. (1979) Studies and critique of Amido Black 10B, Coomassie Blue R and Fast Green FCF as stains for proteins after polyacrylamide gel electrophoresis. *Anal. Biochem.* **96**, 263–278.

6. Spector, T. (1978) Refinement of the Coomassie Blue method of protein quantitation. A simple and linear spectrophotometric assay for <0.5 to 50 µg of protein. *Anal. Biochem.* **86**, 142–146.

7. Lea, M.A., Grasso, S.V., Hu, J., and Seidler, N. (1984) Factors affecting the assay of histone H1 and polylysine by binding of Coomassie Blue G. *Anal. Biochem.* **141**, 390–396.

8. Pierce, J. and Suelter, C.H. (1977) An evaluation of the Coomassie Brilliant Blue G-250 dye-binding method for quantitative protein determination. *Anal. Biochem.* **81**, 478–480.

9. Van Kley, H. and Hale, S.M. (1977) Assay for protein by dye binding. *Anal. Biochem.* **81**, 485–487.

Chapter 3

Amino Acid Analysis by Precolumn Derivatization

E. L. V. Harris

1. Introduction

Amino acid analysis is of central importance to many areas of biological and medical research as a tool for characterization and quantitation of both free amino acids present in biological fluids, beverages, or food and proteins/peptides.

Spackman et al., in 1958 (1), described a method for separation of amino acids by ion exchange chromatography and postcolumn derivatization with ninhydrin. Since then improvements in instrumentation and chromatography matrices have increased both the sensitivity and speed of the technique. In recent years techniques using precolumn derivatization prior to separation by reverse-phase HPLC have become widely used as alternatives to the traditional post-column derivatization techniques.

Precolumn derivatization offers several advantages:

- Instrumentation is not dedicated to amino acid analysis;
- increased sensitivity, <10 pmol/amino acid, compared to ≥500 pmol; and
- decreased analysis time, 20–45 min/run compared to ≥1.

Several derivatization reagents have been used, as shown in Table 1.

The most widely used reagent has been o-phthalaldehyde (OPA)/thiol (2,3). This reagent suffers from two disadvantages; the derivatives have a short half-life, and secondary amino acids such as proline are not detected. In order to overcome the instability of the derivatives, autosamplers capable of mixing reagent and sample immediately prior to injection onto the HPLC column are used. Techniques have been described to enable detection of secondary amino acids by oxidation with chloramine T or hypochlorite prior to derivatization with OPA (4,5); these are not widely used, however, because of technical difficulties.

Phenylisothiocyanate (PITC) (6,7) and 9-fluorenylmethyl chloroformate (FMOC) (8) as precolumn derivatization reagents are gaining popularity, since they both allow for determination of secondary amino acids. Automation of derivatization by either of these reagents is not easy, since drying *in vacuo* or solvent extraction is required to remove excess reagent and byproducts. Waters Associates however, sell instrumentation, reagents, solvents, and column for PITC derivatization

Table 1

Comparison of Properties of Amino Acid Derivatives Suitable for Amino Acid Analysis by Precolumn Derivatization

Derivatization reagent	o-Phthalaldehyde (OPA)	Phenylisothiocyanate (PITC)	9-Fluoroenylmethyl chloroformate (FMOC)	Dansyl or dabsyl chloride
Detection of secondary amino acids	No	Yes	Yes	Yes
Stability of derivatives	Poor <10 min	Moderate 1–2 d	Good	Moderate 1–2 d
Detection limits	0.5 pmol	1 pmol	0.5 pmol	10 pmol (dansyl) 0.5 pmol (dabsyl)
Detection method	Fluorescence	UV absorption	Fluorescence[a] or UV absorption	Fluorescence[a] or UV absorption
Interference by excess reagent or byproducts	No	Yes	Yes	Yes
Method for removing excess reagent or byproducts	—	Drying *in vacuo*	Extraction with pentane	None
Ease of automation	Good	Poor/moderate	Moderate	Poor

[a]Fluorescence is more sensitive than absorption.

under the name Pico-Tag™, and Varian Associates sell a package for FMOC derivatization under the name Amino-Tag™. Both of these packages all for automation of amino acid analysis.

In this chapter the techniques using OPA derivatization and PITC derivatization are described. Figure 1 shows the reactions involved in the two techniques.

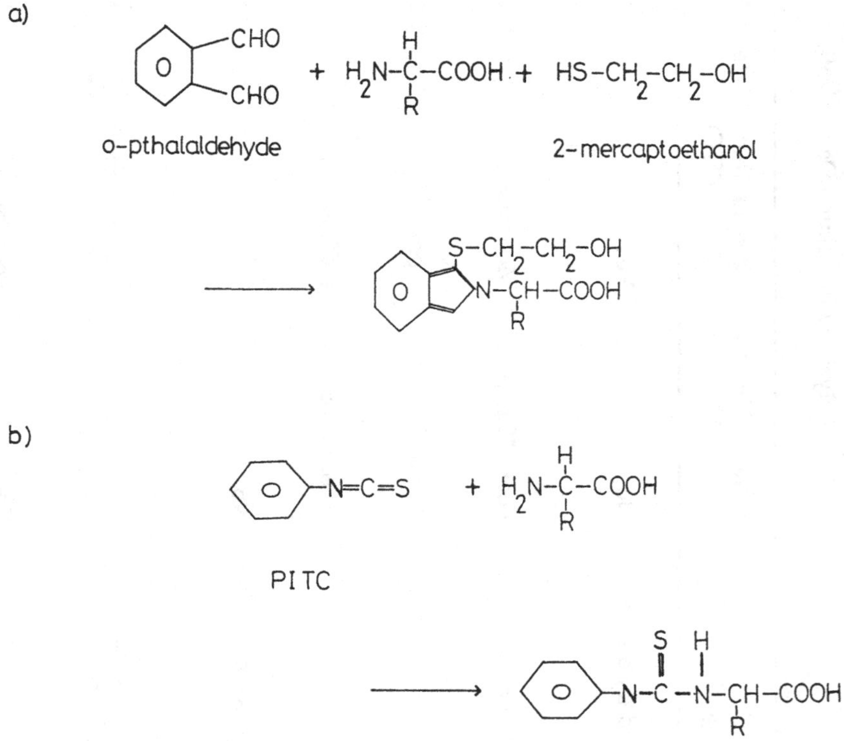

a)

o-pthalaldehyde   2-mercaptoethanol

b)

PITC

Fig. 1. (a) Reaction of *o*-phthalaldehyde (OPA) and a thiol (2-mercaptoethanol) with an amino acid. (b) Reaction of phenylisothiocyanate (PITC) with an amino acid to yield the phenylthiocarbamoyl (PTC) derivative.

2. Materials

2.1. Hydrolysis

1. Constant boiling hydrochloric acid (e.g., Pierce Chemicals) and crystalline phenol.
2. Hydrolysis tubes: Soda glass tubes (about 5 x 0.5 cm) previously treated by baking at 450°C to destroy contaminating proteins/amino acids. Oxygen/butane flame to seal these tubes and diamond tip pen for opening them after hydrolysis.
3. 105°C oven.
4. Freeze-drier or centrifugal concentrator (e.g., Savant Speedvac concentrator) and vacuum pump (e.g., Edwards, to give a vacuum of 50 mtorr; water pumps are not suitable).
5. Amino acid standard. Dissolve the following amino acids together at 5 µM each in HPLC grade water: aspartic acid, glutamic acid, serine, glycine, histidine, arginine, threonine, alanine, proline, tyrosine, valine, methionine, cysteine, isoleucine, leucine, phenylalanine, and lysine.

2.2. PITC Derivatization

1. Small volume, pointed bottom vials with caps, e.g., 0.5 mL polypropylene centrifuge vials from Sarstedt.
2. Ethanol:water:triethylamine (TEA) (2:2:1) (solution 1). Ethanol, TEA, and water must be HPLC grade.
3. Ethanol:TEA:water:phenylisothiocyantate (PITC) (7:1:1:1) (derivatization reagent). This reagent should be made fresh daily. High-purity PITC

should be used and stored at –20°C to prevent formation of breakdown products.

4. HPLC equipment to run a binary solvent gradient including column-heater, autosampler, detector to monitor for absorbance at 254 nm, and an integrator.

5. C18 reverse-phase column with 3 or 4 µm packing (e.g., Millipore Novapak (15 cm x 4 mm id) or LKB 3 µm SuperPac (10 cm x 4 mm id)).

6. 0.14*M* Sodium acetate, 0.05% TEA titrated to pH 6.35 with glacial acetic acid (buffer A).

7. 60% Acetonitrile in water (buffer B).

8. HPLC grade acetonitrile and water should be used for both buffers A and B. Before use, filter both buffers through 0.2 µm filters (cellulose acetate and nitrate filters for buffer A and polyvinyldifluoride filters for buffer B) and degas by bubbling helium through for 5 min.

2.3. OPA Derivatization

1. 400 m*M* Sodium borate buffer. Dissolve 2.47 g of boric acid in HPLC grade water, adjust the pH to 9.5 with sodium hydroxide, and adjust the volume to 100 mL with water.

2. OPA/MCE reagent. Dissolve 50 mg of OPA (e.g., Sepramar grade, BDH Chemicals Ltd.) in 1 mL of methanol (HPLC grade) and dilute to 10 mL with sodium borate buffer. Add 40 µL of 2-mercaptoethanol. Store this reagent in the dark at room temperature for up to 1 wk, adding an additional 4 µL of 2-mercaptoethanol every 3 d. Dilute this reagent 1-in-10 in sodium borate buffer prior to use.

3. Iodoacetic acid reagent. Dissolve 0.74 g of iodoacetic acid and 0.62 g of boric acid in approximately

50 mL of HPLC grade water. Adjust the pH to 9.5 with sodium hydroxide and make up to 100 mL with water.

4. HPLC equipment to run a binary solvent gradient, including autosampler, fluorescence detector with excitation wavelength set at 335 nm and emission wavelength at 455 nm, and an integrator.

5. C-8 or C-18 reverse-phase HPLC column with 5 μm or smaller packing (e.g., Rainin Microsorb, 15 cm x 4.5 mm id, or LKB SuperPac 3 μm, 10 cm x 4.5 mm id).

6. 54.3 mM propionic acid/76 mM disodium hydrogen phosphate adjusted to pH 6.5 with sodium hydroxide. Buffer A is made by mixing the propionate/phosphate buffer with acetonitrile in the ratio 1840:160 (v/v), respectively.

7. Buffer B is water:methanol:acetonitrile:dimethyl sulfoxide (840:600:500:60, v/v/v/v). HPLC grade solvents and water should be used for both buffers A and B. Prior to use, filter both buffers through 0.2 μm filters and degas by bubbling with helium for 5 min.

3. Methods

3.1. Hydrolysis

1. Place the protein samples to be analyzed (0.5–10 μg) in hydrolysis tubes and dry down in the centrifugal concentrator or by freeze-drying.

2. When the samples are dry, add 100 μL of constant boiling HCL and a small crystal of phenol (to give a final concentration of approximately 1%).

3. Seal the tubes by heating in the oxygen/butane flame about 1–1.5 cm from the open end of the tube and twist and draw out the end until sealed.
4. Place in a 105°C oven for 24 h.
5. Crack open the tubes be prescoring with the diamond tip pen.
6. Dry down the samples *in vacuo*, preferably in a centrifugal concentrator to prevent "bumping."

3.2. PITC Derivatization

1. For standards, dry 10 µL of the amino acid mix into a vial.
2. Add 10–20 µL of solution 1 to each of the samples and standards, transfer the samples to vials, and dry *in vacuo*. This step ensures that the pH of the samples is raised to the optimum for derivatization.
3. Add 20 µL of derivatization reagent, cap the vials, and incubate for 20 min at room temperature.
4. Uncap the vials and thoroughly dry the samples and standards *in vacuo*. This should preferably be done overnight to ensure complete removal of excess reagent.

3.3. PTC Separation

1. Redissolve the derivatized samples and standards in a suitable volume of buffer A to allow injection of 5–50 µL onto the HPLC column.
2. The HPLC column should be equilibrated in 90% buffer A:10% buffer B at a flow rate of 1 mL/min at 35°C.
3. The derivatized amino acids are eluted from the HPLC column by a gradient of increasing concen-

tration of buffer B. The exact gradient required will differ from column type to column type; the gradient used with the LKB Super Pac column is shown in Fig. 2.

4. The eluate is monitored for absorbance at 254 nm, and the peaks integrated and quantified by comparison to amino acid standards.

3.4. OPA Derivatization and Separation of Derivatives

1. For standards, dilute the amino acid standard 1:10 in HPLC grade water. Redissolve hydrolysates in 50 µL of HPLC grade water.

2. Equilibrate the HPLC column in buffer A at a flow rate of 1.5 mL/min for 5 µm packing, or 1 mL/min for 3 µm packing.

3. (a) *Manual derivatization.* Mix 20 µL of sample or standard with 90 µL of iodoacetic acid reagent. Add 90 µL of OPA/MCE reagent, mix, and inject onto the column exactly 2 min after addition of the OPA/MCE reagent, or

   (b) *Automated derivatization.* Program the autosampler to mix 20 µL of sample or standard with 90 µL of iodoacetic acid reagent followed by 90 µL of OPA/MCE reagent. The mixing may be done in a separate vial by bubbling air through the solution or may be mixed in-line by passing through a small column of glass beads. The iodoacetic acid reagent and OPA/MCE reagent may be taken from the same vials for a series of samples and standards.

4. The OPA derivatives are eluted by a gradient of buffer B. The exact gradient will vary from column type to column type; that used with the Rainin Microsorb is shown in Fig. 3.

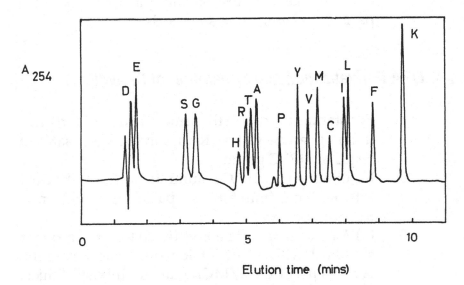

A 254

Elution time (mins)

Fig. 2. Amino acid analysis of a standard mix of amino acids using precolumn derivatization with PITC. The column used is a SuperPac 3μm (10 cm x 4 mm id) at a flow rate of 1 mL/min at 35°C. The composition of the buffers is described in the text. The gradient used was a linear increase from 10% B to 53% B over 10 min, followed by 53% B to 100% B over the next 2 min. The column was re-equilibrated to initial conditions (i.e., 10% B) for 5 min prior to injection of the next sample. The eluate was monitored for absorbance at 254 nm. The amount injected was 5 pmol of each amino acid. The one-letter amino acid code has been used to label the peaks (D, aspartic acid; E, glutamic acid; S, serine; G, glycine; H, histidine; R, arginine; T, threonine; A, alanine; P, proline; Y, tyrosine; V, valine; M, methionine; C, cysteine; I, isoleucine; L, leucine; F, phenylalanine; K, lysine).

5. The eluate is monitored for fluorescence with an excitation wavelength of 335 nm and an emission wavelength of 455 nm. Amino acid peaks are integrated and quantified by comparison with an amino acid standard.

4. Notes

1. In order to achieve maximum sensitivity, all reagents and solvents should be of the highest grade available and all tubes and vials should be scrupulously clean (glass vials should preferably be baked at 450°C, polypropylene tubes can be used without further treatment provided they are kept free from dust). Plastic, talc-free gloves should also be worn to avoid contamination from proteins/amino acids on the skin. Although detection limits of 0.5 pmol, and less, can be achieved, analysis at these levels is not routine because of the levels of amino acids present in reagents and buffers.

2. Samples for amino acid analysis should be free of salt, ammonia, detergents, and buffers containing primary amines such as glycine or Tris. Thus the final purification step prior to analysis should be dialysis or chromatography [e.g., reverse-phase or gel-permeation (*see* Chapters 2 and 5 in vol. 1 of this series)] into water or a volatile buffer (e.g., 0.1% trifluoroacetic acid in water with a gradient of acetonitrile for reverse-phase chromatography or 10 mM ammonium bicarbonate for dialysis or gel-permeation chromatography). The sample is then dried by freeze-drying or drying *in vacuo* [if ammonium bicarbonate is used as the buffer, the sample should be redissolved in water after drying and

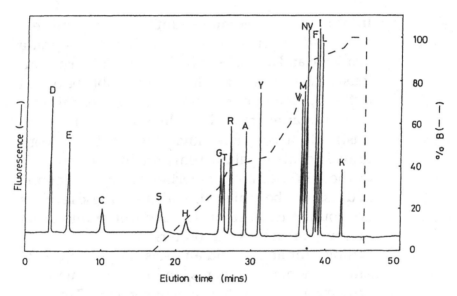

Fig. 3. Amino acid analysis of a standard mix of amino acids using precolumn derivatization with OPA and thiol reagent. The column used is a Rainin Microsorb 5-µm (15 cm x 4 mm id) at a flow rate of 1.5 mL/min at ambient temperature. The composition of the buffers is described in the text, and the gradient of buffer B used is shown in the diagram. The eluate was monitored for fluorescence with an excitation wavelength of 335 nm and an emission wavelength of 455 nm. The amount of each amino acid injected was 10 pmol. The standard one-letter code is used to label the peaks (*see* legend for Fig. 2). NV = norvaline and is included as an internal standard.

then dried a second time to ensure removal of ammonia (ammonium bicarbonate breaks down into ammonia, carbon dioxide, and water *in vacuo*)].

3. Hydrolysis in constant boiling HCl destroys tryptophan and converts asparagine and glutamine to their respective amino acids. Cystine, cysteine, methionine, and tyrosine are partially oxidized, though the effect on tyrosine is minimized by the addition of phenol. Quantification of cysteine is most accurate if the sample is carboxymethylated with iodoacetamide prior to hydrolysis (for method, *see* Chapter 5 in vol. 1 of this series). Hydrolysis for 24-hr results in destruction of approximately 10% of both threonine and serine, and in some cases only partial hydrolysis of bonds between two hydrophobic amino acids (e.g., valine–valine). Thus for accurate quantification of these amino acids the sample is hydrolyzed for different times and the values extrapolated back to zero time to give values for threonine and serine and to infinite time for the hydrophobic amino acids.

4. For analysis of free amino acids in biological fluids, beverages, or food, hydrolysis prior to derivatization is not required. It is preferable, however, to precipitate any proteins present; a suitable protocol is addition of 9 vol of freshly prepared acetontrile containing 0.2% 2-mercaptoethanol. For analysis by PITC derivatization, a longer shallower gradient may be required to separate all the amino acids that occur in biological fluids. The gradient given in Fig. 3 for separation of OPA derivatives should resolve all the amino acids found in biological fluids.

5. Separation of all the amino acid derivatives may require individual tailoring of the eluant buffers to achieve maximum resolution on different columns

and different HPLC systems. This can often be achieved by alteration of the gradient of organic solvent (i.e., decreasing the slope at a particular point in the separation in order to achieve better resolution). Alternatively the ionic strength or pH of buffer A may be altered. Alteration of the ionic strength by up to 10% will effect the separation of arginine or histidine from the other amino acids. The separation of the acidic amino acids can be altered by increasing or decreasing the pH in 0.05-unit steps.

6. The PTC derivatives are stable for several weeks if stored dry at –20°C or for 3 d in solution at 4°C. At room temperature in solution the response obtained for cysteine, valine, and isoleucine decreases by 5% over 6 h; for the derivatives of the other amino acids, the response falls by 5% over 10 h.

7. The PTC derivatives of the standard amino acids obtained from protein hydrolysates are eluted in the first 12 min of the HPLC separation. Peaks will be observed eluting later than this; these peaks are caused by the reagents and do not interfere with the analysis of the amino acids.

References

1. Spackman, D. H., Stein, W. H., and Moore, S. (1958). Chromatography of amino acids on sulphonated polystyrene resins. An improved system. *Anal. Chem.* **30**, 1185–1189.

2. Lindroth, P. and Mopper, K. (1979) High performance liquid chromatographic determination of subpicomole amounts of amino acids by precolumn fluorescence derivatization with *o*-phthalaldehyde. *Anal. Chem.* **51**, 1667–1674.

3. Turnell, D. C. and Cooper, J. D. M. (1982) Rapid assay for amino acids in serum or urine by pre-column derivatization

and reverse-phase liquid chromatography. *Clin. Chem.* **28**, 527–531.

4. Cooper, J. D. H., Lewis, M. T., and Turnell, D. G.(1984) Precolumn o-phthalaldehyde derivatization of amino acids and separation using reverse-phase high performance liquid chromatography.   1. Detection of the imino acids, hydroxyproline and proline. *J. Chromatogr.* **285**, 484–489.

5. Bohlen, P. (1983) Analysis for Imino Acids with o-Phthalaldehyde, in *Methods in Enzymology* vol. 91 (Hirs, C. H. W. and Timasheff, S. N., ds). Academic, New York, London.

6. Koop, D. R., Morgan, E. T., Tarr, G. E., and Coon, M. J. (1982) Purification and characterization of a unique isozyme of cytochrome P-450 from liver microsomes of ethanol-treated rabbits. *J. Biol. Chem.* **257**, 8472–8480.

7. Bidlingmeyer, B. A., Cohen, S. A., and Tarvin, T. L. (1984) Rapid analysis of amino acids using pre-column derivatization. *J. Chromatogr.* **336**, 93–104.

8. Einarsson, S., Josefsson, B., and Lagerkvist, S. (1983) Determination of amino acids with 9-fluorenylmethyl chloroformate and reversed-phase high performance liquid chromatography. *J. Chromatogr.* **282**, 609–618.

Chapter 4

Identification of N-Terminal Amino Acids by High-Performance Liquid Chromatography

E. L. V. Harris

1. Introduction

1-Dimethylaminonaphthalene-5-sulfonyl chloride (dansyl chloride, Dns-Cl) has been widely used in protein chemistry for determination of the N-terminal amino acid of peptides. The dansyl chloride reacts with both α- and Σ-amino groups (Fig. 1). After derivatization the peptide is acid hydrolyzed releasing free amino acids, amino acids such as lysine and histidine with the dansyl group attached to their side chains, together with the derivatized N-terminal amino acid. Traditionally

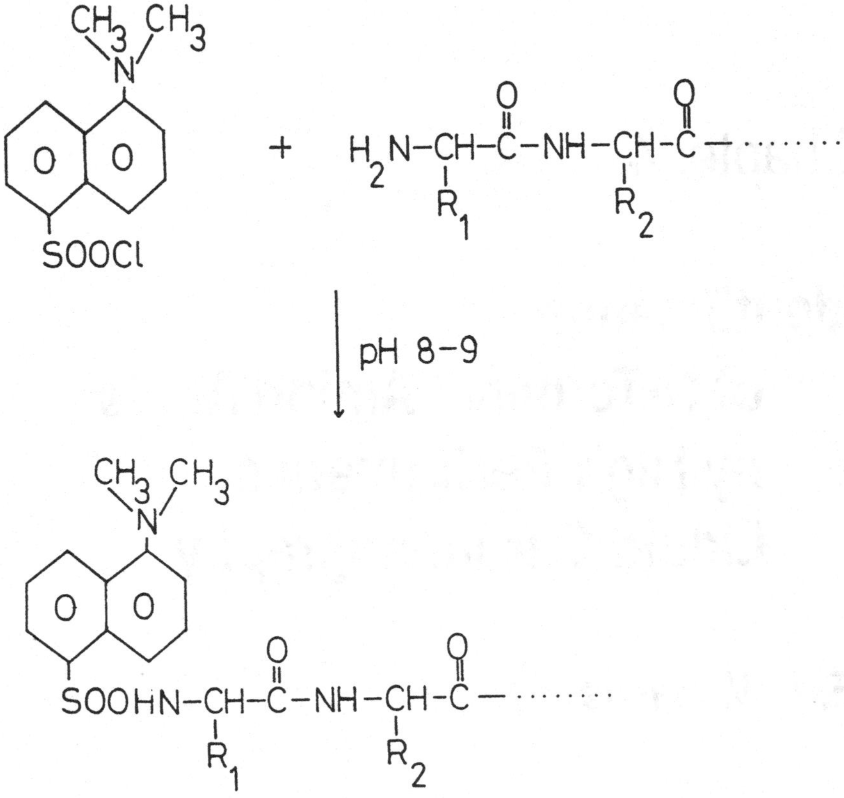

Fig. 1. Reaction of dansyl chloride with the amino terminus of a peptide.

the Dns-amino acids have been separated by thin layer chromatography and viewed by fluorescence in UV light (Chapter 23 in vol. 1 of this series). This procedure is time-consuming, however, and is not easily quantified. More recently, methods have been developed for separation of the Dns-amino acids by reverse-phase HPLC (1,2), thus allowing for automation of the separation and quantitation of the derivatives. In order to minimize the interference by reagent byproducts, the reaction conditions have been modified from those of the traditional method.

2. Materials

1. 2.3 mg/mL of Dansyl chloride in acetonitrile are prepared immediately prior to use from a 100 mg/mL stock in acetone. 100 mg/mL solutions of dansyl chloride in acetone are commercially available (e.g., Pierce Chemicals) and should be stored at 4°C in the dark. Care should be taken when removing aliquots from the stock not to introduce water that hydrolyzes the reagent.
2. 40 mM Lithium carbonate, pH 9.5.
3. 0.1% Ethylammonium chloride.
4. Constant boiling hydrochloric acid (e.g., from Pierce Chemicals).
5. HPLC grade methanol (water-free).
6. Soda glass tubes (approximately 5 x 0.5 cm) with pointed bottoms pretreated by baking at 450°C for 4 h to destroy contaminating amino acids and proteins.
7. Oxygen/butane flame for sealing hydrolysis tubes and a diamond tip for opening them.
8. 105°C Oven.
9. Centrifugal concentrator (e. g., Savant Speedvac) and vacuum pump (e. g., Edwards).
10. HPLC equipment to run a binary solvent gradient with column heater, autosampler, detector for absorption at 254 nm, and integrator. For more sensitive determination, the UV monitor should be substituted with a fluorescence detector with the excitation wavelength at 330 nm and the emission wavelength at 470 nm.
11. Buffer A: 25 mM Trifluoroacetic acid in HPLC-grade water adjusted to pH 7.6 with sodium hydroxide. This sodium trifluoroacetate solution is mixed with HPLC-grade acetonitrile in a ratio of 90:10 (v/v).

12. Buffer B: 30/70 (v/v) sodium trifluoroacetate pH 7.6/acetonitrile. Prior to use, filter both buffers A and B through 0.2-µm filters and degas by bubbling with helium for 5 min.

13. C-18 reverse-phase HPLC column with 5-µm, or smaller, packing (e. g., Ultrasphere ODS column, 25 cm x 4.6 mm id, or LKB SuperPac 3-µm column, 10 cm x 4.6 mm id).

14. Amino acid standard made by dissolving each of the following amino acids at 20 µM in HPLC grade water: aspartic acid, glutamic acid, cysteine, serine, threonine, glycine, alanine, proline, arginine, valine, methionine, leucine, isoleucine, phenylalanine, lysine, histidine, and tyrosine.

3. Methods

3.1. Derivatization and Hydrolysis

1. For standards, pipet 5 µL of the standard mix into a tube. For samples, pipet 5–10 µL of a solution containing 100–200 pmol of peptide into a tube.

2. Dry the standards and samples in the centrifugal concentrator, or if not available, just under vacuum.

3. Redissolve in 5 µL of HPLC grade water and redry. Repeat this procedure once more.

4. Add 4 µL of lithium carbonate solution and 2 µL of a suitable dilution of the 8.7 mM Dns-Cl in acetonitrile to give a 5–10-fold excess of reagent over amino groups (i. e., for the standards, the reagent is not diluted, whereas for 200 pmol of peptide, the reagent is diluted 1 in 10).

5. Centrifuge briefly to ensure that reagents are at the bottom of tube.
6. Cover the tubes with parafilm to minimize evaporation, and incubate at approximately 20°C for 35 min in the dark.
7. Add 2 µL of ethylammonium chloride solution, and centrifuge briefly.
8. After 1 min incubation, dry in the centrifugal concentrator.
9. Add 50 µL of HCl and seal the tube in a oxygen/butane flame by twisting and drawing out when hot.
10. Incubate at 105°C for 8 h. Crack open the tubes by prescoring with the diamond tip pen, and dry off the HCl under vacuum.

3.2. Separation of Derivatives

1. Equilibrate the HPLC column at 39°C in 93% buffer A/7% buffer B at a flow rate of 1.2 mL/min for 5 µm packing or 1.0 mL/min for 3 µm packings.
2. Add 4 µL of dry methanol to each of the sample and standard tubes and dry in the centrifugal concentrator. Repeat this procedure once more. This step ensures removal of all volatile contaminants.
3. Redissolve the samples and standards in a suitable volume to enable injection of 5 µL of solution containing 50–200 pmol of Dns-amino acid(s).
4. The derivatives are eluted with a gradient of buffer B. The exact gradient required will vary from column type to column type; that used for the LKB SuperPac column is shown in Fig. 2.
5. The eluate is monitored for absorbance at 254 nm, or for more sensitive detection for fluorescence

with an excitation wavelength of 330 nm and an emission wavelength of 470 nm. Amino acid peaks are integrated and quantified by comparison with standards.

4. Notes

1. Samples for N-terminal determination should contain no salts, ammonia, or buffers with primary amines such as glycine or Tris. Thus, a preferred final purification step for peptides destined for N-terminal analysis is reverse-phase chromatography in a volatile buffer system such as 0.1% trifluoroacetic acid in water with a gradient of acetonitrile (*see* Chapter 5 in vol. 1 of this series).
2. The procedure described here is suitable for use with peptides of up to 30 amino acids.
3. Hydrolysis in HCl destroys tryptophan and converts asparagine and glutamine to their respective acids. Thus peptides with an N-terminal tryptophan do not yield a Dns derivative, whereas those with an N-terminal asparagine or glutamine will yield Dns-aspartic acid or Dns-glutamic acid, respectively. Several Dns-amino acids are partially destroyed by hydrolysis, e.g., methionine and proline; therefore, for better quantification the amounts obtained should be corrected for losses by comparison to a hydrolyzed Dns-amino acid standard mix. A 4–10 h hydrolysis time will be sufficient for most peptides. The time required for hydrolysis can be shortened further by use of a mix of 5.7M HCl and trifluoroacetic acid (2/1 v/v) at 166°C. Under these conditions, the optimum time of hydrolysis is 50 min.

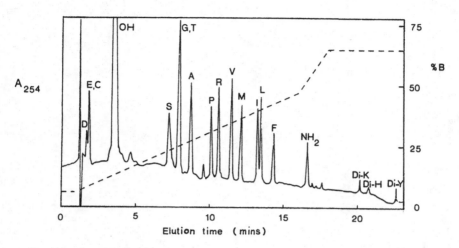

Fig. 2. Separation of a standard mix of amino acids derivatized with dansyl chloride prior to injection onto the HPLC column (LKB SuperPac 3 μm) (10 cm x 4 mm id). The flow rate was 1.0 mL/min with a column temperature of 39°C. The compositions of the buffers used are described in the text, and the gradient of buffer B used to elute the derivatives is shown in the figure. The eluate was monitored for absorbance at 254 nm. 50 Picomoles of each amino acid was injected. The standard one-letter code is used to label the amino acid peaks (*see* Chapter 3, Fig. 2: C, cysteic acid; OH, dansyl sulfonic acid; NH₂, dansyl amide; Di, di-dansyl derivative). For the separation achieved on an Ultrasphere ODS 5 μm column, *see* refs. 1 and 2.

References

1. Levina, N. B. and Nazimov, I. V. (1984) High performance liquid chromatography of Dns-amino acids in the purity control of peptides. *J. Chromatogr.* **286,** 207–216.
2. Nazimov, I. V. and Levina, N. B. (1986) Unified System for High Speed Analysis of Dns- and PTH-Amino Acids, in *Chemistry of Peptides and Proteins* (Voelter, W., Bayer, E., Ovchinnikov, Y. A., and Ivanov, V. T., eds.) Walter de Gruyter, Berlin, New York.

Chapter 5

Enzymatic Methods
for Cleaving Proteins

Bryan John Smith

1. Introduction

There could be various reasons the protein chemist may want to break a protein of interest into fragments, but foremost among them must be the purposes of peptide mapping, primary structure determination, and preparation of particular fragments for studies relating structure and function. Proteinases (or more specifically endo-proteinases) are commonly used to generate peptides for these purposes, but the number of particularly useful and commercially available proteinases is not great. This is because enzyme specificity is an important consideration—the laboratory worker generally requires good yields of clean, pure peptides, but significantly less than 100% cleavage at some or all sites of action gives rise to a complex mixture of a large number of polypeptides.

One of the best characterized and widely used proteinases is trypsin (E.C.3.4.21.4), and it is the use of this enzyme that is described below. Other useful enzymes for cleaving proteins are described in section 4. Trypsin is synthesized as trypsinogen by the vertebrate pancreas, but rapidly becomes converted by the removal of the amino-terminal hexapeptide to the active enzyme of molecular weight approximately 23,500. It has a serine at its active site, and therefore belongs to the serine proteinase family. Trypsin displays good specificity, catalyzing the hydrolysis of the peptide bond to the COOH side of lysyl and arginyl residues. If an acidic residue occurs to either side of the basic residue, hydrolysis is slower, and if the residue to the COOH side is proline, hydrolysis is very slow. Polylysine sequences may also be fully cleaved with difficulty. Trypsin is optimally active at about pH 8.

2. Materials (for Trypsin)

1. Trypsin (E.C.3.4.21.4) is available from various commercial suppliers. It is stable for periods of years as a dry solid at –20°C.
2. Stock solutions:
   a. Trypsin, 1 mg/mL in distilled water. Use fresh or divide into aliquots and store frozen. A stock solution may be thawed and refrozen several times, but for consistent results, thaw only once.
   b. Ammonium bicarbonate (0.4M) in distilled water, pH ~8.5. May be stored refrigerated or frozen for long periods.

3. Method (for Trypsin)

1. Dissolve the substrate in water to 2 mg/mL. Add an equal volume of 0.4*M* ammonium bicarbonate solution. Add trypsin solution to an enzyme/ substrate ratio of 1/50 (w/w), i.e., to 1 mL of 1 mg/ mL substrate solution add 20 μL of 1 mg/mL of trypsin solution.
2. Incubate at 37°C for 24 h.
3. Terminate digestion by immediate submission to peptide mapping or isolation techniques, acidification, freeze-drying, or addition of specific inhibitor (e.g., *N*α-tosyl-L-lysyl chloromethyl ketone, or "TLCK," in molar excess to the trypsin used). As with other serine proteinases, trypsin may be inactivated by reaction of the serine residue at the active site with phenylmethylsulfonyl fluoride (PMSF). Prepare this agent just before use by dissolving in propan-2-ol to a 1*M* solution. Dilute 1000-fold to 1 m*M* in the reaction mixture. It is effective within a few minutes. Beware of the toxic nature of PMSF.

4. Notes

1. Various grades of trypsin are available commercially. Most are from bovine pancreas. That which has been treated with L-1-chloro-3-tosylamido-4-phenylbutan-2-one (or "TPCK"), or similar, is recommended since this treatment specifically inhibits chymotrypsin, which may contaminate trypsin preparations. Enzyme activity may vary, in detail, from source to source or batch to batch. Attention to this is recommended if reproducibility is important (e.g., for peptide mapping).

2. During incubation with substrate, trypsin will also undergo autolysis to produce (among other products) ψ-trypsin, which has chymotrypsin-like activity. This may contribute low-frequency "non-tryptic" cleavages at some tyrosyl, phenylalanyl, or tryptophanyl bonds. The literature claims that low levels of Ca^{2+} (0.1 mM $CaCl_2$) added to digestion buffers may reduce production of ψ-trypsin by autolysis.

3. The buffer described above is a simple volatile buffer that gives an appropriate pH of approximately 8. Other buffers of pH 8 may be readily substituted, e.g., 0.05M Tris-HCl, pH 8.

4. The method described above is the basic procedure and is subject to great change, according to the requirements of the worker and protein in question. Any one set of experimental conditions will give different results with different substrates, and the reader is encouraged to discover optimal conditions for digestion of a substrate empirically.

    For reproducible peptide mapping, when complete digestion is desirable, prolonged digestion may be required (e.g., 48 h at 37°C, pH 8, with a second addition of trypsin, similar to the first at 24 h). On the other hand, for preferential cleavage of particularly sensitive bonds or for production of partial cleavage products (e.g., overlapping peptides for sequencing purposes or production of folded domains), digestion can be limited by use of a low enzyme/substrate ratio, shorter digestion times, and lower incubation temperatures, and/or the buffer may be adjusted to suboptimal pH. High salt conditions (e.g., 0.5M NaCl, in buffer), which favor compact folding of a structured polypeptide chain, may also be used. Such a tightly folded sequence will be more resistant to proteolytic attack than nonstructured regions, though not necessarily completely or indefinitely resistant. Thus, by careful adjustment of buffer and incubation conditions, these partially resistant regions may be prepared and used, for example, in studies on their function.

5. The condition of the substrate is important. First, the substrate should be soluble, or as finely divided as possible, in the digestion buffer. If a sample is not readily soluble in water or ammonium bicarbonate solutions, suitable solvents can be used initially and then adjusted by dilution or titration of pH to allow for trypsin action. If the polypeptide remains insoluble, the precipitate should be kept in suspension by stirring. Thus, 8M urea may be used to solubilize a protein or disrupt a tightly folded structure, and then diluted to 2M urea for digestion by trypsin. Trypsin will also function in 2M guanidinium chloride, in the presence of sodium dodecyl sulfate (SDS, e.g., 0.1% w/v), or in the presence of acetonitrile (up to about 50% v/v), so that fractions from reverse-phase HPLC (acetonitrile gradients in water/trifluoroacetic acid, 0.1% v/v) may be readily digested after simple addition of ammonium bicarbonate or other neutralization, and dilution (if necessary to lower the acetonitrile concentration).

Second, a native protein may be tightly folded such as to markedly slow up or inhibit proteolytic attack. To remedy this, the substrate may be denatured and the structure opened out to allow for access of the proteinase. This may be done by boiling in neutral pH solution, or by use of such agenda as urea, SDS, or organic solvent, as described above. Low concentrations of these agents (e.g., 5–10% v/v acetonitrile) may give more rapid digestion than will a buffer without them, but high concentrations (e.g., 50% v/v acetonitrile) will slow the digestion.

An additional, and very common, technique is reduction and carboxymethylation, i.e., permanent disruption of disulfide bonds. This treatment opens out the protein structure to allow for ready digestion and to minimize complications in peptide separation that are caused by pairs (or greater) of peptides remaining connected by S—S bonds. This treatment is carried out as described in Chapter 7. Alternatively, if a sample has already been digested, S—S bonds may be reduced by simple addition of small amounts of solid dithiothreitol and incubation at room temperature (and pH 8) for 30 min or so. This treatment is followed immediately by HPLC to separate the various peptides. The amount of dithiothreitol required (i.e., slight molar excess over S—S bonds) may be calculated accurately if the cystine content of the protein concerned is known.

6. The result of proteolytic cleavage may be monitored, and the resulting peptides purified, by various forms of electrophoresis and chromatography, as described in other chapters in this series. Typical HPLC conditions might be, for example, reversephase chromatography on a 4 mm id C-18 or C-8 column, using a gradient of 0–100% v/v acetonitrile in water, 0.1% v/v trifluoroacetic acid, at a flow rate of 1 mL/min at room temperature.

7. During digestion, autolysis produces a low background of peptides from the proteinase itself. If this is to prove a problem, enzyme-only controls are recommended. The primary structure of trypsin is known (e.g., *see* the structure of pig trypsinogen

given in ref. *1*), so the products of autolysis should not prove misleading to the protein sequencing fraternity!

8. Lysine is a fairly common constituent of proteins, and digestion with trypsin can generate a large number of peptides of small average size. This is a good point when peptide mapping, but some purposes, such as sequencing, may require longer peptides. The action of trypsin may be modified in an attempt to achieve this. It is done by modification of the side chains of lysyl or arginyl residues as described in ref. *2* and Chapter 7, such that cleavage only occurs at unmodified residues. Perhaps the commonest such method is succinylation of lysyl side chains, leading to tryptic cleavage at arginyl (and any remaining unmodified lysyl) residues (*see* Chapter 7).

    Introduction of additional sites of cleavage by trypsin may be achieved by conversion of cysteinyl residues to aminoethyl cysteinyl residues by reaction with ethyleneimine as described in Chapter 7.

9. The need for substrate modification as described above has been reduced by the commercial advent of proteinases with good specificity for either lysyl or arginyl residues, namely endoproteinases lys C and arg C (*see* Table 1).

    Clostripain is also reported to cleave to the COOH side of arginyl residues, but also may cleave (less frequently) to the COOH side of lysyl residues. It will, however, cleave Arg-Pro bonds.

10. There are a large number of proteinases available commercially. Data on some of the more useful ones are given in Table 1, summarized from personal observation, published data, and suppliers' information. They may be used in much the same

Table 1

Summarized Data on Some Commercially Available Endoproteinases Suitable for the Generation of Peptides from Proteins

Endoproteinase	E.C. No.	Type of proteinase	Apparent molecular weight	pH optimum	Specificity[a]
Chymotrypsin	3.4.21.1	Serine	25,000	7–9	-X↓Y - (X = aromatic)
Clostripain	3.4.22.8	-SH	50,000	7.7	-Arg↓Y- (Lys↓Y-, less frequently)
Elastase	3.4.21.11	Serine	25,900	7–9	-X↓Y- (X = uncharged aliphatic)
Endoproteinase Arg-C ("submaxillary gland protease")	3.4.21—	Serine	30,000	7.5–8.5	-Arg↓Y-; -Lys-Lys↓Y-
Endoproteinase Glu-C	3.4.21.19	Serine	27,000	4 7.8	-Glu↓Y- -Glu↓Y-; -Asp↓Y- (not if Glu within two or three residues of the N-terminus)
Endoproteinase Lys-C (nonreduced) (reduced)	—	Serine	30,000 33,000	8.5–8.8	-Lys↓Y-
Kallikrein (α Form) (β Form)	3.4.21.8	Serine	27,000 28,900	7–8	-X-Arg↓Y- (preferentially if X = Phe, Leu)
Papain	3.4.22.2	-SH	23,000	6–7	-X↓Y- (preferentially if X = Arg, Lys, Gln, His, Gly,Tyr)
Pepsin	3.4.23.1	Acid	34,500	1.8–2.2	-X↓Y- (Y = Hydrophobic; preferentially if X = Phe, met, Leu, Tryp)
Plasmin	3.4.21.7	Serine	85,000	8.9	-Arg↓Y-; -Lys↓Y-
Thermolysin	3.4.24.4	Metallo-(Ca^{2+})	37,500	7–9	-Y↓X- (X = Hydrophobic; also less frequently if Gly, Ser, Thre; Not if -Y-X-Pro-)
Thrombin (α Form) (Factor IIa) (β Form)	3.4.21.5	Serine	36,000 38,000	8.2–9	-Arg↓Y
Trypsin	3.4.21.4	Serine	23,500	8	-Arg↓Y-; -Lys↓Y-; Amino ethylcysteine↓Y-

[a]Y, unspecific. NB, Frequently, susceptibility to proteolysis is reduced or lost if either the potentially labile bond is to proline (e.g., -x-Pro- for chymotrypsin) or is between two like residues (e.g.,-Glu-Glu- for endoproteinase Glu-C).

manner as is trypsin, with due attention paid to buffer (pH) conditions and methods of termination of digestion.

Thus, for pepsin (active in the range pH 1–5), the volatile buffer 5% v/v acetic acid in water may be used. To stop digestion, raise the pH beyond pH 5 and/or freeze dry.

Proteinases such as clostripain containing a sulfydryl group in the active site require activation by inclusion in the buffer of a sulfhydryl agent, such as dithothreitol (1–10 mM) and may be inactivated by addition of iodoacetic acid in molar excess to the sulfhydryl agent present.

Metalloproteinases require the presence of metal ions. Thus, thermolysin requires Ca^{2+}. Preparations of this enzyme, or of a substrate, may themselves contain sufficient metals ions, but otherwise the Ca^{2+} should be added to the buffer to a concentration of at least 5 mM. Reaction may be ended by addition of EDTA in excess to the metal ion concentration (remembering that sample and enzyme may contribute metal also).

Most of the common proteinases are serine proteinases (*see* Table 1) and may be inactivated by reaction with PMSF as described above for trypsin. In addition to these general methods for inhibiting digestion, inhibitors specific for various proteinases are available from commercial sources (*see* suppliers' literature).

11. The specificities of various enzymes are indicated in Table 1, but in addition, two common (but not generally applicable) rules may be applied:

a. Bonds involving prolyl residues may not be cleaved, e.g., endoproteinases Glu-C will commonly not cleave Glu-Pro bonds.

b.   Potentially labile bonds in a multiple sequence
may not each be cleaved in good yield, e.g., Glu-
Glu-X for endoproteinase Glu-C.

12.   One of the most useful proteinases is endoproteinase Glu-C (or "V8 protease") isolated from *Staphylococcus aureus* strain V8, since like trypsin and
endoproteinases Lys-C and Arg-C, it shows good
specificity. It cuts to the COOH side of glutamyl
residue. A lower frequency of cleavage to the
COOH side of aspartyl residues may also occur at
neutral pH, although at pH 4 this may not occur.
Endoproteinase Glu-C functions in buffers containing 0.2% w/v SDS or 4M urea. Its sequence is
known (3).

13.   Other proteinases are of broader specificity and
may be affected by surrounding sequences. Their
action is therefore difficult to predict. In particular instances, however, their observed action may
prove beneficial—perhaps cleaving at one or a few
particularly sensitive sites when incubated in
suboptimal conditions (e.g., short duration digestion or nondenatured substrate).

Good examples of this come from work on
preparation of F(ab)$_2$, antigen-binding fragments of
immunoglobulin IgG, that are bivalent and lack the
constant, Fc region of the molecule. Incubation of
nondenatured IgG molecules with a proteinase of
broad specificity can lead to proteolytic cleavage at
a few sites or a single site in good yield. Pepsin has
been used for this purpose [e.g., pH 4.2–4.5; en-
zyme/substrate, 1/33 w/w; 37°C (4)]. Different
subclasses of mouse IgG were found to be digested
at different rates, however, in the order IgG3 >
IgG2a > IgG1. Different antibodies of the same

subclass may also be degraded differently, some rapidly and without formation of F(ab')$_2$(4). Papain has been used to prepare F(ab)$_2$ fragments from the IgG1 subclass, which is the subclass that is most resistant to pepsin. The method described by Parham et al. (5) uses papain (which has been activated just before use by reaction with cysteine), at pH 5.5 (0.1M acetate, 3 mM EDTA) 37°C, with an IgG concentration of about 10 mg/mL. The enzyme is added at time 0, and again later (e.g., at 9 h) to an enzyme/substrate ratio of 1/20 w/w. Digestion can be halted by addition of iodoacetamide (30 mM) (5). Rousseaux et al. (6) also describe conditions for generating rat F(ab')$_2$, using papain (in the presence of 10 mM cysteine), pepsin, or endoproteinase Glu-C (V8 protease). Incubation of the IgGl and 2a subclasses at pH 2.8 prior to digestion with pepsin improved the yields of F(ab')$_2$ fragments, presumably because the proteins thus denatured were effectively better substrates.

14. Two enzymes of broader specificities are worthy of further mention. The first is thermolysin, for its good thermostability, which may prove useful when keeping awkward substrates in solution. Thermolysin remains active at 80°C or in 8 M urea. The second enzyme is pepsin, which acts at low pH. Disulfide bonds rearrange less frequently in acid than in alkaline conditions, so use of low pH buffers may not only help solubilize a substrate, but may also help preserve naturally disulfide-bonded pairs of peptides. Endoproteinase Glu-C may also be used at low pH, having an optimum at pH 4.

15. For more information on the various proteinases available, the reader is guided to works such as

those by Allen (7), Wilkinson (8) (and references therein), as well as to the commercial suppliers of the enzymes.

16. Exoproteinases are not discussed here, since their use generally does not generate clean peptides. As an example of use of an exoproteinase, the reader is directed to the description of carboxypeptidase Y in vol. 1 of this series (9).

Acknowledgments

I would like to thank my colleagues for helpful discussions.

References

1. Dayoff, M.O. (1976) *Atlas of Protein Sequence and Structure,* vol. 5, suppl. 2. National Biomedical Research Foundation, Washington, DC.
2. Glazer, A.N., Delange, R.J., and Sigman, D.S. (1975) *Chemical Modification of Proteins.* Elsevier/North Holland, Amsterdam.
3. Drapeau, G. (1978) The primary structure of staphylococcal protease. *Can. J. Biochem.* **56,** 534–544.
4. Lamoyi, E. and Nisonoff A. (1983) Preparation of F(ab')₂ Fragments form Mouse IgG of Various Subclasses. *J. Immunol. Meth.* **56,** 235–243.
5. Parham, P., Androlewicz, M.J., Brodsky, F.M., Holmes, N.J., and Ways, J.P. (1982) monoclonal antibodies: Purification, fragmentation and application to structural and functional studies of class I MHC antigen. *J. Immunol. Meth.* **53,** 133–173.

6. Rousseaux, J., Rousseaux-Prevost, R., and Bazin, H. (1983) Optimal conditions for the preparation of Fab and F(ab')₂ fragments from monoclonal IgG of different rat IgG subclasses. *J. Immunol. Meth.* **64**, 141–146.

7. Allen, G. (1981) *Sequencing of Proteins and Peptides.* Elsevier/North Holland, Amsterdam

8. Wilkinson, J.M. (1986) Fragmentation of Polypeptides by Enzymic Methods, in *Practical Protein Chemistry—A Handbook* (Darbe A., ed.). Wiley, Chichester.

9. Klemm, P. (1984) Carboxy-Terminal Sequence Determination of Proteins and Peptides with Carboxy Peptidase Y, in *Methods in Molecular Biology*, vol. 1, *Proteins*, (Walker, J.M., ed.). Humana, Clifton, New Jersey.

Chapter 6

Chemical Cleavage of Proteins

Bryan John Smith

1. Introduction

There is a variety of chemical reactions known to result in the cleavage of the peptide bond. Some are nonspecific—for instance, $6M$ hydrochloric acid at 110°C for 24 h hydrolyzes a polypeptide to a mixture of single amino acids. Others show some discrimination, however, as to the precise nature of the amino acid residues around the bond to be broken. Some of these methods are sufficiently specific to be of use, for instance, for generating peptides for primary structure determination. These methods usefully augment those that use proteolytic enzymes (*see* Chapter 5), especially since they tend to act at positions occupied by less common amino acids and so generate large peptides.

The effeciency of the cleavage reaction is an important factor when considering which method to use for generating peptides from proteins. It is an unfortunate aspect of chemical methods of proteolysis that, in the main, they afford significantly less than 100% yields and commonly exhibit undesirable side reactions. One of the best current methods is cleavage of the bond to the COOH-side of methionyl residues by treatment with cyanogen bromide; yields are 90–100% with few side reactions. It is a well-established and well-used method. A newer technique for cleavage of the bond to the COOH side of tryptophanyl residues also involves treatment with cyanogen bromide, after an oxidation step using dimethylsulfoxide in acid. This method too is highly specific and efficient. Most other methods are less efficient, but still may be of some use for cleavage of small substrates in which problems of poor yields and side reactions are minimized. One such method is limited hydrolysis in dilute acid. This method, together with the two mentioned above, has the advantage of requiring reagents that are stable and pure enough as supplied commercially. The methods are also relatively simple and adaptable to large- and small-scale cleavages. These three methods are described below in some detail.

2. Materials

2.1. Cleavage at Met-X Bonds

1. Ammonium bicarbonate (0.4M) solution in distilled water. Stable for weeks in refrigerated stoppered bottle.

2. 2-Mercaptoethanol. Stable for months in dark, stoppered, refrigerated bottle.
3. Formic acid, minimum assay 98%. Aristar grade.
4. Cyanogen bromide. Stable for months in dry, dark, refrigerated storage. Warm to room temperature before opening. Use only white crystals, not yellow ones. *Beware of the toxic nature of this reagent.*
5. Sodium hypochlorite solution (domestic bleach).
6. Equipment includes a nitrogen supply, fume hood, and suitably sized and capped tubes (e.g., Eppendorf microcentrifuge tubes).

2.2. Cleavage at Tryp-X Bonds

1. Oxidizing solution:
   Glacial acetic acid, 30 vol
   9M HCl, 15 vol
   Dimethylsulfoxide, 4 vol
   Use best grade reagents. Though each of the constituents is stable separately, mix and use the oxidizing solution when fresh.
2. Ammonium hydroxide (15M).
3. Cyanogen bromide solution in formic acid (60%):
   Formic acid (minimum assay 98%, Aristar grade) 6 mL. Make to 10 mL with distilled water. Use fresh. Add cyanogen bromide to a concentration of 0.3 g/mL. Use fresh.
   Store cyanogen bromide refrigerated in the dry and dark, where it is stable for months. Use only white crystals. *Beware of the toxic nature of this reagent.*
4. Sodium hypochlorite solution (domestic bleach).
5. Equipment includes a fume hood and suitably sized capped tubes (e.g., Eppendorf microcentrifuge tubes).

2.3. Cleavage of Asp-X Bonds

1. Dilute hydrochloric acid (approximately 0.013*M*) pH 2 ± 0.04: Dilute constant boiling (6*M*) HCl (220 μL) to 100 mL with distilled water.
2. Pyrex glass hydrolysis tubes.
3. Equipment includes a blowtorch suitable for sealing the hydrolysis tubes, a vacuum line, and an oven for incubation of samples at 108°C.

3. Method

3.1. Cleavage at Met-X Bonds

1. Reduction
a. Dissolve the polypeptide in water to between 1 and 5 mg/mL, in a suitable tube. Add 1 vol of ammonium bicarbonate solution, and add 2-mercaptoethanol to between 1 and 5% v/v.
b. Blow nitrogen over the solution to displace oxygen, seal the tube, and incubate at room temperature for approximately 18 h.
2. Cleavage
a. Dry down the sample under vacuum, warming if necessary to help drive off all of the bicarbonate. Any remaining ammonium bicarbonate will form a nonvolatile salt on subsequent reaction with formic acid.
b. Redissolve the dried sample in formic acid (98%) to 1–5 mg/mL. Add water to make the acid 70% v/v finally.
c. Add excess white crystalline cyanogen bromide to the sample solution, to between 20- and 100-fold molar excess over methionyl residues. Practically,

this amounts to approximately equal weights of protein and cyanogen bromide. To very small amounts of protein, add one small crystal of reagent. Carry out this stage in the fume hood.

d. Seal the tube and incubate at room temperature for 24 h.

e. Terminate the reaction by drying down under vacuum. Store samples at –10°C or use immediately.

f. Immediately after use, decontaminate spatulas, tubes, and so on, that have contacted cyanogen bromide, by immersion in hypochlorite solution (bleach) until effervescence stops (a few minutes).

3.2. Cleavage at Tryp-X Bonds

1. Oxidation: Dissolve the sample to approximately 0.5 nmol/µL in oxidizing solution (e.g., 2–3 nmol in 4.9 µL oxidizing solution). Incubate at 4°C for 2 h.

2. Partial neutralization. To the cold sample, add 0.9 vol of ice-cold NH_4OH (e.g., 4.4 µL of NH_4OH to 4.9 µL oxidized sample solution). Make this addition carefully so as to maintain a low temperature.

3. Cleavage. Add 8 vol of cyanogen bromide solution. Incubate at 4°C for 30 h in the dark. Carry out this step in a fume hood.

4. To terminate the reaction, lyophilize the sample (all reagents are volatile).

5. Decontaminate equipment, such as spatulas, that have contacted cyanogen bromide, by immersion in bleach until the effervescence stops (a few minutes).

3.2. Cleavage at Asp-X Bonds

1. Dissolve the protein or peptide in the dilute acid to a concentration of 1–2 mg/mL in a hydrolysis tube.

2. Seal the hydrolysis tube under vacuum—i.e., with the hydrolysis (sample) tube connected to a vacuum line, using a suitably hot flame, draw out and finally seal the neck of the tube.
3. Incubate at 108°C for 2 h.
4. To terminate the reaction, cool and open the hydrolysis tube, dilute the sample with water, and lyophilize.

4. Notes

4.1. Cleavage at Met-X Bonds

1. The mechanism of the action of cyanogen bromide on methionine-containing peptides is shown in Fig. 1. For further details, see the review by Fontana and Gross (1). Methionine sulfoxide does not take part in this reaction. An acid environment is required to protonate basic groups and so prevent reaction there and maintain a high degree of specificity. Met-Ser and Met-Thr bonds may give less than 100% yields of cleavage because of the involvement of the β-hydroxyl groups of seryl and threonyl residues in alternative reactions, which do not result in cleavage (1).
2. Although the specificity of this reaction is excellent, some side reactions may occur. This is particularly so if colored (yellow or orange) cyanogen bromide crystals are used, when destruction of tyrosyl and tryptophanyl residues may occur.
   The acid conditions employed for the reaction may lead to small degrees of deamidation of side chains and cleavage of acid-labile bonds, e.g., Asp-Pro. A small amount of oxidation of cysteine to cysteic

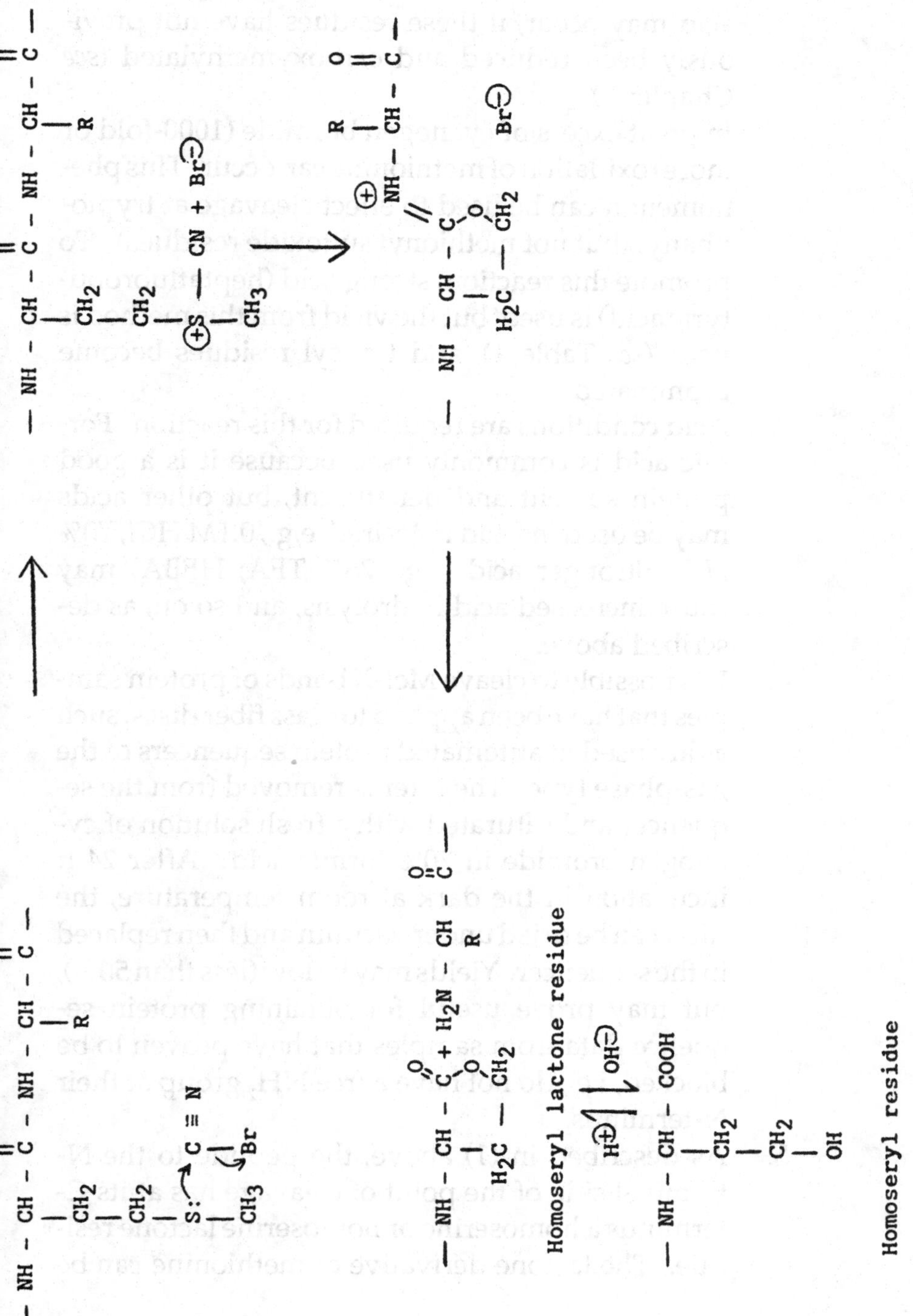

Homoseryl lactone residue

Homoseryl residue

Fig. 1. Mechanism of cleavage of Met-X bonds by cyanogen bromide.

acid may occur, if these residues have not previously been reduced and carboxymethylated (see
Chapter 7).

In great excess of cyanogen bromide (1000-fold or
more) oxidation of methionine can occur. This phenomenon can be used to effect cleavage at tryptophanyl (but not methionyl sulfoxide residues). To
promote this reaction, strong acid (heptafluorobu-
tyric acid) is used, but the yield from this method is
poor (see Table 1), and tyrosyl residues become
brominated.

3. Acid conditions are required for this reaction. Formic acid is commonly used because it is a good
   protein solvent and denaturant, but other acids
   may be used instead if desired, e.g., 0.1M HCl; 70%
   HF. Stronger acid (e.g., 75% TFA; HFBA) may
   cause increased acid hydrolysis, and so on, as described above.

4. It is possible to cleave Met-X bonds of protein samples that have been applied to glass fiber discs, such
   as are used in automated protein sequencers of the
   gas-phase type. The filter is removed from the sequencer and saturated with a fresh solution of cyanogen bromide in 70% formic acid. After 24 h
   incubation in the dark at room temperature, the
   filter can be dried under vacuum and then replaced
   in the sequencer. Yields may be low (less than 50%),
   but may prove useful for obtaining protein sequence data from samples that have proven to be
   blocked, i.e., do not have a free NH_2 group at their
   N-terminus.

5. As described in (1) above, the peptide to the N-
   terminal side of the point of cleavage has at its C-
   terminus a homoserine or homoserine lactone residue. The lactone derivative of methionine can be

coupled selectively and in good yield (2) to solid supports of the amino type, e.g., 3-amino propyl glass. This is a useful technique for sequencing peptides on solid supports. The peptide from the C-terminus of the cleaved protein will, of course, *not* end in homoserine lactone (unless the C-terminal residue was methionine!) and so cannot be so readily coupled.

6. The reagents used are removed by lyophilization, unless salt has formed following failure to remove all of the ammonium bicarbonate. The products of cleavage may be fractionated by the various forms of electrophoresis and chromatography currently available, as described elsewhere in volumes of this series. It should be remembered, however, that since methionyl residues are among the less common residues, peptides resulting from cleavage at Met-X may be large. Accordingly, in HPLC, for instance, use of wide-pore column materials may be advisable (e.g., 30 μm pore size reverse-phase columns, using gradients of acetonitrile in 0.1% TFA in water).

4.2. Cleavage at Tryp-X Bonds

1. The method described is that of Huang et al. (3). Although full details of the mechanism of this reaction are not clear, it is apparent that tryptophanyl residues are converted to oxindolylalanyl residues in the oxidation step, and the bond to the COOH side at each of these is readily cleaved, in excellent yield (approaching 100% in ref. 3) by the subsequent cyanogen bromide treatment. The result is seemingly unaffected by the nature of the residues surrounding the cleavage site.

Table 1

Summary of Chemical Methods for Specific Cleavage of Peptide Bonds

Site of cleavage	Brief method details	Comments	Example ref.
Asp↓X	Incubation in dilute acid (pH 2), high temperature (e.g., 108°C), short duration	Some X-Asp bonds cleaved in lesser yield. Rate of cleavage dependent on various factors (e.g., identity of X). Side reactions include deamidation. Yield moderate to good, but unpredictable.	4
Asp↓Pro	Incubation of 7M guanidine HCl, dilute acid (e.g., 10% acetic acid adjusted to 2.5 with pyridine), moderate temperature (e.g., 37°C), prolonged incubation (e.g., 24–120 h)	Some Asp-Pro may remain resistant despite inclusion of guanidine HCl. Deamidation may occur. Yields moderate to good.	6
Asn↓Gly	Incubation of 2M hydroxylamine, 6M guanidine HCl (brought to pH 9 with LiOH) 45°C, 4 h	Some Asn-Gly bonds may remain resistant despite inclusion of guanidine HCl. Specificity generally good, although some Asn-X bonds may cleave.	7
X↓Cys	(a) Incubation in 6M guanidine HCl, 0.2M Tris-acetate, pH 8, with excess dithiothreitol (b) Add 2-nitro-5-thiocyanobenzoic acid (5-fold molar excess over thiols). Adjust pH to 8 (c) Acidify to pH 4. Dialyze. Lyophilize (d) To cleave, dissolve in 6M guanidine HCl, 0.1M sodium borate, pH 9, 12–16 h, 37°C	For cleavage at cysteine only, not cystine, delete DTT from (a). Cysteinyl residue converted to iminothiazolidinyl residue, which blocks sequencing unless converted to alanyl residues by treatment as in ref. 9. Yield good.	8
His↓X	Incubation in N-bromosuccinimide (3-fold molar excess over His, Tryp, and Tyr) pH 3–4 (e.g., pyridine acetic acid, pH 3.3) 1 h, 100°C	Cleavage at Tryp-X and Tyr-X occurs preferentially (q. v.). Met and Cys may be oxidized. N-Bromosuccinimide unstable in storage. Yields moderate to poor. His converted to lactone derivative.	10
Met↓X	Incubation in cyanogen bromide (20–100-fold molar excess over Met) acid (e.g., 70% formic acid), 18–20 h, room temperature in the dark	Specificity and yield excellent. Cys may be slowly oxidized, and Asp-Pro bonds cleaved. Excess reagent can cause degradation of Tryp and Tyr side chains. Cyanogen bromide is toxic.	1
X↓Ser	(a) Incubation in conc. anhydrous acid (e.g., H_2SO_4, HF or HCOOH) room temperature, few days (b) Controlled acid hydrolysis (e.g., 6M HCl, 18°C, 6 h)	Mechanism involves (reversible) N-O peptidyl rearrangement in Ser. Yield moderate. Thre-X bonds cleaved at lower yield, and others non-specifically at low yield. Tyr sulfonation (in H_2SO_4) and Tryp destruction may occur.	11
X↓Thr	As for X-Ser (N→O shift mechanism)—q. v.	See Ser-X (N → O shift mechanism). Yields moderate to poor.	11

Tyr↓X	As for His-X (N-bromosuccinimide)—q.v.	Tyr-X cleaved more quickly than His-X, more slowly than Tryp-X (q.v.). Yields moderate to poor, or zero, if Tyr is at the N-terminus. Met and Cys may undergo oxidation. Tyr converted to lactone derivative.	10
Tyr↓X	Electrolytic oxidation at Pt electrodes in pH 2.2–2.4 (e.g., 10% acetic acid), 4–5 h	Oxidation of other groups may occur. Yields moderate to poor.	12
Tryp↓X	Incubation in BNPS-skatole (100-fold molar excess over Tryp) in 50% acetic acid (v/v) room temperature, 48 h, in the dark	Reagent is unstable—fresh reagent required to minimize side reactions. Some reaction with Tyr may occur. Met and Cys may be oxidized. Yields moderate to good.	13
Tryp–X↓	As for His-X (N-bromosuccinimide)—q.v.	Also less rapid cleavage at His-X and Tyr-X (q.v.). Yields moderate to poor. Tryp converted to lactone derivative.	10
Tryp↓X	Incubation in very large (up to 10,000-fold) molar excess of cyanogen bromide over Met, in heptafluorobutyric acid 88% H COOH (1/1 v/v), room temperature, 24 h, in the dark	To inhibit Met-X cleavage, Met is photoxidized irreversibly to Met sulfone. Yield poor.	14
Tryp↓X	(a) Incubation in glacial acetic acid: 12M HCl (2/1, v/v) with phenol and dimethylsulfoxide, room temperature, 30 min (b) Add HBr and dimethylsulfoxide, room temperature, 30 min	Cys and Met oxidized. Some hydrolysis of acid-labile bonds around Asp may occur as may deamidation. Fresh (colorless) HBr required. Yields moderate. Tryp converted to di-oxindolalanyl lactone.	15
Tryp↓X	(a) Incubation in glacial acetic acid: 9M HCl (2/1, v/v) with DMSO, room temperature, 30 min (b) Neutralization (c) Incubation with cyanogen bromide in acid (e.g., 60% formic acid) 4°C, 30 h in the dark	Yields and specificity excellent. Met oxidized to sulfoxide.	3
Tryp↓X	Incubation in 80% acetic acid containing 4M guanidine HCl, 13 mg/mL iodosobenzoic acid, 20 µL/mL p-cresol, for 20 h room temperature, in the dark	Specificity and yields good. The p-cresol is used to prevent cleavage Tyr. Tryp converted to lactone derivative.	16

During the oxidation step, methionyl residues are converted to sulfoxides, so bonds at these residues are *not* cleaved by the cyanogen bromide treatment. Cysteinyl residues will also suffer oxidation if they have not been reduced and alkylated beforehand (*see* Chapter 7).

The peptide to the C-terminal side of the cleavage point has a free N-terminus and so is suitable for sequencing.

2. Methionyl sulfoxide residues in the peptides produced may be converted back to methionyl residues by the action, in aqueous solution, of thiols (e.g., dithiothreitol, as described in ref. 3, or see use of 2-mercaptoethanol above).

3. The acid conditions used for this reaction seem to cause little deamidation (3), but one side reaction that can occur is hydrolysis of acid-labile bonds. The use of low temperature minimizes this problem. If a greater degree of such acid hydrolysis is not unacceptable, speedier and warmer alternatives to the reaction conditions described above can be used as follows:

    a. Oxidation—at room temperature for 30 min, but cool to 4°C before neutralization.

    b. Cleavage at room temperature for 12–15 h.

4. As alternatives to the volatile base, NH_4OH, other bases may be used (e.g., the nonvolatile potassium hydroxide or Tris base).

5. As an alternative to formic acid (used in the cleavage step), $5M$ acetic acid may be used.

6. Samples eluted from sodium dodecyl sulfate (SDS) gels may be treated as described, but for good yields of cleavage, Huang et al. (3) recommend that the sample solutions are acidified to pH 1.5 before lyophilization in preparation for dissolution in the

oxidizing solution. Any SDS present may help to solubilize the substrate and, in small amounts at least, does not interfere with the reaction. However, nonionic detergents that are phenolic or contain unsaturated hydrocarbon chains (e.g., Triton, nonidet P-40), and reducing agents, are to be avoided.

7. The method is suitable for large-scale protein cleavage, this requiring simple scaling up. Huang et al. (3) make two points, however:

a. The neutralization reaction generates heat. Since this might lead to protein or peptide aggregation, cooling is important at this stage. Ensure that the reagents are cold and are mixed together slowly and with cooling. A *transient* precipitate is seen at this stage. If the precipitate is insoluble, addition of SDS may solubilize it (but will not interfere with the subsequent treatment.)

b. The neutralization reaction generates gases. Allow for this when choosing a reaction vessel.

8. The results of cleavage may be inspected by polyacrylamide gel electrophoresis, HPLC, and so on, as after cleavage at Met-X bonds (see above). All reagents in the described method are readily removed during lyophilization.

4.3. Cleavage at Asp-X Bonds

1. The bond most readily cleaved in dilute acid is the Asp-X bond, by the mechanism outlined in Fig. 2(i). The bond X-Asp may also be cleaved, in lesser yields—*see* Fig. 2(ii). Thus, either of the peptides resulting from any one cleavage may keep the aspartyl residue at the point of cleavage, or neither

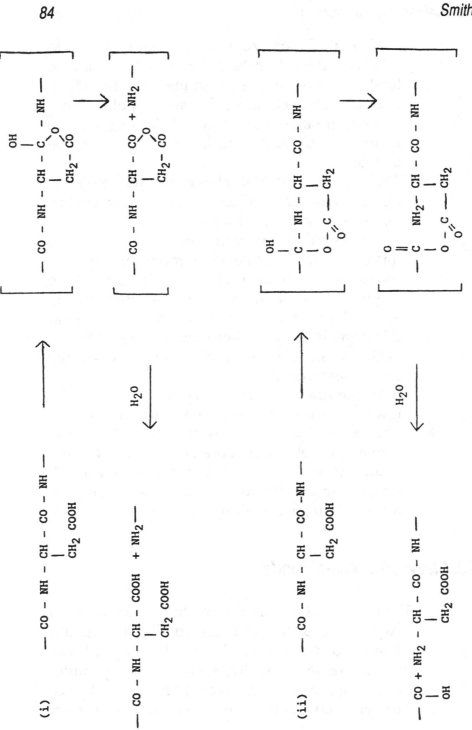

Fig. 2. Mechanisms of the cleavage of bonds to the COOH side (scheme i) and to the NH₂ side (scheme ii) of aspartyl residues in dilute acid.

might, if free aspartic acid is generated by a double cleavage event. Any of these peptides is suitable for sequencing.

The amino acid sequence of the protein can affect the lability of the affected bond. Thus, the Asp-Pro bond is particularly labile in acid conditions (*see* Table 1). Again, ionic interaction between the aspartic acid side chains and basic residue side chains elsewhere in the molecule can adversely affect the rate of cleavage at the labile bond. Such problems as these make prediction of cleavage points somewhat problematical, particularly if the protein is folded up (e.g., a native protein). The method may well prove suitable, however, for use in cleaving small proteins or peptides.

2. The conditions of low pH can be expected to cause a number of side reactions: cleavage at glutamyl residues; deamidation of (and subsequent cleavage at) glutaminyl and asparginyl residues; cyclization of N-terminal glutaminyl residues to residues of pyrolidone carboxylic acid; α-β shift at aspartyl residues, creating a blockage to Edman degradation; partial destruction of tryptophan. The short reaction time of 2 h is intended to minimize these side reactions. Loss of formyl and acetyl groups from N-termini (4) are other possible side reactions, but these may be advantageous in deblocking otherwise unsequenceable polypeptides.

3. A substrate that is insoluble in cold dilute HCl may dissolve during incubation at 108°C. Alternatively, formic acid, a good protein denaturant and solvent, may be used, as follows: dissolve the sample in formic acid, 98%, then dilute 50-fold to pH 2.

4. The comments above concerning the effect of the amino acid sequence and of the environment

around potentially labile bonds, and the various side reactions that can occur, indicate that the consequences of incubation of a protein in dilute acid are difficult to predict—they are best investigated empirically by monitoring production of peptides by electrophoresis, HPLC, and so on (as suggested for cleavage of Met-X, above).

Yields of cleavage of Asp-X bonds of up to about 70% have been reported (4). Such partial hydrolysis at just a few sites can give rise to a complicated mixture of peptides. For this reason this method may not be the one of choice for proteins, but may be suitable for small proteins or peptides containing few aspartyl residues, while also having the advantage of speed and simplicity. It is also readily adapted for large or small (subnanomolar) amounts of substrate.

5. Other Methods of Chemical Cleavage

1. A summary of various of common chemical methods of protein cleavage is given in Table 1. The literature contains descriptions of many other, less frequently used techniques (e.g., *see* ref. 1). Reaction mechanisms are not known for all methods. Yields are generally less than 100% and often vary according to the nature of the substrate, surrounding amino acid sequence, exact conditions of reaction, and so on, but for guidance are put into broad classes in Table 1.

2. The most useful and widely used techniques provide good specificity at good yield, but some of the less likely techniques may also prove useful in particular instances.

The simplicity of a method is also worthy of consideration. Thus cleavage at Met-X can be conducted in one vessel, and losses experienced in manipulation of the sample are minimized. On the other hand, cleavage at X-Cys bonds, as outlined in Table 1, involves dialysis and further treatment to deblock peptides for sequencing. Accumulative losses may limit the minimum size of samples required for this method.

3. Various of the techniques generate a lactone derivative at the C-terminus of the peptide to the NH_2 side of the cleavage point (*see* Table 1). This may be useful for the selective attachment of these peptides to amino supports such as 3-amino propyl glass (5), for the purposes of solid phase protein sequencing.

4. Because of the variety of chemical techniques available, it is not easy to generate general rules. One point for consideration, however, is denaturation of the substrate. It is as well to reduce and alkylate cystinyl residues (*see* Chapter 7) so as to minimize the problems caused by aggregation of proteins or peptides. This is obviously not recommended if positions of disulfide bonds are of interest. If this is the case, be wary of strong acid (e.g., concentrated HCl) or alkaline conditions in which disulfide interchange may occur. Again, folding of the polypeptide may affect reactivity by allowing interaction of side chains. Denaturation by inclusion of agents such as guanidine HCl can help to overcome this and to solubilize the substrate and peptides from it.

Acknowledgments

I wish to thank my colleagues for useful discussions.

References

1. Fontana, A. and Gross, E. (1986) Fragmentation of Polypeptides by Chemical Methods, in *Practical Protein Chemistry—A Handbook*, Darbre, A., (ed.) Wiley, Chichester.

2. Horn, M. and Laursen, R.A. (1973) Solid-phase Edman degradation. Attachment of carboxyl-terminal homoserine peptides to an insoluble resin. *FEBS Lett.* **36**, 285–288.

3. Huang, H.V., Bond, M.W., Hunkapillar, M.W., and Hood, L.E. (1983) Cleavage at tryptophanyl residues with dimethyl sulfoxide–hydrochloric acid and cyanogen bromide. *Meth. Enzymol.* **91**, 318–324.

4. Ingris, A.S. (1983) Cleavage at aspartic acid. *Meth. Enzymol.* **91**, 324–332.

5. Wachter, E. and Werhahn, R. (1980) Cleavage of Trp-,Tyr-, and His-bonds Suited for Attachment to Amino-Supports and Subsequent Edman Degradation, in *Methods in Peptide and Protein Sequence Analysis* (Birr, C., ed.) Elsevier/North Holland, Amsterdam.

6. Landon, M. (1977) Cleavage at aspartyl-prolyl bonds. *Meth. Enzymol.* **47**, 145–149.

7. Bornstein, P. and Balian, G. (1977) Cleavage at cysteine at Asn-Gly bonds with hydroxylamine. *Meth. Enzymol.* **47**, 132–145.

8. Stark, G.R. (1977) Cleavage at cysteine after cyanylation. *Meth. Enzymol.* **47**, 129–132.

9. Otieno, S. (1978) Generation of a free α-amino group by Raney Nickel after 2-nitro-5-thiocyanobenzoic acid cleavage at cysteine residues: Application to automated sequencing. *Biochemistry* **17**, 5468–5474.

10. Ramachandran, L.K. and Witkop, B. (1976) *N*-Bromosuccinimide cleavage of peptides. *Meth. Enzymol.* **11**, 283–299.

11. Iwai, K. and Ando, T. (1976) $N \rightarrow O$ acyl rearrangement. *Meth. Enzymol.* **11**, 236–282.

12. Cohen, L.A. and Farber, L. (1967) Cleavage of tyrosyl-peptide bonds by electrolytic oxidation. *Meth. Enzymol.* **11**, 299–308.

13. Fontana, A. (1972) Modification of tryptophan with BNPS-skatole (2-(2-nitrophenylsulfenyl)-3-methyl-3-bromoindolenine). *Meth. Enzymol.* **25**, 419–423.

14. Ozols, J. and Gerard, C. (1977) Covalent structure of the membranous segment of horse cytochrome b5. Chemical cleavage of the native hemprotein. *J. Biol. Chem.* **252**, 8549–8553.

15. Savige, W.E. and Fontana, A. (1977) Cleavage of the tryptophanyl peptide bond by dimethyl sulfoxide–hydrobromic acid. *Meth. Enzymol.* **47**, 469–468.

16. Fontana, A., Dalzoppo, D., Grandi, C., and Zambonin, M. (1983) Cleavage at tryptophan with o-iodosobenzoic acid. *Meth. Enzymol.* **91**, 311–318.

Chapter 7

Chemical Modification of Proteins

Alex F. Carne

1. Introduction

The two most widely used applications for chemical modification are in primary structure analysis and in the identification of essential groups involved in the binding and catalytic sites of proteins. The methods discussed here are those used frequently in primary structure analysis. Chemical modifications involving protein "active center" identification are the subject of a review by Pfleiderer (1).

With the advent of the gas-phase sequencer (2), the amounts of protein necessary for sequence analysis are in the low picomole range, and it is therefore necessary to modify some of the more well-known methods to suit the smaller amounts of protein.

Cysteine presents a problem both in amino acid and sequence analysis. It is subject to autooxidation, leading to a variety of products such as mixed disulfides (3), and is involved in the formation of protein disulfide bridges. The presence of disulfide bridges within a polypeptide chain leads to difficulties in peptide isolation after enzymic digestion and in sequence determination. Reductive alkylation with iodoacetic acid yields the more stable carboxymethyl derivative and also introduces a charged group that tends to enhance the solubility of the polypeptide in aqueous buffers at alkaline pH. This makes the protein more susceptible to digestion by proteases such as trypsin, chymotrypsin, and so on. Also, using [14C]-iodoacetic acid as the alkylating agent enables a radioactive label to be introduced into the polypeptide chain.

For quantitative amino acid analysis involving cystine or cysteine residues, carboxymethylation is not ideal since it is difficult to assess the completeness of the reaction. A better method is to oxidize the S-S or SH groups using performic acid (3), hydrolyze the polypeptide, and identify the cystine/cysteine as cysteic acid.

A further useful modification is the succinylation of the ε-amino lysine groups. This is an irreversible reaction and hinders tryptic attack at lysine residues, thus giving rise to peptides with arginine at the carboxy-terminus. As with carboxymethylation, the use of [14C]-succinic anhydride allows for the incorporation of a radiolabel, and proteins modified by dicarboxylic anhydrides are generally more soluble at pH 8, and the polypeptide chains are unfolded thus allowing for ready digestion with proteases.

Aminoethylation of cysteine residues, to the *S*-β-aminoethyl derivatives, can be used to introduce new points of tryptic cleavage into a polypeptide chain. Trypsin cleaves peptide bonds involving the aminoethyl derivative since it is an analog of lysine. Cleavage is much slower, however, than that at lysine or arginine residues (4). The positively charged derivative is prepared by reaction of the reduced thiol containing protein with ethyleneimine.

2. Materials

2.1. Reductive Alkylation

1. Denaturing buffer: 6*M* guanidinium chloride, 0.1*M* Tris-HCl, pH 8.5.
2. A source of oxygen-free nitrogen.
3. Dithiothreitol (DTT): 4 m*M* in distilled water. This solution should be freshly made.
4. Iodoacetic acid: 500 m*M* in distilled water, pH adjusted to 8.5 with sodium hydroxide. Iodoacetic acid is light-sensitive, and it is preferable to make up a fresh solution each time, though the solution can be stored in the dark at –20°C.
5. Ammonium bicarbonate: 50 m*M* in distilled water, pH approximately 8.0.
6. Microdialysis equipment (as supplied by BRL).
7. HPLC system with a Brownlee RP-300 "cartridge"-type column fitted (30 x 2.1 mm).
8. 0.1% TFA (HPLC grade).
9. Acetonitrile (HPLC, far UV grade).

2.2. Performic Acid Oxidation

1. Hydrogen peroxide: 30% w/v. Use care! This is a strong oxidizing agent.

2. Formic acid: 88% w/v.
3. Hydrobromic acid: 48% w/v. This acid gives off a caustic, irritating vapor, and its use should be confined to a fume hood.

2.3. Succinylation

1. Guanidinium chloride: 6M in distilled water.
2. Succinic anhydride.
3. Sodium hydroxide: 1M in distilled water.
4. Micro-pH probe.
5. Ammonium bicarbonate: 50 mM in distilled water, pH approximately 8.0.
6. Microdialysis or HPLC equipment.

2.4. Aminoethylation

1. Denaturing buffer: 6M guanidinium chloride, 1.0M Tris-HCl, 0.025% β-mercaptoethanol, pH 8.9.
2. Source of oxygen-free nitrogen.
3. Ethyleneimine (Use care—suspected carcinogen).
4. Ammonium bicarbonate: 50 mM in distilled water, pH approximately 8.0.
5. Microdialysis equipment.

3. Method

3.1. Reductive Carboxymethylation

1. Dissolve the protein in 50 μL of denaturing buffer in a 1.5-mL Eppendorf tube.
2. Gently blow N_2 over the top of the solution for 15 min.
3. Add an equal volume of DTT solution to a concentration of 2 mM.

4. Blow N_2 over the solution for 60 min.
5. Wrap the Eppendorf tube in aluminium foil before the addition of iodoacetic acid.
6. 40 µL of iodoacetic acid solution is added dropwise using a micropipet. During this addition the solution should be stirred. (It is possible to obtain microstirrer bars that will fit Eppendorf tubes.)
7. N_2 is again blown over the surface of the solution and the reaction tube sealed.
8. Incubate in the dark at 37°C for 30 min. The reaction is carried out in the dark to prevent the formation of iodine, from the iodoacetate, thus preventing the iodination of tyrosine residues.
9. Remove excess reagents. This may be performed in one of two ways:

    a. Microdialysis: Here, the solution is pipeted into a 1–2 mL perspex chamber under which there is a sheet of dialysis membrane (mol. wt. cut-off, 6000–8000). The underside of the membrane is in contact with the dialysis buffer, usually 50 mM ammonium bicarbonate. Dialysis is carried out for about 24 h.

    b. HPLC: The protein can be separated from the reaction byproducts by reverse-phase HPLC. To minimize product loss, a 30 x 2, 1-mm microbore "cartridge" column is employed. For proteins, a C3 or C8 matrix with a 10-µm particle diameter and a 30-nm pore size can be used (e.g., Aquapore octyl RP-300). The column is equilibrated in 0.1% aqueous trifluoroacetic acid. The sample is injected onto the column and eluted using an acetonitrile gradient (1%/min, flow rate 0.2 mL/min). The excess reagents and reaction byproducts all appear in the breakthrough, i.e., they are not retained by the column packing material.

3.2. Performic Acid Oxidation

1. Add 100 µL of hydrogen peroxide to 900 µL of formic acid and allow to stand at room temperature for 1 h. This produces performic acid (HCOOOH).
2. Cool the performic acid on ice to approximately 0°C.
3. Dissolve the protein in 50 µL of performic acid (about 400–500 µg/mL), in a precooled tube.
4. Keep at 0°C for 4 h.
5. Add 7.5 µL of *cold* HBr to neutralize the performic acid. Care must be taken at this stage since bromine will be liberated, albeit in small amounts.
6. Dry the sample throughly *in vacuo* over NaOH to remove bromine and formic acid.
7. The sample can then be hydrolyzed and subjected to amino acid analysis.

3.3. Succinylation

1. Dissolve the protein or protein derivative in 200 µL of 6M guanidinium chloride, and using a micro-pH probe adjust the pH to 9.0 with the sodium hydroxide solution.
2. Crush 1–2 crystals (1–2 mg) of solid succinic anhydride and add to the stirred solution over a period of about 15 min, while maintaining the pH of the solution at 9.0 with 1.0M NaOH.
3. Excess reagents are removed by microdialysis or by HPLC.

3.4. Animoethylation

This method is similar to that used by Slobin and Singer for γ-globulin (5).

1. Dissolve the protein (~0.5 mg) in 100 μL of denaturing buffer in a 1.5-mL Eppendorf tube.
2. Flush with N_2, cap, and leave for 2 h.
3. Add 3 μL of ethyleneimine.
4. Flush with N_2, cap, and allow to alkylate for 2 h at room temperature.
5. To remove excess reagents, dialyze against 50 mM ammonium bicarbonate and then distilled water, using the microdialysis apparatus.

4. Notes

1. Reagents for all the methods should be of the highest quality available.
2. The iodoacetic acid used for reductive alkylation must be colorless; any iodine present, revealed by a yellow color, causes rapid oxidation of thiol groups, preventing alkylation and possibly modifying tyrosine residues.
3. If good quality iodoacetic acid is not available, it may be recrystallized from hexane.
4. The method described for reductive alkylation does not include the use of iodo 2-[^{14}C]-acetic acid. The radiolabeled iodoacetic acid should be diluted before use with carrier iodoacetic acid to the required final specific activity, according to the requirements of the worker.
5. Continuous stirring during the addition of the iodoacetic acid is necessary to keep the pH constant throughout the solution, especially when dealing with small volumes.
6. Dialysis tubing should be avoided since the losses involved when handling small amounts of protein can be high. The microdialysis system that uses a

sheet of dialysis membrane is preferable, but losses can still occur. If the reader has access to an HPLC system this should be the method of choice.

7. During performic acid oxidation, the temperature of all the reactants must be at or near 0°C to minimize any side reactions, such as the oxidation of phenolic groups and the hydroxyl groups of serine and threonine.

8. Oxidation of methionine occurs during performic acid oxidation producing methionine sulfone, which can be quantified by amino acid analysis.

9. The addition of HBr in the performic acid method causes bromination of tyrosine residues, producing a mixture of mono- and di-substituted bromotyrosines. These derivatives have lower color values than tyrosine, and quantitation is generally not good after acid hydrolysis.

10. Since several residues are modified, performic acid oxidation is best used in conjuction with amino acid analysis. It is not recommended to use performic acid-oxidized samples for sequence analysis.

11. Succinylation is best carried out in proteins that have been carboxymethylated to prevent S-S or SH groups from reacting. This gives complete selective modification.

12. The addition of small amounts of succinic, anhydride over a 15-min time period enables the worker to control the pH, so that it can be maintained at or near pH 9.0.

13. For the removal of excess reagents following succinylation, the same criteria apply as for carboxymethylation.

14. Succinylation is a better method than maleylation or citraconylation for the modification of lysine residues, since it prevents problems arising from

the still-reactive ethylenic groups. If there is a need for the removal of the blocking groups, however, for subsequent tryptic cleavage at lysine residues, maleylation or citraconylation shoudbe used since the groups introduced by citraconic or maleic anhydride can be removed by treatment with dilute acid.

15. Both clostripain (E.C. 3.4.22.8) and endoproteinase Arg-C (6) can be used for cleavage at arginine residues, instead of succinylation followed by trypsin digestion. Clostripain may cleave at other sites (7), however, and endoproteinase Arg-C, although highly specific for the C-terminal side of arginine residues, does cleave some arginine peptide bonds slowly (8). Thus, succinylation is still of use to the protein chemist trying to achieve specific cleavages coupled with the proper of enhanced solubility.

16. If the aminoethylation reaction is allowed to proceed for a long time with an excess of ethyleneimine, alkylation of methionine (9) or of the α-amino group of the peptide (10) can occur.

17. S-β-Aminoethylcysteine can be detected by amino acid analysis, eluting with the basic amino acids from a cation-exchange column (4).

References

1. Pfleiderer, G. (1985) Chemical Modification of Proteins in *Modern Methods in Protein Chemistry, Review Articles* vol. 2 (Tschesche, H., ed.) Walter de Gruyter, Berlin.

2. Hewick, R.M., Hunkapiller, M.W., Hood, L.E., and Dreyer, W. J. (1981) A gas-liquid-solid-phase peptide and protein sequenator. *J. Biol. Chem.* **256**, 7990–7997.

3. Glazer, A.N., Delange, R.J., and Sigman, D.S. (1975) Chemical Characterization of Proteins and Their De-

rivatives, in *Chemical Modification of Proteins* (Work, T.S., and Work, E., eds.) North Holland/American Elsevier, Amsterdam.

4. Wang, S.S. and Carpenter, F.H. (1968) Kinetic studies at high pH of the trypsin-catalyzed hydrolysis of N-alpha-benzoyl derivatives of L-arginamide, L-lysinamide and S-2-aminoethyl-L-cyteinamide and related compounds. *J. Biol. Chem.* **243**, 3702–3710.

5. Slobin, L. I. and Singer, S.J. (1968) The specific cleavage of immunoglobulin polypeptide chains at cysteinyl residues. *J. Biol. Chem.* **243**, 1777–1786.

6. Levy, M., Fishman, L., and Schenkein, I. (1970) Mouse Submaxillary Gland Proteases, in *Methods in Enzymology* vol. XIX *Proteolytic Enzymes* (Perlmann, G.E. and Lorand, L., eds.) Academic, New York.

7. Mitchell, W.M. and Harrington, W.F. (1970) Clostripain, im *Methods in Enzymology* vol. XIX *Proteolytic Enzymes* (Perlmann, G.E. and Lorand, L., eds.) Academic, New York.

8. Thomas, K.A. and Bradshaw, R.A. (1981) γ-Subunit of Mouse Submaxillary Gland 7S Nerve Growth Factor: An Endopeptidase of the Serine Family, in *Methods in Enzymology*, vol. 80, *Proteolytic Enzymes*, part C (Lorand, L., ed.) Academic, New York.

9. Schroeder, W.A., Shelton, J.R., and Robberson, B. (1967) Modification of methionyl residues during amino-ethylation. *Biochem. Biophys. Acta* **147**, 590–592.

10. Shotton, D. and Hartley, B.S. (1973) Evidence for the amino acid sequence of porcine pancreatic elastase. *Biochem. J.* **131**, 643–675.

Chapter 8

The Design, Preparation, and Use of Immunopurification Reagents

*Andrew Kenney, Lynne Goulding,
and Christopher Hill*

1. Introduction

Immunoaffinity chromatography using immobilized monoclonal antibodies has become one of the most powerful macromolecular separation methods available. In many cases, simple one-step immunoaffinity separations will produce a several thousand-fold purification with yields of better than 80%. Monoclonal antibodies are now available with specificities for many different antigens, and the technology for producing new monoclonal antibodies has become widespread througout the world since its invention (*1; see* vol. 1 of this series). It has been estimated that a new monoclonal antibody is produced every 40 min worldwide.

The wide-ranging use of monoclonal antibodies for immunopurification (IMP) is best illustrated by way of example (Table 1).

Table 1
Examples of Antigens Purified Using Immobilized MAbs

Antigen	Example	Reference
Interferons	Interferon-α	2
Lymphokines	Interleukin-2	3
Enzymes	Tissue plasminogen activator	4
Plasma proteins	Factor I	5
Peptide hormones	Prolactin	6
Immunoglobulins	Rat MAb	7
Membrane-bound proteins	Murine MHC antigens	8

An examination of these examples will demonstrate that IMP is a versatile and powerful technique. It will also be apparent that a number of different strategies have been used for different systems. It is not possible here to explore all aspects of the development of an efficient immunopurification method; rather we have attempted to provide a method in detail that will allow investigators to discover the potential that immunopurification may have for their own application.

In order to exploit this technology successfully, it is necessary to know how best to prepare the reagent. Many different matrices and coupling chemistries have been developed that are suitable for the immobilization of monoclonal antibodies (Table 2). The choice for research laboratory applications will normally be dictated by the availability and ease of preparation, and in most cases this means cyanogen bromide-activated agarose. The method described here will in most cases prove reli-

Table 2

Matrices Available for Immobilization of Antibodies for Affinity Chromatography

Class	Commercial name	Support material	Coupling chemistry	Supplier	Pore size exclusion
Natural polymers	CNBr-Sepharose 4B	Agarose	CNBr	Pharmacia	20×10^6
	CM Bio gel A	Agarose	EDAC	Bio Rad	—
	Affigel 102/202	Agarose	EDAC	Bio Rad	—
	CNBr-Sepharose CL4B	Crosslinked agarose	CNBr	Pharmacia	20×10^6
	Epoxy-Sepharose CL4B	Crosslinked agarose	Epoxide	Pharmacia	4×10^6
	Affigel 10/15	Crosslinked agarose	N-Hydroxysuccinimde	Bio Rad	5×10^6
	Reacti gel (6x)	Crosslinked agarose	CDI	Pierce	4×10^6
	Reacti gel (25 DF)	Crosslinked agarose	CDI	Pierce	5×10^3
Synthetic polymers	Reacti gel (HW 65F)	Fractogel	CDI	Pierce	5×10^6
	Avid gel	Trisacryl	FMP	Bio Probe	20×10^6
	Eupergit C	Acrylic beads	Epoxide	RhonePharma	100–2500 nm
	Amino ethyl Bio gel P2/P150	Polyacrylamide	EDAC	Bio Rad	—
Inorganic	Glycophase	Controlled pore glass	CDI	Pierce	20 nm
	HP beads	Controlled pore glass	—	Chemicon	—
	Biomag	Magnetic oxide	Glutaraldehyde	Advanced Magnetics, Inc.	—
	Various	Silica	Various	Boehringer	200 nm
				Beckman	—
				Chemapol	—
				Diagen	50–1000 nm
Composite	Acrobeads	Agarose/polyacrolein	Glutaraldehyde	Galil	—
	Ultrogel AcA 22	Agarose/acrylamide	Glutaraldehyde	LKB	—
Membrane	Biodyne	Nylon membrane		PALL	—

able and reproducible and is easy and convenient to use. In some cases, however, other matrices and chemistries will produce much better results. When the antigen to be purified is of a very high molecular weight, the choice of a matrix with a very large pore size will enable more efficient use of the antibody. This is because antibody that becomes immobilized well within the matrix bead will not be available to an antigen that is too large to diffuse into the pores of the bead. In addition, for applications in which leakage of the antibody from the matrix into the product is undesirable, it may be advantageous to choose an alternative coupling chemistry to CNBr. It is not possible to produce rules for use in choosing the matrix or chemsitry, but the method described here will allow investigators to begin studying the performance of their antibody for immunopurification. Later on it may be necessary to investigate alternatives in order to enhance the performance in individual applications.

The development of a method for using the prepared immunosorbent is usually an empirical process, especially with regard to the choice of eluant. The immunosorbent must first be characterized in terms of its capacity for antigen under the conditions to be employed. Failure to do this will either result in a marked under-utilization of the immunosorbent or in a loss of antigen in the unbound protein fraction. Depending on the affinity of the antibody for antigen, the eluant will either be a simple change in pH and ionic strength or may need to make use of denaturants and chaotropes. A list of eluants that have been found useful is given in Table 3. These have been presented in approximately the order in which they should be tried. Individual circumstances, however, particularly in respect to the stability of antibody and antigen in these solutions, will mitigate against the use of some of them.

Table 3
Typical Elution Condition Used
for Immunoaffinity Chromatography

Eluant	Reference
Extremes of pH	
0.1M Glycine/HCl, pH 2.5	9
1.0M Propionic acid	10
0.15M NH$_4$OH, pH 10.5	11
0.1M Na CAPS, pH 10.7	12
Chaotrophic agents	
4.0M MgCl$_2$, pH 7.0	13
2.5M NaI, pH 7.5	14
3.0M NaSCN, pH 7.4	15
Protein-deforming agents	
8M Urea, pH 7.0	16
6M Guanidine/HCl	17
Organic solvents	
50% v/v Ethanediol, pH 11.5	18,19
Dioxane/acetic acid	

2. Materials

1. Monoclonal antibody at greater than 80% purity.
2. CNBr activated Sepharose (Pharmacia).
3. Coupling buffer: 0.1M sodium hydrogen carbonate (NaHCO$_3$), pH 8.3, containing 0.5M sodium chloride—to be made fresh daily.
4. 1mM Hydrochloric acid.
5. Acetate buffer, pH 4: 0.1M sodium acetate containing 0.5M sodium chloride adjusted to pH 4.0 with glacial acetic acid—to be made fresh daily.
6. Ethanolamine solution: 1M ethanolamine adjusted to pH 8 with concentrated HCl. Note: Make up ethanolamine to about half the desired volume and

add conc. HCl dropwise. This evolves a considerable amount of heat and requires quite a large volume of HCl to bring the pH to 8. Ethanolamine is
also an irritant, which should not be inhaled.

7. Phosphate-buffered saline (PBS): 20 mM sodium
    phosphate buffer pH 7.2 containing 0.9% sodium
    chloride.
8. Phosphate-buffered saline with preservative: PBS
    containing 0.1% sodium azide.
9. pH 2.5 buffer: 0.1M glycine/HCl pH 2.5.

3. Methods

All procedures carried out at room temperature
unless otherwise stated.

3.1. *Immobilization of Monoclonal Antibody*

1. Prepare the purified monoclonal antibody (MAb)
    in coupling buffer at a concentration of 1–5 mg/
    mL. An estimate of mouse MAb concentration can
    be obtained by measuring absorbance at 280 nm
    and by calculation as follows:

$$(MAb) = A_{280}/1.4 \text{ mg/mL}$$

The MAb solution should be prepared and chilled
to 4°C before proceeding to the next step.

2. Weigh desired amount of CNBr-activated Sepha-
    rose 4B: As a guide 0.3 g swells to produce approxi-
    mately 1.0 mL gel. Calculate the volume of gel
    required on the basis of 1.0 mL of gel for every 5 mg
    of antibody to be coupled.
3. Swell and wash the CNBr-activated Sepharose
    with 1mM HCl, using 200 mL of HCl/g of Sepha-

rose. Sprinkle the Sepharose onto a small volume of 1 mM HCl contained in a sintered glass funnel (G3), and very gently stir the gel to ensure even swelling. Alow the gel to stand for 15 min at room temperature in 1 mM HCl, ensuring that the gel does not dry out. Then wash through the remaining 1 mM HCl under gentle suction.

4. Wash the swollen gel with 10 gel vol of cold coupling buffer (4°C) to chill the gel, and at the same time equilibrate in coupling buffer. Preparation of the gel for coupling should take approximately 30 min (from weighing to adding the antibody). Coupling capacity of the gel can be reduced if it is allowed to stand for longer at room temperature.

5. Add the swollen, washed, and chilled CNBr-activated Sepharose to 5 vol of monoclonal antibody solution (*see* 1 above) also cooled to 4°C, and allow the mixture to roll gently overnight at 4°C (*see* Note 6 in section 4).

6. Filter the coupling mixture through a sintered glass funnel (G3) using a gentle suction. Add an equal volume of cold coupling buffer to the gel slurry and mix gently to wash away unbound antibody. Drain this through the sintered glass funnel under gentle suction and combine the washes to determine the total amount of unbound antibody.

7. Return the gel slurry to the mixing vessel and add 5 vol of 1M ethanolamine pH 8 to block the unreacted CNBr-activated sites. Roll the slurry for 2 h at room temperature.

8. Filter the blocking agent away from the gel and wash the gel with tree cycles of pH 4 and pH 8 buffers to remove noncovalently bound proteins as follows:

   a. Mix the blocked gel with 10 gel vol of pH 4 acetate

buffer. Check that the pH of the mixture is close
to 4 and then allow the mixture to stand for at
least 5 min before removing the buffer under
gentle suction.
b. Add 10 gel vol of pH 8.3 coupling buffer to the
slurry. Check that the pH is close to 8 and allow
this to stand for at least 5 min and then remove
using gentle suction.
c. Repeat steps (a) and (b) twice.

9. Finally wash the gel with pH 8.3 coupling buffer
containing 0.1% sodium azide and store the gel in
this solution at 2–4°C. *Do Not Freeze.*

10. To facilitate measuring known volumes of gel, it is
recommended that the gel be stored as a 50% (v/v)
gel buffer mixture. This is best done by allowing
the gel to settle in a suitable container (at least 2x
larger than gel volume) at 4°C for 24 h. When the
gel has settled, the volume of the storage buffer can
then be corrected to the same volume as the gel.
When taking samples of gel from the vessel, the
slurry should be gently mixed to an even consisten-
cy before removing twice the desired gel volume.

3.2. Affinity Chromatography Using the Immobilized Monoclonal Antibody

1. The 50% (v/v) gel slurry should be allowed to
warm to room temperature.

2. Mix the gel slurry gently to form an even suspension, and then remove twice the desired gel volume.

3. Pack the MAb-Sepharose gel into a suitable column
using an appropriate running buffer. We suggest
PBS (*see* Note 8 in section 4) using a flowrate of 10
mL/h.

4. Pass through the column 5 column vol of elution buffer (Table 3). In this protocol, we suggest a buffer of pH 2.5. This step will substantially reduce the amount of antibody that might otherwise leak into the eluted product.

5. Equilibrate the column with 10 column vol of running buffer.

6. Apply the crude sample to the column. Ideally the sample should be equilibrated in running buffer.

7. Wash the sample onto the column with 5 column vol of running buffer. If the Ag:Ab interaction is shown to be strong enough to withstand high salt (e.g., up to 1 M KCl), then this can be included in the washing step to improve the purity of the eluted material.

   Before applying elution buffer, check that the A_{280} has returned to base line. Elute the bound antigen with 2 column vol of elution buffer (Table 3). If using an extreme condition, then the eluate should be neutralized upon collection (e.g., collect pH 2.5 eluate into 2 M Tris-HCl pH 8.3).

8. The column can be regenerated by washing with 10 column vol of PBS/0.1% sodium azide. If all the antigen does not elute, then it is necessary to remove this before re-equilibration using more extreme eluant (e.g., pH 2).

9. The column should be stored in a preservative at 4°C between runs, e.g., it should be left packed if possible. Avoid handling the gel any more than necessary.

4. Notes

1. The size of antigen can influence the nature of the matrix chosen. Here we have illustrated the use of

Sepharose, which is suitable for the majority of proteins, but if large proteins or protein complexes are being purified, then pore size must be a consideration. The nature of the matrix can affect the flow properties.

2. Coupling chemistries: Many different matrix types are available commercially with a variety of available coupling chemistries (*see* Table 2).

3. CNBr Sepharose CL 4B is available from Pharmacia and is preferable to noncrosslinked material.

4. *Never* allow the gel to dry out, especially when under suction.

5. The activated gel is incubated at 4°C prior to adding antibody because at higher temperatures the antibody would couple more quickly and tend to be mopped up by CNBr sites on the outside of the beads. This wastes sites on the inside of the beads and could crowd the outside sites enough to reduce efficiency.

6. Agitation machines come in various forms. We recommend a roller that imparts a continuous movement in one direction, e.g., shakers that change direction can inflict unnecessary shear forces.

7. The blocking reagent should contain an amine group attached to a moiety that will not impart ion-exchange properties or a hydrophobic character to the support, which could result in enhanced non-specific binding of proteins. Ehanolamine is commonly used because the moiety that becomes immobilized is small and is hydrophilic without carrying a charge.

8. Most generally available biochemical buffers are suitable for use with immunopurification. We have found phosphate or Tris to be useful in many situ-

ations. In general the buffer should be chosen to maintain a pH at which the ligand to be purified is stable. Buffers that can interfere with subsequent analysis should be avoided when practicable.

References

1. Kohler, G. and Milstein, C.(1975) Continuous culture of fused cells secreting antibody of predefined specificity. *Nature* (Lond.) **256**, 495–497.

2. Secher, D.S. and Burke, D. C. (1980) A monoclonal antibody for large scale purification of human leucocyte interferon. Nature **235**, 446–450.

3. Robb, R.J., Kutny, R.M., and Chowdhry, V. (1983) Purification and partial sequence analysis of human T-cell growth factor. *Proc. Natl. Acad. Sci. USA* **80**, 5990–5994.

4. Andreasson, P.A., Nielson, L.S., Grondahl-Hanson, J., Skriver, L., Zeuthen, J., Stephens, R.W., and Dano, K. (1984) Inactive proenzyme to tissue-type plasminogen activator from human melanoma cells, identified after affinity purification with a monoclonal antibody. *EMBO J*. **3**, 51–56.

5. Hsuing, L-M., Barclay, A.M., Brandon, M.R., Sim, E., and Porter, R.R. (1982) Purification of human C3b inactivator by monoclonal antibody affinity chromatography. *Biochem. J*. **203**, 293–299.

6. Stuart, M.C., Boscato, L.M., and Underwood, P.A. (1983) Use of immunoaffinity chromatography for purification of ^{125}I-labelled human prolactin. *Clin. Chem*. **29**, 241–245.

7. Bazin, H., Xhurdebise, L-M., Burtonboy, G., Lebacq, A-M., Clercq, L.D., and Cormont, F. (1984) Rat monoclonal antibodies. Rapid purification from in vitro culture supernatants. *J. Immunol. Meth*. **66**, 261–269.

8. Stallcup, K.C., Springer, T.A., and Mescher, M.F. (1981) Characterization of an anti-H2 monoclonal antibody and its use in large scale antigen purification. *J. Immunol*. **127**, 923–950.

9. Hudson, L. and Hay, F.C. (1980) in *Practical Immunology*, Blackwell Scientific, Oxford.

10. Kristianson, T. (1978), in *Affinity Chromatography* (Hoffmann-Osteuhof, O., Breitenbach, M., Koller, F., Kraft, D., and Scheiner, O., eds.) Pergamon, New York.

11. Chidlow, J.W., Bourne, A.J., and Bailey, A.J. (1974) Production of hyperimmune serum against collagen and its use for the isolation of specific collagen peptides on immunosorbent columns. *FEBS Lett.* **41**, 248–252.

12. Hill, C.R., Birch, J.R., and Benton, C. (1986) *Affinity Chromatography using Immobilised Monoclonal Antibodies. Bioactive Microbial Products* III. *Downstream Processing Special Publication of the Society for General Microbiology.* Academic, London.

13. Mains, R.E. and Eipper, B.A. (1976) Biosynthesis of adrenocaricotrophic hormone in mouse pituitary tumour cells. *J. Biol. Chem.* **251**, 4115–4420.

14. Avrameas, S. and Ternynck, T. (1967) Use of iodide salts in the isolation of antibodies and the dissolution of specific immune precipitates. *Biochem. J.* **102**, 37C.

15. Zoller, M. and Matzku, S. (1976) Antigen and antibody purification by immunoadsorption elimination of non-biospecifically bound proteins. *J. Immunol. Meth.* **11**, 287–295.

16. Melchers, F. and Messer, W. (1970) The activation of mutant β-Galactosidase by specific antibodies. Purification of eleven antibody activatable mutant proteins and their subunits on Sepharose immunoadsorbents. Determination of the molecular weights by sedimentation analysis and acrylamide gel electrophoresis. *Eur. J. Biochem.* **17**, 267–272.

17. Weintraub, B.D. (1970) Concentration and purification of human chorionic somato-mammotropin (HCS) by affinity chromatography. Application to radioimmunoassay. *Biochem. Biophys. Res. Commun.* **39**, 83–89.

18. Hill, R.J. (1972) Elution of antibodies from immunoadsorbents: Effects of dioxane in promoting release of antibody. *J. Immunol. Meth.* **1**, 231–245.

19. Andersson, K.K., Benjamin, Y., Douzov, P., and Bolny, C. (1979) The effects of organic solvent and temperature on desorbtion of yeast 3-phosphoglycerate kinase from immunoadsorbent. *J. Immunol. Meth.* **25**, 375–381.

Chapter 9

Dye–Ligand Chromatography

Sarojani Angal

1. Introduction

The affinity of two kinases for Blue Dextran was discovered in 1968 during gel permeation chromatography (1) and agarose electrophoresis (2). The observed interaction was mediated by the chromophore of Blue Dextran, Cibacron Blue F3G-A (3), which was postulated to bind to the supersecondary structure of certain enzymes known as the dinucleotide fold (4). Immobilized Cibacron Blue F3G-A replaced more expensive biological group-specific ligands in the affinity chromatographic purification of a wide variety of nucleotide-binding enzymes (5). The adsorbent was easier to synthesize, biochemically more stable, and had much higher capacities than the traditional nucleotide-based affinity matrices, making it more suitable for large-scale purifications.

Although the dinucleotide fold hypothesis was under scrutiny in many laboratories worldwide, a large number of other fiber-reactive dyes became available and proved their usefulness as "general" ligands in enzyme and protein purification (6,7). It has now become evident that the special place afforded to Cibacron Blue in dye ligand chromatography is merely historical.

The dyes appear to interact with ligand binding pockets or grooves on proteins through both hydrophobic and ionic forces complementary to those of the protein. The structures of most of the dyes are undisclosed, and the three-dimensional structures of most of the proteins of interest to the purification chemist are unknown. Thus it is not possible at present to predict which dye will provide the "best fit" for a given protein. The selection of a dye for protein purification must of necessity be empirical. The inexpensiveness of the dyes makes this possible even for the most modestly equipped laboratory. This chapter details methods for the synthesis and systematic screening (8,9) of a range of dye-adsorbents, together with guidance for the design of a purification scheme and for the optimization of operational parameters.

2. Materials

1. A selection of reactive dyes (e.g., Procion MX, H, HE, and P series available from ICI, Organics Division, Blackley, Manchester, UK. For other manufacturers, consult ref. 10). Some dyes are available from suppliers such as Sigma. Table 1 shows a list of suitable dyes.
2. Beaded agarose (e.g., Sepharose 4B from Pharmacia). About 40 mL of slurry per dye (see section 3).
3. Sodium chloride, 20% w/v.

Table 1
Examples of Some Triazine Dyes Suitable
for Making a Screening Kit[a]

Monochlorotriazines	Dichlorotriazines
Procion Blue H-B	Procion Brown MX-5BR
Procion Blue HE-RD	Procion Blue MX-4GD
Procion Red H-8BN	Procion Blue MX-R
Procion Green H4G	Procion Rubine MX-B
Procion Green HE-4BD	Procion Scarlet MX-G
Procion Brown H-2G	Procion Orange MX-G
Procion Red HE-3B	Procion Yellow MX-8G
Procion Violet H-3R	
Procion Turquoise H-A	
Procion Yellow H-A	

[a]Dyes near the top of the lists tend to bind more protein than those from the bottom of the lists. Consult refs. *9* and *10* for additional information.

4. Sodium carbonate, 5% w/v stock.
5. 10 mM Sodium phosphate (PB), pH 7.4.
6. 6M Urea.
7. Disposable plastic columns, 6 mm id, 1 mL bed volume and 5 mL reservoir volume (from, e.g., BioRad, Amicon).
8. 1M KCl in PB.
9. Crude extract containing protein of interest (e.g., human serum).
10. Materials required for assay of protein of interest.

3. Method

3.1. Preparation of Dye-Ligand Adsorbents

1. Wash the Sepharose repeatedly with a total of 10 vol of distilled water on a sintered glass funnel (porosity 3) under gentle suction pressure from a

water pump. Each time, allow surplus water to be removed to the point at which the surface of the gel is just beginning to "crack." This is referred to as moist gel.

2. Weigh out 20 g of moist gel into separate containers (one for each dye and one for undyed control). The container should have lids and wide necks for ease of transfer (e.g., 150 mL Sterilin plastic containers). Suspend in distilled water to a final volume of 50 mL.

3. Weigh ≥ 200 mg of each dye into a 25-mL container. Great care must be taken in weighing, since most of the dyes are extremely fine. Place a large sheet of white paper under the balance and moisten with water. Wear gloves and weigh into containers with lids. Wash the balance pan afterward, and wipe the surrounding area carefully.

4. Dissolve the dyes in distilled water to a concentration of 10 mg/mL.

5. Add 20 mL of the dye solution to the suspension of Sepharose. Add 20 mL of distilled water to the "control." Mix and leave at room temperature for 5 min.

6. Add 10 mL of the sodium chloride solution to each suspension. Incubate at room temperature for 30 min with occasional mixing.

7. Add 20 mL of the sodium carbonate stock solution to each suspension. Incubate the gels in a gently shaking water bath set at 45°C. Do not use magnetic stirrer bars, since these will damage bead structure. Dichlorotriazine-containing dyes (MX series) need to be incubated for 1 h; therefore the incubations should be staggered to allow for sufficient gel-washing time between preparation. Monochlorotriazines need to be incubated for 40 h.

8. Wash each dyed adsorbent with several liters of warm (45°C) distilled water followed by 10 vol of 6*M* urea and finally by distilled water alone until the washings are colorless. Some gel may get trapped in the sinter, particularly if the sinter is old, in which case it should be cleaned with a few milliliters of warm 6*M* HCl (or with chromic acid if unsuccessful).

9. Suspend the dichlorotriazine adsorbents in an equal volume of 1% (w/v) sodium carbonate and incubate at 45°C for 24–30 h to hydrolyze the second chlorine of the triazine ring. Wash once on a sintered glass funnel and suspend in 0.1% (w/v) sodium carbonate. Alternatively, decant the higher concentration of sodium carbonate and replace with water. Suspend the monochlorotriazine adsorbents in 0.1% (w/v) sodium carbonate.

10. Store the adsorbents at 4°C in the presence of 0.02% (w/v) sodium azide to inhibit microbial growth. Storage in mild alkali prevents acid hydrolysis of the dye from the matrix.

3.2. Screening of Dye Ligand Adsorbents

1. Pack 1 mL of adsorbent into each disposable column. It is best to code each adsorbent at this stage to save writing the full dye name on fractions later. The screening experiment is run at room temperature under gravity. Rest the "lip" or reservoir of the column above a 15-mL test tube, such that the surface of the gel is below the top of the tube. This arrangement can be engineered to prevent columns running dry. Place the test tubes in racks (LKB fraction collector racks from older models are best.

Each rack can accommodate 10 columns and the rack holders allow 20 columns to be run at once.) It is best to run all the selected columns at the same time, but since they will not all have identical flow properties, good organization is needed.

2. Prewash all columns with $1M$ KCl in PB (5 mL) and equilibrate with PB (10 mL). If any dye leaches from the columns, the prewash process will need to be repeated.

3. Adjust the protein concentration of the crude protein extract to about 20 mg/mL with equilibration buffer (previously dialyzed and clarified, if necessary). Apply 0.5 mL to each column, allow to run through, then collect the effluent. Add 0.5 mL of PB to the top of each column, allow to run through, and collect. This procedure allows proteins to reside in columns for similar amounts of time. Wash with a further 4 mL of PB and collect into the same tube.

4. Transfer the column to a new tube and elute with $1M$ KCl in PB (5 mL). Re-equilibrate in PB containing 0.02% sodium azide for storage.

5. Measure the absorbance at 280 nm of all the flow through and eluate fractions. Similarly measure the amount of the protein of interest. Compare both values with those of the "control" column, which represents the extent of nonspecific adsorption of proteins to underivatized Sepharose.

6. Calculate the purification factor offered by each column, as well as the yield. A good "positive" column is one that binds the protein of interest and little else, and gives a good yield on elution with KCl. A good "negative" column is one that allows the protein of interest to flow through, but binds many of the impurities.

4. Notes

1. All the fiber-reactive dyes are destined for the textile industry, and the crude preparations available from manufacturers contain various amounts of other solids (e.g., buffer salts, sodium chloride antidusting agents, surfactants). Thus the amount of chromophore in a given dry weight may vary from batch to batch or from one manufacturer to another. This implies that the ligand concentration of all the adsorbents prepared at the same time may not be comparable. Although the consequences of this variation may be minimal in a preliminary screening experiment, the possibility should be noted when synthesizing further batches (*see* Notes 5 and 10 in this section). Alternatively, the selected dyes can be purified before use (*10*), and dry weights standardized to give similar numbers of moles in a reaction mix.

2. Dye preparations are usually heterogeneous in composition and may contain other chromophores as minor impurities. The same dye (i.e., having the same Color Index number) when synthesized by two different manufacturers may contain different isomers. Whether these factors influence dye–ligand chromatography is not yet documented. It is therefore preferable to use the same manufacturer for a particular dye. It should be noted that some dyes are no longer being manufactured, and therefore availability should be checked before planning a scale-up.

3. Additional care needs to be taken with dichlorotriazinyl dyes, since the second chlorine may be available for reaction with proteins during chromatography. This complication is minimized by incubat-

ing the dyed adsorbents in alkaline solutions as described above, but quantitative hydrolysis of the second chlorine may not occur even under these conditions. It is therefore preferable to use monochlorotriazinyl dyes if possible, particularly as "positive" adsorbents.

4.   Although agarose is the most commonly used matrix, other supports have been reported (11). Crosslinked agarose polymers support lower ligand concentrations, but the molar effectiveness (moles of protein bound per mole of dye) is comparable to uncrosslinked agarose. Copolymers of agarose and acrylamide exhibit lower molar effectiveness, whereas cellulose-based adsorbents are particularly ineffective in binding proteins, although they bind much more dye than agaroses. Dextran polymers are comparable to agarose. The use of microparticulate silica has also been successful, but it is unlikely to be useful for screening experiments.

The choice of matrix can be crucial to the success of dye–ligand chromatography. It is recommended that initial screening experiments are carried out using agarose-based adsorbents. Once a dye is selected, further exploration can be carried out on matrices that may be more suitable to specific applications for reasons of physicochemical strength or stability.

5.   The amount of dye attached to a matrix can be measured by several methods:

a.   Spectrophotometric measurement of the unbound dye after coupling and comparison with total amount of dye at the start of synthesis. Dye absorbances are nonlinear at high concentrations because of ring-stacking effects, and solutions need to be well diluted. If the molar absorption coefficients are not known, an estimate can be based on dry weight (*see* Note 1 in this section).

b. Direct spectrophotometric measurement of dye released by acid hydrolysis (e.g., boiling for 10 min in 50% acetic acid or incubation at 37°C for 5 min in 6N HCl) of a known weight or settled volume of gel. Suitable for agaroses.

c. Direct spectrophotometric measurement of beads suspended in 50% v/v glycerol. This method is suitable for celluloses or acrylamide crosslinked gels that are not easily hydrolyzed by acid.

6. The number of dyes available at any one time may be 60–70. It is recommended that as many as possible be tried in an initial screen, particularly if the protein or source has not been evaluated before. Since dye adsorbents are simple to prepare and stable for many years, this is not a difficult task. If for some reason (e.g., difficulty of assay) this is not possible, however, it is advisable to start screening with dyes known to bind substantial amounts of protein and attempt selective elution.

   Some dyes bind more proteins from crude extracts than others in a consistent manner, whereas the reverse is true of others (9). It therefore follows that the best positive and negative dyes may be found by screening dyes in the low-binding and high-binding groups, respectively.

7. The best outcome of a screening experiment is the discovery of a positive dye column that can generate the protein of interest in pure form with little or no modification of the elution conditions. This is likely to be rare, particularly if the initial abundance of the protein is low.

8. The second best outcome is the discovery of a pair of adsorbents, one negative and one positive, that

can be operated sequentially to give substantially pure protein. When scaling up, it is preferable to run the negative column first, since it then protects the positive column from fouling, and the two columns can be run in tandem with almost the ease of a single step (8). This reasoning applies to an even greater extent if a good positive column is not found and a bioaffinity step is used instead.

9. A more common outcome of a screening experiment is that several positive and negative columns generate reasonable purification factors and must be investigated further before a decision can be made on selection. This can be done using the fractions already available by use of analytical techniques such as polyacrylamide gel electrophoresis. The qualitative examination of contaminants, together with the known physical properties of the protein of interest (e.g., molecular size) may make the choice of the positive adsorbent obvious. This in turn should allow a logical choice to be made for the best negative adsorbent (i.e., the one that binds most of the contaminant proteins bound by the positive dye).

10. Occasionally, the binding of a desired protein may be very tight, and the protein is not found in the flow-through or eluate. Under these conditions, stronger chaotrophic elution (e.g., 0.5M NaSCN) or affinity elution (e.g., substrate of effector) may be successful. Alternatively, a lower immobilized dye concentration can lower the affinity of the dye for the protein. This should not be done by dilution with blank Sepharose, but by variation of incubation time during synthesis.

11. If a qualitative examination after an initial screening does not give a clear result, further screening

should be carried out to modulate dye–protein interactions. The choice of buffer composition and pH has a significant effect on the binding of proteins to dye columns (12). A greater number of proteins will bind when the pH is lowered to 5.5. Thus adsorption at lower pH values can be used to improve the purification obtained on negative columns. Conversely, increasing pH to 8 and adding 100 mM NaCl to the adsorption buffer will minimize ionic interactions of positive dyes with proteins and result in improved purification factors. Another variable parameter is the inclusion of divalent cations that promote binding of proteins to dye matrices (10). Investigation of these parameters will allow for selection of the best positive and negative dyes from the secondary screening.

12. Once a decision is made about the number and identity of the dye-ligands to be used and the most significant parameters are optimized, a model purification should be carried out at the laboratory scale to fine tune the process.

References

1. Haekel, R., Hess, B., Lauterborn, W., and Winsler, K.H. (1968) Purification and allosteric poroperties of yeast pyruvate kinase. *Hoppe Seylers Z. Physiol. Chem.* **349**, 699.
2. Kopperschlaeger, G., Freyer, R., Diezel, W., and Hofmann, F. (1968) Some kinetics and molecular properties of yeast phosphofructokinase. *FEBS Lett.* **1**, 137.
3. Kopperschlaeger, G., Diezel, W., Freyer, R., Liebe, S., and Hofmann, F. (1971) Interaction of yeast phosphofructokinase with Blue Dextran 2000. *Eur. J. Biochem.* **22**, 40–45.
4. Stellwagen, E. (1977) Use of Blue Dextran as a probe for the nicotinamide adenine dinucleotide domain in proteins. *Acct. Chem. Res.* **10**, 92–98.

5. Dean, P.D.G. and Watson, D. H. (1979) Protein purification using immobilised triazine dyes. *J. Chromatogr.* **165**, 301–320.

6. Bonnerjea, J., Oh, S., Hoare, M., and Dunnill, P. (1986) Protein purification: The right step at the right time. *Bio/Technology* **4**, 954–958.

7. Qadri, F. (1985) The reactive triazine dyes: Their usefulness and limitations in protein purifications. *Trends Biotechnol.* **3**, 7–12.

8. Hey, Y. and Dean, P.D.G. (1983) Tandem dye-ligand chromatography and biospecific elution applied to the purification of glucose-6-phosphate dehydrogenase from *Leuconostoc mesenteroides. Biochem. J.* **209**, 363–371.

9. Scopes, R.K. (1986) Strategies for enzyme isolation using dye-ligand and related adsorbents. *J. Chromatogr.* **376**, 131-140.

10. Lowe, C.R. and Pearson, J.C. (1984) Affinity chromatography on immobilised dyes. *Meth. Enzymol.* **104**, 97–113.

11. Angal, S. and Dean, P.D.G. (1977) The effect of matrix on the binding of albumin to immobilized Cibacron Blue. *Biochem. J.* **167**, 301–303.

12. Angal, S. and Dean, P.D.G. (1978) The use of immobilised Cibacron Blue in plasma fractionation. *FEBS Lett.* **96**, 346–348.

Chapter 10

Aminohexyl-Sepharose Affinity Chromatography

Purification of an Auxin Receptor

R. D. J. Barker and H. M. Bailey

1. Introduction

Affinity chromatography is a widely used technique that is based on specific binding, usually between proteins and low-molecular weight ligands. It is often found that the presence of a spacer arm between the supporting matrix and the affinity ligand is beneficial to binding capacity. A convenient starting material for affinity chromatography involving carboxyl-containing ligands is aminohexyl-Sepharose, which is commercially available and contains a spacer arm; any carboxyl-containing ligand can easily be coupled to the amino group of the resin. As an example of the use of ami-

nohexyl-Sepharose, this chapter describes the purification of a soluble auxin receptor using an affinity resin
produced by derivatization of aminohexyl-Sepharose.

Plant-growth regulators, such as the auxin indole-
3-acetic acid (IAA), exert considerable influence on the
growth of higher plants (1). It is generally held that such
effects are likely to be mediated by receptors whose
function is to produce a cellular response upon percep-
tion of the plant-growth regulator. Such receptors may
be expected to show the majority of the following properties: they are likely to be proteins, in that proteins
possess the necessary specificity of binding; they must
display reversible binding; they must possess an asso-
ciation constant that would result in saturation of the
receptor at the low concentrations of plant-growth regulator present within cells; they are likely to be present in
very low concentrations; finally, they must cause
transduction of the plant growth regulator "signal": one
such effect is regulation of expression of particular
genes at the level of transcription (2). Plant hormone
receptors have recently been extensively reviewed (3).

To fully characterize plant hormone receptors, purification is necessary. Since cytoplasmic receptors are
present at extremely low concentrations [typically between 0 and 0.5 pmol binding sites per mg protein, (4,5)]
this is a difficult task. The possession of a low dissociation constant (typical K_d 10^{-6} to 10^{-9} M) suggests that
affinity chromatography could be successfully applied.
Various attempts have been made to use this method for
purification of IAA receptors. Venis (6) synthesized a
derivative of the synthetic auxin, 2, 4-dichlorophenoxy-
acetic acid (2,4-D); this derivative (2,4-D-ε-L-Lysine),
when coupled to Sepharose, retained proteins from pea
and wheat leaf extracts. Although a fraction, eluted
with 2 mM KOH (pH 11.2), was found to stimulate

Escherichia coli RNA polymerase-dependent incorporation of ATP into 10% trichloroacetic acid (TCA) precipitable material, no binding assays were conducted. Further studies (7), using the same adsorbent and elution conditions, confirmed stimulation of UTP incorporation in a TCA-insoluble form, but no binding assays were carried out.

Tappeser et al. (8) synthesized a number of affinity adsorbents by coupling different auxins and anti-auxin analogs to epoxy-activated Sepharose 6B. The most useful adsorbent, when tested with solubilized auxin-binding proteins from corn membranes, was found to be 2-hydroxy-3,5-diiodobenzoic acid. Although $0.15M$ NaCl eluted all auxin-binding proteins, they concluded that elution with $10^{-4}M$ naphthalene-1-acetic acid (NAA) gave greater purification: unfortunately no binding assays were conducted to confirm this. Auxin binding proteins have also been purified from etiolated mung bean seedlings using 2,4-D-ε-L-Lysine coupled to CNBr-activated Sepharose 4B (9). These have been shown to possess high-affinity binding for 2,4-D (10) and to stimulate RNA synthesis (11). A soluble auxin-binding protein from tobacco callus has been partially purified by gel filtration: this showed high-affinity binding and stimulated RNA synthesis in vitro (4).

The procedure described here has been successfully applied (5) to an IAA-binding protein extracted from suspension cultured cells of tobacco (*Nicotiana tabacum*). The adsorbent used for affinity chromatography is prepared by coupling 2,4-D to aminohexyl-Sepharose (AH-Sepharose). When extracts of tobacco cells possessing high-affinity IAA binding, as demonstrated by Scatchard analysis (12), are passed through a column of this resin, the high-affinity binding is retained on the column. Desorption is achieved by addition of IAA to the eluting buffer; after removal of IAA by gel filtration, high-affinity IAA binding may again be demonstrated. In the presence of IAA, this purified material stimulated transcription, as measured by the incorporation of UTP into TCA-insoluble material by isolated nuclei from tobacco cells (5).

2. Materials

2.1. Preparation of 2,4-D-AH-Sepharose

1. AH-Sepharose 4B may be obtained from Pharmacia.
2. 0.5M NaCl; 2 L is sufficient.
3. 900 mg of 2, 4-dichlorophenoxyacetic acid (2,4-D) is dissolved in approximately 30 mL of distilled water; to facilitate this, 1M KOH is added dropwise to maintain pH in the range of 8–11, while stirring in a fume hood. It does not matter if the final volume exceeds 30 mL. *Caution: 2,4-D is toxic; appropriate precautions must be taken, including use of gloves and mask; whenever possible, procedures involving this reagent should be conducted in a fume hood.*
4. 900 mg of 1-ethyl-3-(3-dimethyldiaminopropyl) carbodiimide (EDC) is dissolved in approximately 30 mL of distilled water. EDC is toxic; take appropriate precautions, as described for 2,4-D.
5. 1M HCl.
6. 0.05M sodium acetate buffer, pH 4.0, containing 1.0M NaCl; 1 L is adequate.
7. 0.05M Tris-HCl, pH 9.0, containing 1.0M NaCl; 1 L is sufficient.
8. Sodium azide.
9. Sephadex G-50 (Pharmacia). A 2.4 x 38 cm column of Sephadex G-50, equilibrated in BB + 0.1M NaCl

(see below), is suitable for removal of IAA following desorption of proteins from the 2,4-D-AH-Sepharose column.

2.2. Extraction and Binding Assay

1. The buffer used for extraction of the soluble IAA receptor from tobacco cells contains:
   $0.2M$ H_3BO_4
   $0.13M$ KCl
   2 mM $Na_2B_4O_7$
   4 mM Sodium diethyldithiocarbamate (Sigma) (added just before use)
   This is adjusted to pH 6.8 with KOH (*see* Note 1 in section 4).

2. Ultrafiltration cell and filters. A capacity of 100 mL is required: membranes with a molecular weight cut-off of 10,000 are appropriate (e.g., Ulvac ultrafiltration membrane type G1OT, Chemlab).

3. Binding assays are carried out in binding buffer (BB), which contains: 20 mM Tris, 50 mM H_3BO_4, 1.5 mM $Na_2B_4O_7$, 100 mM KCl, 10 mM $MgCl_2$, 25 mM $NaMoO_4$, 0.1 mM Na_2 EDTA, and 0.1 mM dithiothreitol (DTT). The pH is adjusted to 7.5 with $1M$ HCl. For chromatography this buffer is supplemented as follows:

   a. For equilibration of columns, by addition of NaCl to $0.1M$ (BB + $0.1M$ NaCl);

   b. For elution of proteins retained by the affinity column, by addition of NaCl to $0.05M$ and IAA to $0.04M$ (BB + $0.05M$ NaCl + $0.04M$ IAA; IAA dissolves more readily if the pH is raised with $1M$ KOH; once in solution the pH should be adjusted to 7.5). The increase in ionic strength is to minimize nonspecific binding to the affinity column.

4. High specific activity IAA is required: ^{3}H-IAA (777 G Bq/mmol) from Amersham International is suitable. For use ^{3}H-IAA is diluted with BB. Appropriate safety precautions must be taken while using radioisotopes, and local rules must be observed.

5. Nonradiolabeled IAA is initially dissolved in methanol. A $3 \times 10^{-2}M$ stock solution is suitable and may be stored at $-20°C$ for up to 1 mo. For use the stock solution is diluted with BB. IAA solutions are light-sensitive and should be protected from light.

6. Dextran-coated charcoal (DCC) suspension: 0.6% (w/v) charcoal and 0.06% (w/v) dextran (Sigma, average mol. wt., 80,000) in distilled water.

7. Protein assay: a means of estimating protein concentration is required. The method of Bradford (13), as modified by Spector (14), is suitable (*see* Chapter 1).

3. Method

Unless otherwise specified, all procedures are carried out at 4°C.

3.1. Preparation of 2,4-D-AH-Sepharose

The method used for the preparation of the affinity resin is based on the general procedure for coupling carboxyl-containing ligands to AH-Sepharose provided by the manufacturers (Pharmacia, Sweden). Coupling is carried out at room temperature.

1. Six grams of AH-Sepharose 4B is allowed to swell in an excess (1200 mL) of 0.5M NaCl and is then washed with 4 x 200 mL aliquots of 0.5M NaCl, followed by 4 x 200 mL distilled water. These washes are carried out on a sintered glass funnel

over a vacuum (a water pump is adequate). The resin is transferred to a round-bottom quickfit flask.

2. The 2,4-D and EDC solutions are added to the resin, and the pH is adjusted to pH 5.0. The pH is maintained between 4.5 and 5.5 for 90 min by the addition of drops of 1M HCl. The flask is gently swirled (by hand) during this period.

3. The slurry is then stirred overnight by rotation on a slowly moving rotavap (without vacuum; a Buchi Rotavapor is suitable); magnetic stirrers, which may destroy the bead structure, should be avoided.

4. After 16–20 h, the resin is transferred to a glass sinter and washed alternately with buffered solutions of alkaline and acidic pH (0.05M sodium acetate buffer, pH 4.0, containing 1.0M NaCl, and 0.05M Tris-HCl, pH 9.0, containing 1.0M NaCl). Four washes with each buffer (200 mL per wash) are adequate; the resin is finally washed with 200 mL of distilled water.

5. The resin is stored at 4°C in 0.05M sodium acetate buffer, pH 4.0, contining 1M NaCl and 0.02% (w/v) sodium azide as a preservative (stable for several weeks at 4°C.) For the protocol described here, a 6 x 1 cm column of 2,4-D-AH-Sepharose is adequate.

3.2. Binding Assay for IAA Receptor

A variety of strategies are available (*see* Note 2 in section 4). Assays utilizing Dextran-coated charcoal (DCC) to adsorb free IAA, leaving receptor-bound IAA in the supernatant, have been found satisfactory with the IAA receptor from tobacco cells. This procedure, based on that of Oostrom et al. (*15*), is described here, but it is recommended that workers use whatever assay they find effective and convenient for soluble receptors from any given tissue.

1. Twelve tubes are set up as shown:

Tube	^{3}H-IAA, pmol	Nonradioactive IAA, pmol
1	2.5	0
2	2.5	2.5
3	2.5	5.0
4	2.5	10
5	2.5	25
6	2.5	40
7	2.5	80
8	2.5	120
9	2.5	180
10	2.5	240
11	2.5	2400
12	2.5	0

All tubes are made up to 0.5 mL with BB.

2. One half milliliter of sample is added to each tube 12: This tube is made up to 1 mL with BB and is used to check on the efficiency of binding of free IAA by the DCC.
3. The tubes are incubated at 25°C for 20 min and are then cooled in an ice bath for 20 min.
4. One milliliter of DCC suspension is added to each tube and the tubes are shaken for 20 min at 4°C and centrifuged at 2000g for 10 min.
5. An aliquot of the supernatant is removed for counting in a liquid scintillation counter. Dissociation constants and numbers of binding sites per mL of sample are calculated by Scatchard analysis (12).

3.3. Affinity Chromatography Using 2,4-D-AH Sepharose

1. Thirty grams (fresh weight) of cells are harvested and suspended in 50 mL of extraction buffer and

homogenized in a Potter-Elvehjam glass vessel (5 strokes at 1500 rpm) (*see* Note 3 in section 4).

2. The homogenate is centrifuged at 100,000g for 45 min and the supernatant concentrated in an ultrafiltration cell to approximately 15 mL (*see* Note 4 in section 4).

3. The concentrated extract is adjusted to pH 7.5 with 1M KOH, and solid NaCl is added to a final concentration of 0.1M; 6 mL is used in a prepurification binding assay (*see* Note 5 in section 4).

4. The remainder is applied to a 6 x 1 cm column of 2, 4-D-AH-Sepharose, pre-equilibrated by the passage of 40–50 mL of BB + 0.1M NaCl. A flow rate of 0.25–0.3 mL/min is used throughout, and 3-mL fractions are collected. After the sample has been applied, the column is washed with 30 mL BB + 0.1M NaCl. The nonretained fraction, identified by absorbance at 280 nm, may be collected for assay of high-affinity IAA binding (*see* Note 6 in section 4).

5. Desorption of protein bound to the 2,4-D-AH-Sepharose column is achieved with BB + 0.05M NaCl + 0.04M IAA (*see* note 7 in section 4). The first 4–5 fractions that contain IAA (detected by A_{280}) are pooled (12–15 mL), and a small sample is taken for protein assay.

6. The desorbed fraction is applied to the Sephadex G-50 column to remove IAA from the protein. To ease identification of protein-containing fractions after gel filtration, an inert protein, such as hemoglobin, may be added (0.33 mg/mL) prior to application to the Sephadex G-50 column (*see* note 8 in section 4).

7. The protein peak (void volume) eluted from the Sephadex G-50 column is pooled and assayed for high affinity IAA binding in a binding assay. This partially purified material is suitable for use in studies of receptor function (*see* Note 9 in section 4).

4. Notes

1. The extraction buffer described is based on that of van der Linde et al. (4). They found that the use of Tris buffer at pH 7.5 for extraction of soluble receptors from tobacco callus resulted in high levels of contamination by polyphenols. The use of borate buffers at pH 6.8 appreciably reduced this contamination. When polyphenols are not a problem, a 50 mM Tris-HCl buffer, pH 7.5, containing 2 mM Na$_2$ EDTA and 2 mM DTT, can be used.

2. Various methods are available for assessment of dissociation constants and numbers of binding sites. Equilibrium dialysis is the most reliable method, but requires relatively large numbers of binding sites; the same drawback applies to the use of gel-filtration columns. Dextran-coated charcoal has been widely used to adsorb free hormone; following centrifugation, receptor-bound hormone remains in the supernatant and may be quantified. Although this is not an equilibrium method, it can be used, providing the rate constant for dissociation of the receptor–hormone complex is slow. Alternative assays involve filtration through a nitrocellulose filter, the use of ultrafiltration (or centrifugal ultrafiltration), or precipitating with ammonium sulfate to separate receptor-bound hormone from free hormone. Methods of quantifying and analyzing plant hormone receptors have been reviewed (3, 16).

3. Any homogenization method that results in the majority of cells being broken, while avoiding denaturation of proteins, is acceptable.

4. This step is to ensure that receptor concentration is sufficient to be detected in the binding assay; binding assays with extracts of low protein concentra-

tions are often unsuccessful. Concentration also serves to reduce the time taken to load the extract onto the 2,4-D-AH-Sepharose column.

5. Affinity chromatography has only been successful when high-affinity binding has been detectable in the concentrated extract.

6. The nonretained fraction routinely contains more than 90% of the protein applied to the 2,4-D-AH-Sepharose column; no high-affinity binding is found in this fraction.

7. The high concentration of IAA is to ensure complete desorption of bound receptor; the lower ionic strength of this buffer, compared to that of the equilibration buffer (also used to wash the column prior to desorption), is to minimize nonspecific desorption. If the $0.04M$ IAA is replaced by $0.04M$ acetate, high-affinity IAA binding is not recovered.

8. The protein concentration of the fraction desorbed by BB + $0.05M$ NaCl + $0.04M$ IAA is usually in the range of $0.02–0.04$ mg/mL; this fraction contains 1–2% of the protein initially applied.

9. The procedure described typically results in a 35–40-fold purification of IAA receptor, with recovery exceeding 40% of that initially present. In all cases in which high-affinity binding has been demonstrable, the purified fraction has stimulated transcription in vitro, as measured by incorporation of UTP into TCA-insoluble material (5).

References

1. Wareing, P. F. and Phillips, I. D. J. (1981) *Growth and Differentiation in Plants* Pergamon, Oxford.
2. Hagen, G., Kleinschmidt, A., and Guilfoyle, T. (1984) Auxin-regulated gene expression in intact soybean hypocotyl and excised hypocotyl sections. *Planta* **162**, 147–153.

3.   Venis, M. (1985) *Hormone Binding Sites in Plants* Longman, New York.

4.   van der Linde, P. C. G., Bouman, H., Mennes, A. M., and Libbenga, K. R. (1984) A soluble auxin-binding protein from cultured tobacco tissues stimulates RNA synthesis *in vitro*. *Planta* **160**, 102–108.

5.   Bailey, H. M., Barker, R. D. J., Libbenga, K. R., van der Linde, P. C. G., Mennes, A. M., and Elliott, M. C. (1985) Auxin binding site in tobacco cells. *Biol. Plant.* **27**, 105–109.

6.   Venis, M. A. (1971) Stimulation of RNA transcription from pea and corn DNA by protein retained on Sepharose coupled to 2,4-dichlorophenoxyacetic acid. *Proc. Natl. Acad. Sci. USA* **68**, 1824–1827.

7.   Rizzo, P. J., Pedersen, K., and Cherry, J. H. (1977) Stimulation of transcription by a soluble factor isolated from soybean hypocotyl by 2,4-D affinity chromatography. *Plant Sci. Lett.* **8**, 205–211.

8.   Tappeser, B., Wellnitz, D., and Klambt, D. (1981) Auxin affinity proteins prepared by affinity chromatography. *Z. Pflanzenphysiol.* **101**, 295–302.

9.   Sakai, S. and Hanagata, T. (1983) Purification of an auxin-binding protein from etiolated mung bean seedlings by affinity chromatography. *Plant Cell Physiol.* **24**, 685–693.

10. Sakai, S. (1985) Auxin-binding protein in etiolated mung bean seedlings: Purification and properties of auxin-binding protein-II. *Plant Cell Physiol.* **26**, 185–192.

11. Sakai, S., Seki, J., and Imaseki, H. (1986) Stimulation of RNA synthesis in isolated nuclei by auxin-binding proteins-I and -II. *Plant Cell Physiol.* **27**, 635–643.

12. Scatchard, G. (1949) The attractions of proteins for small molecules and ions. *Ann. NY Acad. Sci.* **51**, 660–672.

13. Bradford, M. M. (1976) A rapid and sensitive method for the quantitation of microgram quantities of protein utilizing the principle of protein-dye binding. *Anal. Biochem.* **72**, 248–254.

14. Spector, T. (1978) Refinement of the Coomassie Blue method of protein quantitation. *Anal. Biochem.* **86**, 142–146.

15. Oostrom, H., Kulescha, Z., van Vliet, Th. B., and Libbenga, K. R. (1980) Characterization of a cytoplasmic auxin receptor from tobacco pith callus. *Planta* **149**, 44–47.

16. Rubery, P. H. (1981) Auxin receptors. *Ann. Rev. Plant Physiol.* **32**, 569–596.

Chapter 11

Purification of DNA-Dependent RNA Polymerase from Eubacteria

N. W. Scott

1. Introduction

In eubacteria, the structure of DNA-dependent RNA polymerase is conserved; the enzyme is composed of at least four different subunits: beta (β), beta-prime (β'), and alpha (α), which form the core (ββ'α$_2$), and sigma (σ), which is required for efficient promoter selection (1–3).

Most of the procedures available for the isolation of this enzyme follow a set format; they normally make use of a gross fractionation step, such as phase separation, followed by several forms of column chromatography that make use of the protein's affinity for DNA or heparin and its binding to compounds such as DEAE (1,4,5).

Another feature of the enzyme that is utilized for its purification is its size. The absolute sizes of the subunits vary between species, but the M_r values of the beta subunits are usually within the range 140,000–180,000, the M_r values of the alpha subunit are about 40,000, and M_r values of the sigma subunits are in the range 20,000–100,000. This means the holoenzyme can have a M_r of greater than 400,000, so it can be separated from other proteins by molecular sieving or even centrifugation (1–3). The conservation of subunit structure also means that the enzyme can be easily identified on polyacrylamide gels; this is a useful property during the purification procedure.

Despite the similarity in enzyme structures among species, there has been a proliferation of isolation procedures available (see ref. 4 for a list of the 14 procedures used for *Bacillus subtilis* alone). This is because different species have intrinsic properties that can influence RNA polymerase isolation and because of the presence of minor forms of the enzyme. For most organisms, only a single form of the enzyme has been found (1–3), but studies of *B. subtilis* (4), *Escherichia coli* (6), and *Streptomyces coelicolor* (7) have indicated that these organisms contain multiple forms of the enzyme that contain different sigma factors, and as such have different promoter specificities. The difficulty for purification comes from the fact that the alternative forms of the enzyme can represent a very minor amount of the total holoenzyme pool, less than 5% of the total in the case of the four *B. subtilis* minor holoenzymes(8).

The procedure described in this chapter is one used successfully to purify RNA polymerase from several Gram negative bacteria (9). It can be used with other bacteria and should yield enzyme purified to an extent that makes it suitable for transcription studies or for analysis of subunit properties. The method has four steps: (1) cell lysis and removal of cell debris; (2) a series of ammonium sulfate precipitations; (3) Bio-Gel A-5m chromatography; and (4) DNA-cellulose chromatography. The first two steps are modifications of methods first introduced by Burgess (10); the aim of these steps and step 3 is to separate the enzyme from the bulk of the cell protein and particularly to remove contaminating DNA-binding proteins and any remaining nucleic acids prior to affinity chromatography with DNA-cellulose. This column is very effective for the purification of RNA polymerase, particularly when a linear salt gradient is used to elute the protein (4). It has advantages over other materials in that it is easy to make, it allows the user the choice of DNA for optimum purification, it stores well, and it is comparatively inexpensive. The column can also be used for the isolation of DNA binding proteins.

2. Materials

1. The buffer used in the purification is TGED (11): 0.01M Tris-HCl, 5% (v/v) glycerol, 0.1 mM EDTA, 0.1 mM dithiothreitol. TGED containing various concentrations of NaCl are required; 1-L volumes of the following are usually adequate:

    TGED
    TGED + 0.15M NaCl
    TGED + 0.5M NaCl
    TGED + 1.0M NaCl
    TGED + 1.5M NaCl

    The purified enzyme is stored in TGED buffer containing 50% glycerol and 0.1M NaCl (11).
2. Phenylmethyl sulfonyl fluoride (PMSF) (use care, toxic). Make fresh prior to use. It is solubilized in ethanol prior to dilution with TGED + 1M NaCl buffer, and is then added to the lysate when re-

quired (see methods) to give a final concentration of 23 µg/mL.

3. Ammonium sulfate.

4. Bio-Gel A5-m (Bio-Rad). Use a 25 x 2 cm column pre-equilibrated with TGED + 0.5M NaCl.

5. DNA–cellulose. This is prepared by the method of Alberts and Herrick (12). Mix 160 g of cellulose with 500 mL of 2 mg/mL calf thymus DNA (or DNA of choice) in 0.01M Tris-HCl (pH 7.9) and 0.001M EDTA (TE buffer), and dry in large drying dishes (14-cm Petri dishes are suitable) at 50°C for 3–4 d followed by 24 h under vacuum. Powder the paste during the drying period with a glass rod. Suspend the powder in 20 vol of TE buffer and leave at 4°C for 24 h. Wash the DNA–cellulose twice with TE buffer to remove free DNA; use an acid-washed scintered glass funnel over a vacuum for this procedure. Suspend the slurry in TE buffer containing 0.15M NaCl and either divide into 20-mL aliquots and freeze at –20°C until required or load into a column. For the protocol given here, a 30 x 1 cm column is adequate. The amount of DNA absorbed to the cellulose can be ascertained by boiling the suspension at 100°C for 20 min then measuring A_{260nm} of the liquid phase (50 µg mL DNA gives absorbance of 1). Approximately 50% of the DNA will absorb to the cellulose under the conditions described. Before and after use, wash the DNA-cellulose column with 10 mL of TGED + 1.5M NaCl and 60 mL of TGED + 0.15M NaCl at a flow rate of 15 mL/h.

6. The RNA polymerase assay used here has evolved from the assays used by Burgess (10) and Jaehning et al. (13). As with all radioactive assays, safety precautions must be taken and local rules must be

followed. Make the assay buffer as a 10x stock and keep at –20°C until required. 10x assay buffer:

400 m*M* Tris-HCl (pH 7.9)
1.0 mM Dithiothreitol
2.0 mM Spermidine
100 mM MgCl$_2$

Dissolve ribonucleoside triphosphates and DNA in 1x assay buffer. The ribonucleoside triphosphate stock solution contains ATP, CTP, and GTP at 2 m*M*, UTP at 0.5 m*M*, and 200 µCi/mL ³H-UTP. Store the solution at –20°C. Make a 1 mg/mL calf thymus DNA (or DNA of choice) solution fresh for the assay.

The assays are carried out in acid-washed, oven-sterilized glass tubes. The components should be added to the tubes, held in ice, in the following order:

30 µL Column fraction
50 µL 1x Assay buffer
10 µL DNA solution
10 µL Ribonucleoside triphosphate solution

Start RNA synthesis by shifting the assay tubes to 30°C, and continue the incubation for 20 min. Stop the reaction by putting 40 µL duplicate samples onto labeled GF/C filters (Whatman), and dropping them into ice-cold 5% (w/v) trichloroacetic acid (TCA) containing 0.01*M* sodium pyrophosphate (reduces nonspecific binding). After 15 min, wash the filters twice with TCA and twice with absolute ethanol. Dry the filter, then measure the precipitated radioactivity.

7. SDS-Polyacrylamide gel electrophoresis (PAGE) (*see* vol. 1 of this series). For rapid analysis of column fractions use 7.5–10% (w/v) acrylamide gradient gels and stain the gel with Coomassie brilliant

blue, the fractions containing the enzyme can be identified from the presence of core subunits (9).

8. Dialysis tubing.

3. Method

All procedures carried out at 4°C.

3.1. Cell Lysis

1. Lyse 50 g of cells (wet weight) suspended in 200 mL TGED + 1*M* NaCl using a French pressure cell (*see* Note 4 in section 4).
2. Add 23 µg/mL PMSF to the lysate to reduce proteolytic activity (*see* Note 6 in section 4).
3. Centrifuge the suspension at 100,000*g* for 1.5 h to remove cell debris (*see* Note 8 in section 4).

3.2. Ammonium Sulfate Precipitations

1. Measure the volume of supernatant, then put it into a beaker on ice, and stir using a magnetic bar. The stirring rate should be fast enough to provide effective mixing, but not fast enough to cause frothing. Slowly add 23.1 g of solid ammonium sulfate per 100 mL supernatant, making sure there is no build up of undissolved material. Also add 0.05 mL of 1*M* NaOH per 10 g of ammonium sulfate to keep the pH above 7. Continue the stirring for 15 min, then centrifuge at 10,000*g* for 30 min, and discard the pellet.
2. Measure the volume of the supernatant and stir. Slowly add 10.75 g of solid ammonium sulfate per

100 mL of supernatant to give a 50% saturated solution. Again control the pH with NaOH. Centrifuge at 10,000g for 30 min, and discard the supernatant.

3. Suspend the pellet in TGED + 1M NaCl containing 25.6 g of solid ammonium sulfate (42% saturation). Stir for 20 min. Centrifuge at 10,000g for 30 min, discard the supernatant, and resuspend in the same buffer (42% ammonium sulfate), centrifuge as above, then discard the supernatant.

3.3. Bio-Gel A-5m Chromatography

1. Dissolve the pellet in 5 mL of TGED buffer and load onto the Bio-Gel column; use a flow rate of 20 mL/h. Elute the column with TGED + 0.5M NaCl and collect 10-mL fractions. The highest molecular weight proteins elute first, so analyze 10 fractions around 1.8 x the void volume; this can be estimated from the A_{280nm} of the fractions. Use PAGE and the enzyme assay to identify the fractions containing the enzyme.

2. Pool the enzyme containing fractions and measure the volume. Add 35 g of ammonium sulfate per 100 mL of eluate (plus required NaOH) to give a 50% saturated solution. Centrifuge at 10,000g for 30 min. Suspend the pellet in TGED such that the conductivity of the solution is equivalent to TGED buffer containing 0.15M NaCl.

3.4. DNA–Cellulose Chromatography

1. Load the dissolved enzyme onto the DNA–cellulose column; use a flow rate of 15 mL/h. Wash the column with 60 mL of TGED + 0.15M NaCl, then

elute the protein with a linear salt gradient (60 mL TGED + 0.15–1.5M NaCl). Collect 6-mL fractions, and analyze the salt gradient by PAGE and the RNA polymerase assay. The gradient fractions can be identified by $A_{280\,nm}$ and by using a refractometer to measure salt concentration.

2. Pool the fractions containing the enzyme and dialyze into storage buffer. Store at –20°C.

4. Notes

1. If large amounts of DNA cellulose are made, it can either be stored at 4°C for up to 9 mo or frozen at –20°C. To avoid excessive freeze-thaw cycles, store in small aliquots. Commercially prepared DNA, such as calf thymus, can be used for this column. Alternatively, DNA prepared from the organism being studied can be used.

2. RNA polymerase assays may require modification depending upon the properties of the enzyme being studied. A larger volume of sample or a longer incubation time may be required for analysis of column fractions. To optimize assays with purified enzyme, salt concentration, pH, the DNA template, the specific activity of the label, and the type of labeled ribonucleoside can all be changed. The enzyme is particularly sensitive to rapid dilution, so to get maximum activity it is particularly important to add the assay components in the order indicated (14).

   Simple modification of the assay can provide valuable information on the nature of the enzyme. For instance, it can be seen if the extract contains a holoenzyme: prebinding the enzyme to DNA for 10 min at 30°C prior to the addition of ribonucleosides allows for the formation of a rifampicin-resistant complex only if the enzyme used is a holoenzyme (3). If species-specific bacteriophage or plasmids containing defined promoters are available, they can be used to study the promoter specificity of the enzyme. If they are not available, however, analysis of binding to T7 promoters is a simple system that can be used to provide some information (3).

3. The method described can be used for between 10 and 500 g of cells, if modifications are made to the size of the columns. It is not always possible to get such amounts of cells, for instance for stages of sporulation or for synchronized cell cultures. For smaller amounts of cells, the micropurification system of Gross et al. is suitable (15).

4. If a French pressure cell cannot be used for lysis, other methods can be tried. Some experimentation may be necessary to identify the best procedure for the cells being used. Some of the most popular techniques in use involve blending frozen cells in a Waring blender (10). Either blend 50 g of frozen cells with 150 g of Superbrite 100 glass beads and 50 mL of lysis buffer for 5 min at low speed and 10 min at high speed (10), or blend the cells in the presence of lysozyme and deoxycholate instead of the glass beads. For this latter method, blend pieces of frozen cells 2 cm or less in diameter, in grinding buffer (0.05M Tris-HCl) (pH 7.9), 5% (v/v) glycerol, 2 mM EDTA, 0.1 mM dithiothreitol, 1 mM 2-mercaptoethanol, 0.233M NaCl, 130 µg lysozyme/mL, and 23 µg PMSF/mL. For 50 g of cells, use 150 mL of buffer, blend at low speed for 2–3 min to resuspend the cells, then incubate the cells at 4°C for 20 min. Add 2.5 mL 4% (v/v) sodium deoxycholate and blend at

low speed for 30 s, and continue the incubation for another 20 min. Finally, blend at high speed for 30 s to shear the DNA. This method has been used most often with the Polymin P purification technique (Note 7 in this section). Grinding frozen cells with alumina or gentle lysis with deoxycholate and lysozyme (15) are techniques that can also be used.

5. No DNAse is used in this method because it was reported that it can introduce proteases (11). More suitable grades of DNAse may now be available.

6. PMSF is added to the lysate to reduce proteolytic activity. If this is not effective, the lysate may be treated with hemoglobin–Sepharose to absorb out the proteases (4). The effectiveness of enzyme inhibitors may be judged by PAGE analysis of protein degradation in extracts incubated in the presence and absence of inhibitors.

7. 1M NaCl is included in the lysis buffer to dissociate the enzyme from DNA; the core and σ factors have DNA-binding properties (9).

The ammonium sulfate concentrations used here were optimized for *E. coli*. They may require modification for use with other organisms, particularly because the use of high concentrations of NaCl and ammonium sulfate can be detrimental to enzyme activity and can dissociate subunits from the core (4). Avoid using high-salt treatments with purified enzyme, since this can inhibit enzyme activity (4).

Alternative procedures that could be used are PEG6000 (15) or Polymin P (11) precipitations, or it is possible to apply crude lysates directly to affinity columns without using a crude fractionation (7).

8. Centrifugations can be carried out with a cooled bench centrifuge, but longer spin times must be used.

9. The Bio-Gel A-5m column is prewashed with TGED + 0.5M NaCl, and can be reused indefinitely if kept in this buffer.

10. Although Mg^{2+} improves RNA polymerase stability, it is not used in the column buffer since it interferes with RNA polymerase binding to DNA-cellulose (4). The absence of Mg^{2+} also reduces any DNAse activity in the extract and prevents fungal growth on the agaraose column (11).

11. Some methods use higher concentration of glycerol [up to 20% (v/v)] in the buffers since it stabilizes the holoenzyme on DNA–cellulose (4).

12. Both the assay and PAGE are used to locate the enzyme, since some discretion is allowed as to which fractions are to be pooled. This is particularly important for the DNA–cellulose column, since fraction-containing contaminants can either be subjected to further purification steps or discarded (9).

    To decide whether a protein that copurifies with the core is an enzyme subunit or a contaminant is not an easy task. Several different approaches can be used to investigate this. A physical association can be established by centrifugation (10), one-dimensional PAGE (4), or even two-dimensional PAGE (first-dimension, nondenaturing, second-dimension, denaturing) (9). To show a functional relationship involves purification of individual proteins, by either column chromatography (e.g., ref. 5) or excision from acrylamide gels, and addition to in-vitro transcription systems, where the effect of the protein on enzyme function can be evaluated (16).

13. If further purification steps are required, several different approaches are available. A second DNA-cellulose column with a different salt gradient may

resolve the enzyme from the contaminants. Alternatively a different affinity column could be used; Heparin–agarose columns are suitable (5). Ion exchange chromatography using DEAE sephadex (9) or DEAE-cellulose can also be effective (4), and finally glycerol gradient centrifugation can also be used (10).

If different columns are used, avoid changing the salt in the buffer, since this can lead to enzyme inactivation.

14. The purity and amount of enzyme that will be recovered by this method will be species-specific and will relate to the conditions used to grow the bacteria, but the amount of enzyme may well be in the order of 1–10 mg. Enzyme activity is also likely to be variable; enzyme activity can vary between 200 and 2000 units (per mg protein) (9) (where a unit is the amount of enzyme required to incorporate 1 nmol UMP into acid-insoluble material in 10 min).

References

1. Burgess, R.R. (1976) Purification and Physical Properties of E. coli RNA Polymerase, in *RNA Polymerase* (Losick, R. and Chamberlin, M., eds.) Cold Spring Harbor Laboratory, Cold Spring Harbor, New York.
2. Fukuda, R., Ishihams, A., Saitoh, T., and Taketo, M. (1977) Comparative studies of RNA polymerase subunits from various bacteria. *Mol. Gen. Genet.* **154**, 135–144.
3. Wiggs, J.L., Buch, J.W., and Chamberlin, M.J. (1979) Utilization of promoter and terminator sites on bacteriophage T7 DNA by RNA polymerases from a variety of bacterial orders. *Cell* **16**, 97–109.
4. Doi, R.H. (1982) RNA Polymerase of *Bacillus subtilis*, in *The Molecular Biology of the* Bacilli, vol. 1, *Bacillus subtilis* (Dubnau, D.D., ed.) Academic, New York.

5. Sternbach, H., Engelhardt, R., and Lezius, A.G. (1975) Rapid isolation of highly active RNA polymerase from *Escherichia coli* and its subunits by matrix-bound Heparin. *Eur. J. Biochem.* **60**, 51–55.

6. Grossman, A.D., Erickson, J.W., and Gross, C.A. (1984) The *htp* R gene product of E. coli is a sigma factor for heat shock promoters. *Cell* **38**, 383–390.

7. Westpheling, J., Ranes, R., and Losick, R. (1985) RNA polymerase heterogeneity in *Streptomyces coelicolor*. *Nature* **313**, 22–27.

8. Price, C.W. and Doi, R.H. (1985) Genetic mapping of rpoD implicates the major sigma factor of *Bacillus subtilis* RNA polymerase in sporulation initiation. *Mol. Gen. Genet.* **201**, 88–95.

9. Scott, N.W. and Dow, C.S. (1986) Purification and partial characterization of DNA-dependent RNA polymerase from *Rhodomicrobium vannielii*. *J. Gen. Microbiol.* **132**, 1939–1949.

10. Burgess, R.R. (1969) A new method for the large scale purification of *Escherichia coli* deoxyribonucleic acid-dependent ribonucleic acid polymerase. *J. Biol. Chem.* **244**, 6160–6167.

11. Burgess, R.R. and Jendrisak, J.J. (1975) A procedure for the rapid, large scale purification of *Escherichia coli* DNA-dependent RNA polymerase involving Polymin P precepitation and DNA cellulose chromatogaphy. *Biochemistry* **14**, 4634–4638.

12. Alberts, B.M. and Herrick, G. (1971) DNA-cellulose chromatography *Meth. Enzymol.* **21**, 198–217.

13. Jaehning, J.A., Wiggs, J.L., and Chamberlin, M.J. (1979) Altered promoter selection by a novel form of *Bacillus subtilis* RNA polymerase. *Proc. Natl. Acad. Sci. USA* **76**, 5470–5474.

14. Gonzalez, N., Wiggs, J., and Chamberlin, M.J. (1977) A single procedure for resolution of *Escherichia coli* RNA polymerase holoenzyme from core polymerase. *Arch. Biochem. Biophys.* **182**, 404–408.

15. Gross, C., Engbaek, F., Flammang, T., and Burgess, R.R. (1976) Rapid micromethod for the purification of *Escherichia coli* ribonucleic acid polymerase and the preparation of bacterial extracts active in ribonucleic acid synthesis. *J. Bacteriol.* **128**, 382–389.

16. Haldenwang, W.G., Lang, N., and Losick, R. (1981) A sporulation-induced sigma-like regulatory protein from *B. subtilis*. *Cell* **23**, 615–624.

Chapter 12

Direct Immunoprecipitation of Protein

Christopher F. Thurston
and Lucy F. Henley

1. Introduction

Many biochemical experiments depend on the measurement of the amount of an enzyme (or other protein) independently of any enzymatic activity it may have. When working with cell-free extracts or other complex mixtures this is not simple to do, and most procedures that have general application involve the use of antibodies. Two distinct strategies have been used to good effect. The amount of antibody bound can be made to reflect amount of antigen, as in an ELISA assay (ref. 1 and Vol. 1 of this series) or in a "Western blot" (ref. 2 and Chapters 28 and 29 in this volume). Alternatively, the antibody can be used effectively to purify the antigen protein. The method described here is of the latter sort and has advantages of simplicity and rapidity both over ELISA and blotting techniques, on the one hand, and indirect precipitation techniques on the other.

Direct immunoprecipitation is achieved by addition of a specific antibody to a sample at appropriate concentration to form a precipitable matrix with antigen (3). The precipitate so formed is contaminated with nonspecifically bound protein that is "scrubbed" off the precipitate by centrifugation through a viscous sucrose solution (4). The resultant contamination-free precipitate may be counted for radioactivity (of labeled antigen) or subjected to further analysis by electrophoresis or isoelectric focusing (*see* Fig. 1).

If the protein to be detected/quantitated is distinguishable from immunoglobulin polypeptides on the basis of size or charge, the method can be used with nonradioactive samples, but in practice its usefulness is for analysis of radiolabeled mixtures, in which very small amounts of labeled antigen can be isolated for analysis by coprecipitation with a relatively large amount of unlabeled "carrier" antigen.

2. Materials

1.   Antigen: The protein to be analyzed must be available in a highly purified form as the immunogen for antibody production. Purified antigen is also required to titrate the antibody and, when samples containing very small amounts of antigen are to be analyzed, to act as carrier. The pure antigen is conveniently used at 1–2 mg/mL and must be dissolved in a buffer mixture that does not interfere with antibody binding, such as TEC described below.

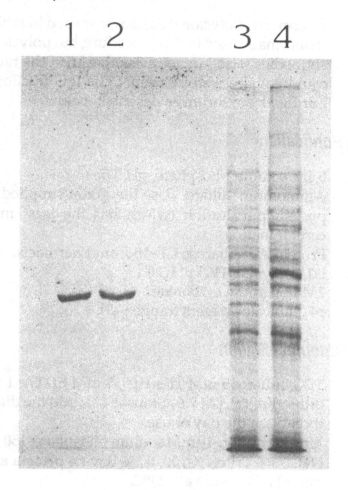

Fig. 1. Analysis of isocitrate lyase synthesis in the green alga *Chlorella fusca*. Fluorograph of an SDS gel showing immunoprecipitated isocitrate lyase in tracks 1 and 2, obtained from the cell-free extracts run in tracks 3 and 4, respectively. Protein was labeled in vivo by ^{35}S-methionine incorporation for 2 h (tracks 1 and 3) and 5 h (tracks 2 and 4) after the removal of inducer (acetate) from the medium. Clearly there was significant residual enzyme synthesis after inducer removal, although this represents a small fraction (1–2%) of total incorporation and does not result in easily distinguishable bands in the total protein tracks. Kinetic analysis of these data are shown in ref. 5.

2. Antiserum: Polyclonal antibodies raised in rabbits are normally used in this procedure, but polyclonal antibodies raised in animals other than the rabbit could equally well be used. The IgG fraction is purified from serum as described below.

2.1. IgG Purification

1. 0.1M sodium phosphate, pH 7.5.
2. Ammonium sulfate. Use the grade supplied for protein purification (which has the least metal contamination).
3. Protein A Sepharose CL-4B from Pharmacis.
4. 0.1M Glycine-HCl pH 3.0.
5. 2M Potassium carbonate.
6. pH indicator papers (range 6–8).

2.2. Immunoprecipitation

1. TEC buffer: 15 mM Tris-HCl, 1 mM EDTA, 1 mM dithiothreitol, pH 7.6. Store at 4°C; add the dithiothreitol on the day of use.
2. Binding buffer: 10 mM sodium phosphate, 150 mM NaCl, 1% Triton X-100, 0.5% bovine plasma albumin pH 7.2. Store at –20°C.
3. 0.9M Sucrose. Make up in sterile distilled water and store at 4°C. Prior to use, centrifuge as 1-mL amounts in microfuge tubes for 15 min at 4°C (*see* Note 6 in section 4).

3. Method

3.1. IgG Purification

1. All steps should be done at 4°C or in containers on ice.

2. Measure 10 mL of serum into a small beaker or flask and stir magnetically while adding 2.26 g of ammonium sulfate (to bring the solution to 40% saturation). Maintain the pH at 7.5 by adding ammonia solution. This should be fairly concentrated, so as not to significantly increase the overall volume. Continue stirring for 10 min after all of the ammonium sulfate has been added.

3. Centrifuge for 10 min at 8000g, and discard the supernatant, allowing the pellet to drain thoroughly.

4. Dissolve the pellet in 10 mL of phosphate buffer.

5. Repeat steps 2 and 3.

6. Dissolve in 5 mL of phosphate buffer, and dialyze against the same buffer, using three successive 1-L amounts.

7. Centrifuge the dialyzed protein solution for 15 min at 12,000g to remove any particulate material. If the supernatant is not homogeneous, filter through a 0.45-µm pore size membrane filter.

8. Load the clarified protein solution onto a 5-mL protein A-Sepharose column at 30 mL/h (*see* Note 1 in section 4). Reapply the eluate twice to ensure maximum retention of IgG.

9. Wash the column with phosphate buffer at 30 mL/h collecting samples from time to time for measurement of absorbance at 280 nm. Continue washing until the absorbance of the eluate is less than 0.05 (*see* Note 2 in section 4).

10. Elute the bound IgG with glycine buffer, collecting 1.5 mL fractions that are immediately neutralized with 2M potassium carbonate (*see* Note 3 in section 4).

11. Measure the absorbance of each fraction at 280 nm, and pool those fractions containing a significant

proportion of the eluted IgG. Store at –20°C in 200–500 µL amounts.

3.2. Immunoprecipitation

1. The amounts for the immunoprecipitation mix are as follows: A standardized amount of antigen must be used, 5 µg is sufficient (*see* Note 4 in section 4). The antigen is made up to 50 µL with TEC in a 400-µL microfuge tube. IgG fraction (diluted to give a twofold excess, *see* Note 5 in section 4) is added in 50 µL, followed by 25 µL of binding buffer.
2. Incubate the mixture for 2 h at 4°C to allow for formation of the immunoprecipitate.
3. Centrifuge for 15 min at 4°C to to pellet the crude precipitate.
4. Carefully aspirate off the supernatant.
5. Add 50 µL of binding buffer to the precipitate and take up and expel with a pipet 5–10 times to disperse the precipitate as a fine suspension.
6. Overlay the suspended precipitate onto 200 µL of 0.9*M* sucrose in a 400-µL microfuge tube, and centrifuge for 15 min at 4°C (*see* Note 6 in section 4).
7. Carefully aspirate the supernatant and resuspend the pellet for analysis (*see* Note 7 in section 4).

4. Notes

1. Protein A Sepharose should be equilibrated and regenerated as recommended by the manufacturer. A 10-mL syringe is a convenient container in which to make the column. Control the system by pumping off the bottom of the column with a peristaltic pump for this and steps 9 and 10.

2. It is more convenient to monitor the absorbance of the column with a flow monitor if available, washing until there is no significant absorbance in the eluate (say less than 0.5% of the absorbance of the sample loaded on the column).

3. The stability of antibody at pH 3 is limited and varies from one antibody to another; there is no doubt that the neutralization step is the most crucial in this part of the procedure. In our experience it is quicker to use pH papers than a pH electrode, but whatever can be done quickest is best.

4. The total amount of antigen must be sufficient to give a visible immunoprecipitate, for which purpose 5 µg is adequate for proteins that bind about twice their own weight of immunoglobulin. If the antigen is an abundant constituent of the sample mixture, e.g., 5% of total protein, then the total protein concentration of the sample must be adjusted to, for example, 2 mg/mL so that 50 µL contains 5 µg of antigen. In this situation it is not uncommon to find that some samples are too diluted and need to be concentrated rapidly. Amicon "Minicon-B" disposable clinical sample concentrators are very useful for this, both because they are quick and because the extent of concentration can be quite accurately estimated by eye at any stage in the process.

   For many other antigens the concentration in cell-free extract or other sample (such as the products of in vitro translation) is much less, and the precipitation is achieved by using unlabeled carrier antigen. In this case use 5 µL of 1 mg/mL carrier and 45 µL of sample extract, containing not more than 1 µg of antigen.

5. For each batch of IgG prepared, a calibration must be done. Set up tubes containing 5 µg antigen, and

add a range of volumes of IgG fraction (say 10–200 µL) making all tubes up to the same volume with phosphate buffer and adding binding buffer to give a 2:1 v:v ratio of phosphate buffer to binding buffer. Stand for 2 h at 4°C, and centrifuge for 15 min. Thoroughly aspirate off the supernatants and dissolve the pellets for protein determination. If a suitable range of IgG amounts is tested, the pellets increase progressively to a plateau value of about 20 µg. The minimum amount of IgG to give the plateau value (i.e., complete precipitation of the antigen) is doubled in the routine procedure, which is what is intended by "a twofold excess" in step 1. If 50 µL of IgG fraction is insufficient to give this amount, concentrate the solution as described for protein samples in Note 4 above. As an example, one recent preparation of anti-isocitrate lyase IgG fraction required 280 µg to give twofold excess for precipitation of 5 µg of antigen protein. If the antigen of interest has an easily assayed activity that is not inactivated by the conditions of the precipitation reaction, the residual activity in supernatants from the precipitation step can be used to determine the amount of IgG required for complete precipitation more easily, and often more precisely, than measurement of precipitated protein.

6. The centrifugation of the sucrose solution prior to its use never produces a visible pellet, but the removal of nonspecific contamination works less well if it is not done. The immunoprecipitate may be colored prior to centrifugation through sucrose (for instance, when precipitating from reticulocyte lysates, the pellets are bright pink at this stage), but after passage through the sucrose, the pellets should be white.

In this and all centrifugation steps except those during the preparation of IgG fraction, in which larger volumes are involved, use a microcentrifuge giving at least 10,000g. Since microcentrifuges are usually not refrigerated, the centrifuge should be used in a cold room or cold cabinet. The standard 1.5-mL capped polythene "Eppendorf" microfuge tubes are suitable for precentrifuging the sucrose solution, but not for the immunoprecipitation reaction itself or the centrifugation of the immunoprecipitate through sucrose. The long, narrow 400-μL tubes are used for these steps since their shape results in a more compact pellet that can be seen relatively easily. The pellets formed using 5 μg of antigen are clearly visible once the procedure has become familiar, but they can be difficult to see in microfuge tubes that have been badly moulded and are uneven at the bottom.

7. The final immunoprecipitates can be dissolved in loading buffers for analysis by electrophoresis or isoelectric focusing, or may be counted for radioactivity without further processing, in which case the bottom 5 mm of each tube can be cut off and dropped directly into a scintillation vial containing a protein-solubilizing agent.

References

1. Engvall, E. and Perlmann, P. (1972) Enzyme-linked immunosorbent assay, ELISA. 111. Quantitation of specific antibodies by enzyme-labeled antiimmunoglobulin in antigen-coated tubes. *J. Immunol.* **109**, 129–135.
2. Vaessen, R. T. M. J., Kreike, J., and Groot, G. S. P. (1981) Protein transfer to nitrocellulose filters. *FEBS Lett.* **124**, 193–196.

3.   Palmiter, R. D., Oka, T., and Schimke, R. T. (1971) Modulation of ovalbumin synthesis by estradiol-17β and actinomycin D as studied in explants of chick oviduct in culture. *J. Biol. Chem.* **246**, 724–737.

4.   Yeung, A. T., Turner, K. J., Bascomb, N. F., and Schmidt, R. R. (1981) Purification of an ammonium-inducible glutamate dehydrogenase and the use of its antigen affinity-column purified antibody in specific immunoprecipitation and immunoadsorption procedures. *Anal. Biochem.* **110**, 216–228.

5.   Henley, L. F. and Thurston, C.F. (1986) The disappearance of isocitrate lyase from the green alga *Chlorella fusca* studied by immunoprecipitation. *Arch. Microbiol.* **145**, 266–271.

Chapter 13

Detection of Proteins
in Polyacrylamide Gels Using
an Ultrasensitive Silver Staining
Technique

Ketan Patel, David J. Easty,
and Michael J. Dunn

1. Introduction

Polyacrylamide gel electrophoresis is a versatile
and powerful tool for the analysis of biological samples
and is capable of good separation and high resolution of
complex protein mixtures. Although Coomassie Brilliant Blue R250 has proved ideal as a general protein
stain for the more traditional applications of this
method, current trends toward thinner gels, decreased
sample loading (to improve resolution), and the recent

developments of two-dimensional gel electrophoresis and peptide mapping techniques have necessitated increasingly sensitive detection methods. Electrophoretic separation of radioactively labeled proteins followed by autoradiography permits the detection of trace proteins (10^{-4} to 10^{-5}% of total protein) in a sample (1). However, problems inherent in radioactive methods include: (a) in vitro labeling may alter physical properties of proteins, and (b) in vivo experiments require excessively large quantities of isotopes that are prohibitively expensive in animals and unethical in human clinical studies. Alternative methods such as fluorescent staining (2) and heavy metal stains (3) are of less than, or at best of equivalent sensitivity to, Coomassie Brilliant Blue R250.

The property of silver to develop images was discovered in the mid-17th century. Subsequently, imaging techniques using silver were first applied to photography, followed closely by their use in histology. Recently, silver staining methods have been developed to detect proteins after polyacrylamide gel electrophoresis; first by employing histological principles (4) and then by photochemical methods (5). Silver staining techniques are reported to be between 20 and 200 times more sensitive than Coomassie Brilliant Blue R250 and represent a significant advantage in the range of protein detection methods in polyacrylamide gels. Silver staining can be used as a detection method for techniques other than protein polyacrylamide gel electrophoresis. Proteins transferred to nitrocellulose membranes can be visualized using silver staining procedures (6), as can DNA (7) and RNA molecules (8) in gels and on membranes.

The precise mechanism leading to the formation of a visible silver/protein complex has not been fully clarified, but both histological and photochemical stains involve the reduction of ionic silver to the metallic form. It has been proposed that silver cations complex with protein amino groups in an alkaline environment and with cysteine and methionine sulfur residues (9). More recently, however, Gersten et al. (10) have implicated three-dimensional protein structure and, therefore, the steric presentation of reactive moieties in three-dimensional space as being of most consequence. They proposed that other factors such as amino acid composition are of secondary importance. Caution should be exercised in the interpretation of protein staining patterns since certain protein species, e. g., calmodulin (11), fail to respond to single-step silver staining.

After testing several current silver staining techniques (12–14), we have found that the most adequate procedure for our purposes in terms of sensitivity, resolution, time required, and expense was the method described by Wray et al. (15). This method has been assessed favorably in comparative methodological studies made by other workers (16). We describe the Wray technique for silver staining of gels, together with modifications and technical advice that will enable an experimenter to optimize results. An example of a 2-D gel stained by this procedure is shown in Fig. 1.

2. Materials

1. All water used must be distilled and deionized.
2. Gel fixation solution: Trichloroacetic acid (TCA) solution, 20% (w/v).
3. Silver diamine solution: 21 mL of 0.36% (w/v) NaOH is added to 1.4 mL of 35% (w/v) ammonia and then 4 mL of 20% (w/v) silver nitrate is added drop-wise with stirring. When the mixture fails to

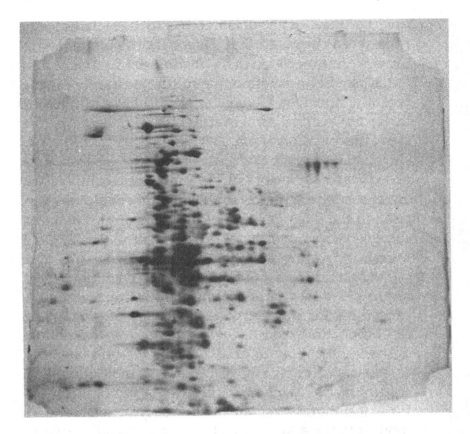

Fig. 1. Two-dimensional gel of bacterial proteins stained with the silver staining procedure. A horizontal, flat-bed pH 3–10 isoelectric focusing gel was used in the first dimension, and an 8–20% gradient polyacrylamide SDS-PAGE gel was employed in the second dimension.

clear with the formation of a brown precipitate, further addition of a minimum amount of ammonia results in the dissolution of the precipitate. The solution is made up to 100 mL with water. The silver diamine solution is unstable and should be used within 5 min.

4. Reducing solution: 2.5 mL of 1% citric acid, 0.26 mL of 36% (w/v) formaldehyde made up to 500 mL with water.
5. Farmer's reducer: 0.3% (w/v) potassium ferricyanide, 0.6% (w/v) sodium thiosulfate, 0.1% (w/v) sodium carbonate.
6. Clearing solution: 100 mL of 1.5M sodium thiosulfate mixed with an equal volume of a solution of 0.15M copper sulfate, 0.6M sodium chloride, and 0.9M ammonium hydroxide.

3. Method

3.1. Fixation

1. After electrophoresis the gel is immediately fixed in 200 mL of TCA for a minimum of 1 h at room temperature. High-percentage polyacrylamide and thick gels require an increased period for fixation, and overnight soaking is recommended.
2. The gel is placed in 200 mL of 50% (v/v) methanol and soaked for 2 x 30 min with agitation.
3. The gel is washed in excess water for 2 x 20 min, facilitating the rehydration of the gel and the removal of methanol. An indication of rehydration is the loss of the hydrophobic nature of the gel.

3.2. Staining

1. The gels are soaked in the silver diamine solution for 15 min with constant agitation. For thick gels (1.5 mm upward), it is necessary to use increased volumes so that the gels are totally immersed. Caution should be exercised in disposal of the

ammoniacal silver reagent, since it decomposes on standing and may become explosive. The ammoniacal silver reagent should be treated with dilute hydrochloric acid (1*N*) prior to disposal.

2. The gel is washed (2 x 5 min) in water.

3. The gel is placed in reducing solution. Proteins are visualized within 10 min, after which the background will gradually increase. It is important to note that this reaction displays inertia, and that staining will continue for 2–4 min after removal of the gel from the reducer. Proteins are visualized from 8 min until 15 min, after which background becomes unacceptable.

4. The gels are rinsed in water for 2 min.

5. Stain development is terminated by immersing the gel in 45% (w/v) methanol, 10% (v/v) acetic acid. It is important to note that staining displays inertia and will continue for a further 2–4 min after removal of the gels from reducing solution.

    If staining is of low sensitivity, it can be intensified by recycling. Stained gels are washed in water (5 min) and returned to fixation in methanol (section 3.1, step 2), and the whole process repeated.

3.3. Destaining

1. Selective destaining: Gels are washed in water (5 min) to remove the stop solution. For the controlled removal of background staining that obscures proper interpretaion of the protein pattern, partial destaining of gels using Farmer's reducing reagent (9) is recommended. The gels are placed in Farmer's reducer for a time dependent upon the intensity

of the background. Destaining is terminated by returning gels to the stop solution.

2. Complete destaining: Gels that are totally unusable because of overstaining can be efficiently destained prior to restaining by immersion in the clearing solution. Once the gel has cleared it must be washed extensively in water (3 x 30 min) prior to return to 50% (v/v) methanol (section 3.1, step 2) and recycled.

4. Notes

1. Gloves should be worn at all stages when handling gels, since silver staining will detect skin and sweat proteins present on fingers.

2. Volumes of the solutions used at all stages should be sufficient such that the gels are totally immersed. If the volume of solution is insufficient for total immersion, staining tends to be uneven and leads to gel surface drying.

3. All solutions for the staining and destaining steps should be freshly prepared, and overnight storage is not recommended. Fixing solutions can be stored.

4. All solutions must be prepared using clean glassware and deionized, distilled water.

5. An inherent problem with the staining of gradient SDS-PAGE gels is uneven staining along the concentration gradient. The less concentrated polyacrylamide region develops background staining prior to the more concentrated region. A partial solution to this problem is to increase the time of staining in silver diamine (*see* section 3.2, step 1).

6. Various chemicals used in one- and two-dimensional procedures will inhibit staining, whereas

others impair resolution or produce artifacts. Acetic acid will inhibit staining and should be completely removed prior to the addition of silver diamine solution (section 3.2, step 1). Glycerol, used to stabilize SDS gradient gels during casting, and urea, used as a denaturing agent in isoelectric focusing, are removed by water washes. Agarose, often used to embed rod IEF gels onto SDS-PAGE gels in 2-D PAGE procedures, contains proteins that are detected by silver staining as diffuse bands and give a strong background. Tris and glycine present in electrophoresis buffers complexes with silver (17) and must be washed out with water. The use of 2-mercaptoethanol as a disulfide bond reducer should be avoided since it leads to the appearance of two artifactual bands at 50 and 67 dalton on the gel (18).

7. Using this method, most proteins appear as dark brown to black spots or bands. Certain species, however, stain to give yellow spots regardless of protein concentration. The phenomenon of color variation in silver staining has been linked to the post-translational modification of the proteins. Glycosylation and the addition of lipid groups have been known to affect the staining intensities of proteins as well as their pI values.

8. Radioactively labeled proteins can be detected by silver staining prior to autoradiography or fluorography for the majority of the commonly used isotopes (^{14}C, ^{35}S, ^{32}P, ^{125}I). In the case of ^{3}H, however, silver deposition will absorb most of the emitted radiation.

References

1. O'Farrell, P.H. (1975) High resolution two-dimensional electrophoresis of proteins. *J. Biol. Chem.* **250**, 4007–4021.
2. Patel, K., Easty, D., and Dunn, M.J. (1986) A Qualitative Study of ANS Fluorescence Staining of Proteins after 1 and 2D Gel Electrophoresis, in *Electrophoresis '86* (Dunn, M.J., ed.) VCH Verlagsgesellschaft, Weinheim.
3. Quitarelli, G., Bellocci, M., and Geremia, R. (1973) On phosphotungstic acid staining. IV. Selectivity of the staining reaction. *J. Histochem. Cytochem.* **21**, 155–160.
4. Kerenyi, L. and Gallyas, F. (1972) A highly sensitive method for demonstrating proteins in electrophoretic, immunoelectrophoretic and immunodiffusion preparations. *Clin. Chim. Acta* **38**, 465–467.
5. Merril, C.R., Dunau, M., and Goldman, D. (1981) A rapid sensitive silver stain for polypeptides in polyacrylamide gels. *Anal. Biochem.* **110**, 201–207.
6. Merril, C.R. and Pyatt, M.E. (1986) A silver stain for the rapid quantitative detection of proteins or nucleic acids on membranes or thin layer plates. *Anal. Biochem.* **156**, 96–110.
7. Marshall, T. (1983) Detection of DNA and lipopolysaccharide in thick polyacrylamide gels using an improved silver stain. *Electrophoresis* **4**, 269–272.
8. Guillemette, J.G. and Lewis, P.N. (1983) Detection of subnanogram quantities of DNA and RNA on native and denaturing polyacrylamide and agarose gels by silver staining. *Electrophoresis* **4**, 92–94.
9. Heukeshoven, J. and Dernick, R. (1985) Simplified method for silver staining of proteins in polacrylamide gels and the mechanism of silver staining. *Electrophoresis* **6**, 103–112.
10. Gersten, D.M., Wolf, P.H., Ledley, R.H., Rodriguez, LV, and Zapolski, E.J. (1986) The relationship of amino acid composition to silver staining of proteins in electrophoresis. *Electrophoresis* **7**, 327–332.
11. Rochette-Egly, C. and Studdi-Garaud, C. (1984) Selective detection of calmodulin in polyacrylamide gels by double staining with Coomassie Blue and silver. *Electrophoresis* **5**, 285–288.
12. Switzer, R.C., Merril, C.R., and Shifrin, S. (1979) A highly sensitive stain for detecting proteins and peptides in polyacrylamide gels. *Anal. Biochem.* **98**, 231–237.

13. Sammons, D.W., Adams, L.D., and Nishizawa, E.E. (1981) Ultrasensitive silver-based color staining of polypeptides in polyacrylamide gels. *Electrophoresis* **2**, 135–141.

14. Oakley, B.R., Kirsch, D.R., and Morris, N.R. (1980) A simplified ultrasensitive silver stain for detecting proteins in polyacrylamide gels. *Anal. Biochem.* **105**, 361–363.

15. Wray, W., Boulikas, T., Wray, V.P., and Hancock, R. (1981) Silver staining of proteins in polyacrylamide gels. *Anal. Biochem.* **118**, 197–203.

16. Ochs, D.L., McConkey, E.H., and Sammons, D.W. (1981) Silver stains for proteins in polyacrylamide gels: A comparison of six methods. *Electrophoresis* **2**, 304–307.

17. Ames, G.F-L. and Nakaido, K. (1976) Two-dimensional gel electrophoresis of membrane proteins. *Biochemistry* **15**, 616–623.

18. Guevara, J., Johnston, D.A., Ramagli, L.S., Martin, B.A., Capitello, S., and Rodriguez, L.V. (1982) Quantitative aspects of silver deposition in proteins resolved in complex polyacrylamide gels. *Electrophoresis* **3**, 197–205.

Chapter 14

Chromatofocusing

Ake Sidén and Paolo Gallo

1. Introduction

Investigations of different protein species in complex biological fluids, e.g., cerebrospinal fluid (CSF) and serum, are important tasks in clinical laboratory work as well as in biochemical research. One crucial part of such investigations is the application of a suitable method in order to separate the individual proteins and/or resolve a certain protein into subtypes or microheterogeneous components. There are two major categories of procedures that are used to achieve these separations. One is represented by electromigration techniques, such as electrophoresis or isoelectric focusing, which separate the sample molecules according to size or charge; an immunoelectrophoretic step may also be used, i.e., a separation according to antigenicity. The second major category is formed by the liquid chromatographies, which give a partition of the sample molecules on the basis of size, charge, hydrophobicity, or binding to a biospecific ligand.

Chromatofocusing is a liquid chromatography that separates biomolecules on the basis of their respective isoelectric point (p*I*), i.e., it is a chromatographic analog of isoelectric focusing. This procedure was initially described in 1976 by Sluyterman and Wijdenes (*1*), and later more detailed basic data were published (*2–4*). The method is based on the interaction of an ion exchanger and a suitable buffer, which results in the formation of a pH gradient without the use of any gradient mixer. Accordingly, proteins that have been bound to the ion exchanger are eluted and focused in order of the p*I* values. The following steps form the basis of the procedure:

1. Choice of the optimal pH gradient for the protein(s) of interest.
2. Selection of start and eluent buffers on the basis of the above decision.
3. Sample preparation and application onto the ion exchanger, which has been equilibrated with start buffer.
4. Performance of the separation by running the eluent buffer through the ion exchanger.
5. Washing and regeneration of the ion exchanger followed by equilibration with start buffer.

The present description of chromatofocusing is based on experiences with the FPLC separation media and instruments. References *5–7* give further basic theoretical and practical data for the methodology, as well as different applications. The chromatofocusing column contains an anion exchanger prepared from a monodisperse beaded hydrophilic resin carrying tertiary and quaternary amines as charged groups. The sample is applied onto the column at conditions giving a negative charge to the molecules of interest, i.e., at a pH value above the pI of these latter components. Consequently, these sample molecules are bound to the positive charges of the ion exchanger. The eluent contains a mixture of amphoteric buffers with different pK_a values and has been adjusted to a pH value below the pI of the molecules that are to be separated. During the elution, these amphoteric species interact with the anion exchanger that results in the formation of a pH gradient within the column and a gradually decreasing pH of the eluate. The lowering of the pH at a certain level of the column is more pronounced than at a level further down, i.e., the pH increases down the length of the column until the whole column is in equilibrium with the eluent buffer. The net result of these conditions will be an elution and focusing of the sample molecules in order of their respective pI values (components with a higher pI being eluted before those with a lower pI).

The following sections will give basic methodological data for the use of chromatofocusing in studies of CSF and serum proteins, predominantly immunoglobulin G (IgG). The results from such separations have already been presented in more detail (8–12). Because of the application for CSF protein separations, these methods were designed to be used for samples with relatively low protein amounts.

2. Materials

1. All chemicals and solvents, including the water, should be of HPLC or analytical grade, and their compatibilities with the system should be checked. The buffers and other solutions run through the column have to be filtered through a sterile 0.2-μm

filter and degased. Furthermore, they should be kept at 3–8°C when not used and prior to use should be carefully inspected with regard to microbial contamination.

2. Buffers:

Start buffer: 0.025M diethanolamine, pH 9.5.

Eluent buffers:

10% (v/v) Polybuffer 96 (Pharmacia, Uppsala, Sweden), pH 6.0.

10% (v/v) Polybuffer 74 (Pharmacia), pH 4.0.

The buffers are titrated to their respective pH values with hydrochloric acid prior to the addition of the final few milliliters of water up to the total volume. The Polybuffer eluents should be protected from light.

3. Solutions for column washing and regeneration:

2M Sodium chloride

70% (v/v) Acetic acid

4. Chromatofocusing columns and equipment (Pharmacia):

1 Mono P HR 5/20 (5 mm x 20 cm) column

1 Small chromatofocusing column[a]

2 Prefilters

1 GP-250 gradient programmer[b]

2 P-500 Pumps

1 0.6-mL Mixer

1 V-7 Valve

2 V-8 Valves[c]

1 UV-1 Single-path monitor

1 HR Flow cell

1 280 nm Filter kit

1 REC-482 Two-channel recorder

1 FRAC-100 Fraction collector

1 PSV-100 Valve[c]

1 pH Monitor[d]

1 Standard pH electrode^d
1 Flow-through pH electrode^c
1 Chromatography rack
1 10-mL Superloop
Sample loops 0.5 and 1.0 mL
Cable and flanging/start up kits

^aNot commercially available; it was prepared by the manufacturer of the equipment according to instructions from the authors: a HR 5/5 (5 mm x 5 cm) glass column packed with 1 mL of the same medium as the Mono P column.

^bAlternatively, a LCC-500 control unit; besides carrying out the same functions as the GP-250, the LCC-500 may also be used to control automated multisample/multidimensional separations. Several further components are necessary, however, in order to make full use of this control unit.

^cOptional; the V-8 valves make it possible to connect several columns to the system, the PSV-100 valve increases the flexibility of the FRAC-100 fraction collector, and the flow-through pH electrode simplifies the registration of the pH gradient.

^dCheck whether suitable components are already available in your laboratory.

3. Method

The method is based on the following procedures: equipment preparation, sample preparation, sample loading and running, column washing and regeneration, and result evaluation. Instrument set-up, calibration, and programming, as well as column preparation for first-time use, are performed according to instructions from the manufacturer. The prefilters are connected to the system in such a way that they protect the column(s). A schematic system design is given in Fig. 1.

Three different chromatofocusing procedures will be described in section 3.3. The first of these methods (section 3.3.1) is based on the use of a commercially available column and a pH gradient suited for most IgG components. The second technique (section 3.3.2) util-

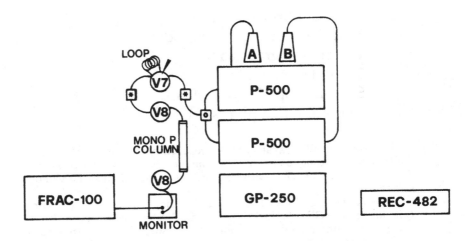

Fig. 1. Instrument set-up when two V-8 valves are used. In systems without such valves, the column is introduced (with the second prefilter) between the V-7 valve and the UV-1 monitor. A and B: reservoirs for the buffers. The sample loop is connected to the V-7 valve and loaded with a syringe via a 0.5-mm outer diameter, blunt-tipped needle (arrowhead).

izes the same type of pH gradient, but the separations are run on a small column. Finally, a method for the use of a gradient covering alkaline, as well as acid pH values, is described (section 3.3.3).

3.1. Equipment Preparation

1. Program the chromatofocusing method(s) as well as the recording and fraction collection modes into the equipment. The basic parameters are specified in section 3.3.

2. Fill the reference cell of the UV-1 single-path monitor with the Polybuffer eluent (10% Polybuffer 96,

pH 6.0, is used also for the broad pH gradient chromatofocusing method).
3. Make sure that the intended connections between the buffer reservoirs and the P-500 pumps will be correct: the start buffer (buffer A) to pump A and the eluent buffer (buffer B) to pump B. If two different eluent buffers are used, the first is connected to pump A and the second to pump B.
4. Perform blank runs in order to check the equipment, programs, pH gradients, pressures, background absorbances, and baselines.

3.2. Sample Preparation

1. Lipid-containing samples should be delipidated. This is recommended for all sera, but can generally be omitted for CSF samples (unless they are known to be significantly contaminated by lipids). Depending on the available amount of material, delipidation may be performed before or after adjustment of the IgG concentration. The sample is mixed with an equal volume of chloroform, thoroughly agitated, and, 30 min later, centrifuged at 1400–1500g for 10 min.
2. The IgG concentration is adjusted to 1 mg/mL by either dilution with 0.15M sodium chloride or concentration, e.g., by vacuum ultrafiltration.
3. The samples are diluted 1:5 in start buffer (giving an IgG concentration of 0.2 mg/mL) and filtered through a sterile 0.2-µm filter. There should not be any turbidity of the final material.

3.3. Sample Loading and Running

The equipment is switched on. It should be noted that the UV-1 single-path monitor will need about 2 h to warm up. The buffer reservoirs are connected to their respective P-500 pumps.

3.3.1. *Chromatofocusing by Use of the Mono P Column and Polybuffer 96*

The basic parameters of the chromatofocusing program (volume based) are the following:

1. The Mono P column has been equilibrated with start buffer.
    Buffers injected into the column after sample loading: 5 mL of 100% buffer A (start buffer) followed by 35–40 mL[a] of 100% buffer B (10% Polybuffer 96, pH 6.0).[b]
    Flow rate 0.25 mL/min.
    UV absorbance of the eluate measured at 280 nm (AU mode) and range selector set at 0.02.
    Input to one channel of the recorder from the UV monitor, the second channel registers % buffer B or pH of the eluate (when a flow-through pH electrode is used); chart speed 0.25 cm/min.
    Collection of consecutive 1-mL fractions.
2. Turn the V-7 valve to the "load" position. Check the positions of the V-8 valves if such are used to connect the column to the system.
3. Inject 0.5 mL of the prepared sample (having an IgG content of 0.1 mg) into the 0.5-mL sample loop.
4. Turn the V-7 valve to the "inject" position.
5. Enter the appropriate method number in instruction "Run method 0" on the display for the GP-250/LCC-500 unit and start the run by pressing "do store."
6. Run the separation.

3.3.2. *Chromatofocusing by Use of the Small (HR 5/5) Column and Polybuffer 96*

1. The same basic parameters are used with the following exceptions: Buffers injected into the column after sample loading: 2 mL of 100% buffer A (start buffer) followed by 15 mL of 100% buffer B (10% Polybuffer 96, pH 6.0).[b]
    Range selector set at 0.05–0.1.
2–6. The same procedures as in section 3.3.1.

3.3.3. *Chromatofocusing by Use of the Mono P Column and a Combination of Polybuffer 96 and 74*

1. The instrument set-up and basic parameters of the program deviate from those of the previous methods as follows:
    A 10-mL Superloop is introduced between the P-500 pumps and the V-7 valve. The superloop is used as start buffer reservoir. Pumps A and B are used for the first and second eluent buffers, respectively. Buffers injected into the column after sample loading: 10 mL of 100% start buffer followed by 100% of the first (10% Polybuffer 96, pH 6.0) and second (10% Polybuffer 74, pH 4.0) eluent buffer for 25 and 45 mL, respectively. Range selector set at 0.02–0.1.
2. Fill the Superloop with 10 mL of start buffer.
3–7. The same procedures as step 2–6 in section 3.3.1.

[a]A buffer B volume of 35 mL brings down the pH of the eluate to 6.5; an increment of the buffer B volume to 40 mL gives a final eluate pH of 6.0.
[b]In our previous applications of this chromatofocusing procedure (8,9,11,12), a final start buffer volume was injected after the eluent buffer in order to get a more well-defined state at the end of the separation. We have found this procedure to be unnecessary, however.

3.4. Column Washing and Regeneration

1. It is recommended to use a reversed-flow direction with these procedures. The 1.0 mL sample loop is used for the solutions.

2. Turn the V-7 valve to the "load" position, and inject 1 mL of 70% (v/v) acetic acid into the sample loop. Turn the V-7 valve to the "inject" position, and run the acetic acid through the column. Totally, three such procedures are performed.

3. Use the same technique to run three 1-mL volumes of 2M sodium chloride through the column.

4. Equilibrate the column with start buffer. This is accomplished by running start buffer through the column until the pH of the eluate is the same as that of the buffer.

3.5. Result Evaluation

1. The pH gradient as well as the reproducibility of the separations should be checked. Registration of the gradient is performed by either pH measurements on the collected fractions or use of a flow-through pH electrode. The reproducibility of the separations is checked by duplicate runs.

2. Evaluations of the standard of the separations are based on investigations of the final product, i. e., the material in the collected fractions, by suitable qualitative or quantitative methods. This step is especially important during the period of method development for a certain application, but should also be performed regularly when the technique is run more routinely.

   Instructions for cleaning of heavily contaminated columns as well as for long-time storage of columns and equipment are available from the manufacturer.

4. Notes

1. As already stated in the introductory section, the present chromatofocusing methods were designed for applications on samples with relatively low protein amounts. Furthermore, in order to be suited for IgG separations, they were based on the use of quite broad pH gradients embracing the alkaline range. Figure 2 exemplifies the results from small-column chromatofocusing (section 3.3.2) of CSF with oligoclonal IgG and the paired serum sample with normal IgG. Figure 3 gives an example from chromatofocusing of normal CSF by use of the Mono P column and two eluent buffers (section 3.3.3). As shown by the figures, these two techniques give gradients covering the interval of pH 9.5–6.0 and 9.5–4.0, respectively. Figure 3 also illustrates the chromatofocusing profile obtained from normal IgG with the method in section 3.3.1. An eluent buffer volume of 35 mL for this latter method gives a gradient between pH 9.5 and 6.5; an increment of the eluent buffer volume from 35 to 40 mL will bring the acidic end of the gradient to pH 6.0.

2. The three chromatofocusing methods have been applied for separations of samples having an IgG content of 0.025–0.5 mg and a total protein amount in the range of 0.3–6 mg. These quantities are substantially lower than the protein capacity of the Mono P column: about 25 mg/column or 5 mg/single peak. The corresponding values for the small column are in the range of 5–10 and 1–2 mg, respectively. It is recommended to use relatively low protein loads, e.g., 0.1–0.2 mg IgG, during optimization of the method. Scaling up can then be performed by gradual increments of the protein load.

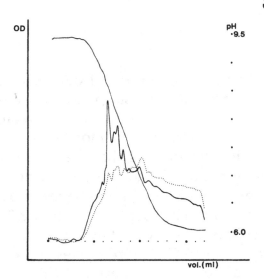

Fig. 2. Chromatofocusing (section 3.3.2) of CSF (continuous line) and serum (dotted line) from a patient with oligoclonal CSF IgG. The X-axis gives the initial start buffer volume (2 mL) followed by 15 mL of the eluent buffer; the Y-axis shows the UV absorbance of the eluate. The pH values are indicated to the right, and the continuous line across the graph shows the pH gradient. There are distinct peaks in the CSF profile that correspond to oligoclonal IgG (reproduced from ref. 14 with kind permission from Pergamon Press).

At the same time, signs of a deteriorating resolution must be looked for, since this will occur with too high-protein loads. Preparative separations for obtaining larger protein amounts are in that case performed by repetitive runs on submaximal protein amounts.

3. It is obvious that the choice of the pH gradient, i.e., the start and eluent buffers, has to be based on the pI of the protein(s) of interest. Naturally, the sample must also be compatible with the pH values and ionic strengths of the buffers. Proteins known to give well-focused bands at isoelectric focusing can be expected to give good results also at chromatofocusing. It may, however, be found that certain

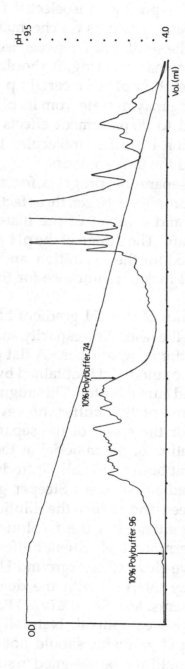

Fig. 3. Chromatofocusing (section 3.3.3.) of normal CSF. In this case, the X-axis gives the initial start buffer volume (10 mL) followed by 25 mL of the first and 45 mL of the second eluent buffer. The left part of the graph shows the IgG profile. The right part of the graph corresponds to proteins with more acidic p*I* values: albumin gives the broad prominent peak (partly outside the graph), and the three narrow peaks to the left correspond to transferrin, including the tau fraction (reproduced from ref. *10*, with kind permission from Pergamon Press).

proteins that focus poorly on isoelectric focusing give more satisfactory results on chromatofocusing. In spite of the similarities between isoelectric focusing and chromatofocusing, it should be observed that the elution pH of a certain protein at chromatofocusing may deviate from its p*I*. This is probably related to displacement effects and/or disproportionation of buffer molecules between matrix pores and the bulk solution.

4.   In principle, a separation program for a certain sample is a compromise between three factors: resolution, speed, and dilution of the material obtained in the eluate. The design of the pH gradient, the flow rate used for the separation, and the column type are of major significance for the aforenamed factors.

5.   The characteristics of the pH gradient have profound effects on the resolution capacity and are defined by the buffer compositions. A flat gradient increases the resolution and is obtained by using a relatively diluted eluent buffer. This augments the necessary volume of this buffer, however, with consequences for the speed of the separation as well as the dilution of the material in the eluate. Too-dilute eluent buffers will also introduce risks of poor pH gradient shapes. Steeper gradients increase the speed and reduce the dilution of the material in the eluate, but the resolution is decreased. Furthermore, high-eluent buffer concentrations will give elevated background UV absorbances that may interfere with the detection of protein components. Mostly, 10% (v/v) Polybuffer solutions have proved optimal. Naturally, unnecessarily broad pH gradients should not be used. The gradient should also be designed in such a way that the components of interest elute in the inter-

mediate section of the pH interval and especially not in the terminal part where an impaired resolution may occur. The initial start buffer volume has to be large enough to eliminate proteins not bound to the column.

6. An increment of the flow rate increases the speed of the separation, but there will be some reduction of the resolution. We have, in spite of the relatively long separation times (about 3 and 5 h for the method in section 3.3.1 and 3.3.3, respectively), mostly used a flow rate of 0.25 mL/min. Flow rates of 0.5 or 1 mL/min have, however, also given quite good results.

7. A reduction of the column size makes it possible to form the gradient with a lower amount of eluent buffer. In turn, this decreases the separation time (about 75 min for the method in section 3.3.2) and the dilution of the material in the eluate. This latter fact also gives a more favorable ratio between sample and background UV absorbances. There is a decrease of the peak separation caused by the reduced length of the column. The resolving power is still quite satisfying, however.

8. Chromatofocusing may be run in analytical or preparative scale. In the first-mentioned case, the method is used to detect heterogeneities within a class or subclass of proteins. It has exhibited a high capacity to detect mono- or oligoclonal components of IgG in body fluids such as CSF and serum. Generally, electromigration techniques performed in gel media are more suitable for analytical purposes, but situations may be encountered in which an alternative method such as chromatofocusing could prove advantageous. The main applications will, however, probably be found in preparative work, e.g., when the technique is used as a first-step

procedure aimed at the isolation of relevant components for complementary investigations. Chromatofocusing has several promising features for such purposes: high resolution, relatively fast and easy performance, comparatively mild separation conditions, and methodological flexibility. Furthermore, the obtained material can be used directly, or after a simple concentration/dilution procedure, for examinations by, for example, isoelectric focusing, two-dimensional gel electrophoresis, or ELISA (*8,9,11,12*).

9. Trouble shooting with respect to components of the FPLC equipment is performed according to instructions in the respective manuals. Problems not referable to these parts are generally to be sought in the sample, the column, or the buffers (*6*). Column hygiene is of utmost importance for obtaining good and reproducible results, as well as for avoiding costs by a reduced life-time of the column. All samples, buffers, and other solutions run through the column should be carefully prepared, filtered, and free of air bubbles. There are three major groups of biomolecules that are prone to contaminate the column: polyions, lipids, and certain proteins (*6*). CSF samples and sera have been found to be well compatible with the column when the material is free of lipids. We have used a simple chloroform extraction for delipidation. This procedure could introduce denaturation of certain proteins. No significant problems have been observed, however, for IgG. Alternative methods for delipidation can also be applied, e.g., techniques based on the use of lipophilic dextran (*13*). A thorough elimination of residual material should be performed after each run, and we have found it advantageous to routinely use an acid as well as a salt solution. It may

be found for certain applications, however, that such a procedure is needed first after 5–10 runs. In that case, the column is washed only with 2*M* sodium chloride after each of the intervening runs. Column contamination is usually manifested by a declination of the reproducibility/resolution or an increment of the backpressure.

10. The present review on chromatofocusing is aimed at giving basic methodological data and indicating some advantages as well as applications of the technique. We have experienced few technical problems when the method is applied in a proper way and its limitations are considered. One draw-back is the high price for the equipment. It should, however, be observed that chromatofocusing can be performed also by using an alternative stock of instruments (perhaps already available in your laboratory). Finally, you may find it advantageous to modify the methodology in order to optimize it for your special needs. Good luck!

Acknowledgments

This work was supported by grants from the Karolinska Institue, the Swedish Medical Research Council (grant no. 3381), the Swedish Multiple Sclerosis Society, the Swedish Society of Medical Sciences, and the Tore Nilson Foundation.

References

1. Sluyterman, L.A.A.E. and Wijdenes, J. (1976) Chromatofocusing: Isoelectric focusing on Ion Exchangers in the Absence of an Externally Applied Potential, in *Proceeding of the International Symposium on Electrofocusing and Isotachophoresis* (Radola, B. J. and Graesslin, D., eds.) Walter de Gruyter, Berlin.

2. Sluyterman, L.A.AE. and Elgersma, O. (1978) Chromatofocusing: Isoelectric focusing on ion-exchange columns. I. General principles. *J. Chromatogr.* **150**, 17–30.

3. Sluyterman, L.A.AE. and Wijdenes, J. (1978) Chromatofocusing: Isoelectric focusing on ion-exchange columns. II. Experimental verification. *J. Chromatogr.* **150**, 31–44.

4. Soderberg, L., Laas, T., and Low, D. (1981) Chromatofocusing: A New High Resolution Method for Protein Fractionation, in *Protides of the Biological Fluids 29* (Peeters, H., ed.) Pergamon, Oxford.

5. Richey, J. (1982) FPLC: A comprehensive separation technique of biopolymers. *Am. Lab.* **October**, 104–129.

6. Pharmacia Laboratory Separation Division (1985) *FPLC Ion Exchnage and Chromatofocusing: Principles and Methods* Pharmacia, Uppsala, Sweden.

7. Chromatofocusing: reference list, Pharmacia, Uppsala, Sweden.

8. Gallo, P. and Siden, A. (1985) Examinations of cerebrospinal fluid immunoglobulin G by chromatofocusing. *J. Neurol.* **232**, 231–235.

9. Gallo, P., Olsson, O., and Siden, A. (1986) Small-column chromatofocusing of cerebrospinal fluid and serum immunoglobulin G. *J. Chromatogr.* **375**, 277–283.

10. Siden, A., Gallo, P., and Olsson, O. (1985) Broad pH-Gradient Chromatofocusing, in *Protides of the Biological Fluids 33* (Peeters, H., ed.) Pergamon, Oxford.

11. Olsson, O., Bergenbrant, S., Gallo, P., Gudmundsson, C., and Siden, A. (1985) Examinations of Monoclonal IgG with Preparative Chromatofocusing Followed by 2-Dimensional Gel Electrophoresis, in *Protides of the Biological Fluids 33* (Peeters, H., ed.) Pergamon, Oxford.

12. Garcia-Merino, A., Gallo, P., Siden, A., and Tavolato, B. (1986) Isoelectric Focusing Combined with Chromatofocusing and ELISA for Investigations of Anti-MBP Antibodies, in *Protides of the Biological Fluids 34* (Peeters, H., ed.) Pergamon, Oxford.

13. Dahlberg, E., Snochowski, M., and Gustafsson, J.-A. (1980) Removal of hydrophobic compounds from biological fluids by a simple method. *Anal. Biochem.* **106**, 380–388.

14. Siden, A., Bergenbrant, S., Gallo, P., and Olsson, O. (1985) Applications of Small-Column Chromatofocusing and 2-Dimensional Gel Electrophoresis in CSF and Serum Protein Examinations, in *Protides of the Biological Fluids 33* (Peeters, H., ed.) Pergamon, Oxford.

Chapter 15

Hybrid Isoelectric Focusing Using Mixed Synthetic-Carrier Ampholyte-Immobilized pH Gradient Gels

Michael J. Dunn and Ketan Patel

1. Introduction

An exciting recent innovation in isoelectric focusing (IEF) has been the development of Immobiline reagents (LKB) (1) for the preparation of polyacrylamide gels containing immobilized pH gradients (IPGs). The Immobilines are a series of seven acrylamide derivatives with the general structure:

$$CH_2=CH-\overset{\overset{\displaystyle O}{\|}}{C}-\overset{\overset{\displaystyle H}{|}}{N}-R$$

where R contains either a carboxyl (acidic Immobilines) or a tertiary amino (basic Immobilines) group, forming a series of buffers with different pK values (3.6, 4.4, 4.6, 6.2, 7.0, 8.5, 9.3). IPG IEF gels are prepared in the same manner as conventional gradient polyacrylamide gels using a concentration gradient, stabilized with a density gradient (e.g., glycerol), of the Immobiline reagents mixed with acrylamide and *N,N'*-methylene bisacrylamide (Bis) under the control of a suitable gradient forming device. Thus, during polymerization the buffering groups forming the pH gradient are covalently attached and immobilized via vinyl bonds to the backbone of the polyacrylamide matrix. This immobilization of the species forming the pH gradient results in the elimination of pH gradient drift (but *not* electroendosmosis), making it possible to use extremely narrow pH gradients spanning from 0.1 to 1 pH unit. Additional advantages of this technology are that the pH gradients are not disturbed by high salt concentrations, and separations can be carried out under conditions of controlled ionic strength and buffering power. Although this is a relatively new methodology, it has already been the subject of several review articles (2–6).

Initial efforts concentrated on the development of narrow and ultranarrow pH gradients to exploit the extremely high resolving power of such systems. It is claimed that a difference in pI of only 0.002 pH units can be detected. These gradients are generated using only two Immobilines, one acting as a buffer and the other as a titrant. Detailed formulations for a series of 58 such 1 pH unit-wide gradients are documented in refs. 6 and 7, and narrower pH gradients can be readily derived by linear interpolation.

Narrow and ultranarrow pH gradients have an excellent resolution capacity, but have limited application to samples containing proteins of widely varying pI values. Wider pH gradients are essential for the analysis of such samples, but the formation of extended pH gradients, spanning greater than 1 pH unit, is complicated by the necessity to mix several buffering Immobiline species. This problem has been resolved by the creation of computer programs designed to calculate the mixtures required for this purpose (8,9). Using this approach, formulations have been derived for the generation of pH gradients spanning 2–6 pH units (9–11). Gradients extending from pH 4–10 can now be readily constructed, but IPGs cannot be extended to pH 3 without using two nonbuffering species, very acidic (pK < 1) and alkaline (pK, 12) Immobilines, as titrants. These reagents are not yet available commercially from LKB. It has also proved possible to generate nonlinear IPGs of any desired shape (12,13). This technique of pH gradient engineering can be used to optimize the shape of the pH gradient to obtain maximal resolution of the proteins in a particular type of sample.

Immobiline gels are best cast in standard vertical cassettes on GelBond PAG supports. Gels should generally be 1 mm or less in thickness. It is possible to form sample application wells in the gel at the time of gel casting using strips of adhesive embossing tape (e.g., Dymo) (6). Such depressions can accommodate 10–12 µL of sample, but it should be remembered that the depressions in the gel can disrupt the electric field during IEF. Once the appropriate recipe for the required pH gradient has been selected, the IPG gel can be formed using a simple two-chamber gradient-mixing device (6,7). Alternatively, a computer-controlled stepmotor-driven buret apparatus, commercially available from Desaga, can be used (14,15). This apparatus is expensive, but substantially improves reproducibility, flexibility, and documentation of the pH gradients. It is especially useful for the preparation of wide-range IPG gels in which pH gradient engineering, for example, localized gradient flattening, will be used to enhance resolution.

In the standard recommended procedure (6,7) IPG gels are polymerized by heating at 50°C for 1 h. This procedure is used because at this temperature all the Immobiline reagents have been found to react at the same rate (16,17), with 84–88% incorporation of all seven Immobilines being obtained. These conditions are potentially disastrous for wide-range pH gradients, however, because of the instability of the pK 8.5 and pK 9.3 species at elevated temperatures (85% destruction of the former in less than 1 h at 60°C) (18). To overcome this problem, Righetti's group advises careful control of the pH of the mixture used for gel preparation, by suitable titration with acids or bases, to around pH 7.5–8 (18). An alternative strategy has been devised by Altland that permits IPG gels to be polymerized at room temperature (19). In this procedure the stock Immobiline reagents are titrated with Tris buffer to near neutrality and stored in aliquots at –20 to –70°C. Polymerization in neutral buffer at ambient temperature is a controlled and convenient procedure, but these advantages are gained at the expense of the necessity for removing the buffer from the gels by washing and prerunning steps prior to sample application (19).

After polymerization, the gel on its plastic support is removed from the cassette. IPG gels cannot be used immediately for IEF, but must first be washed to remove catalysts and unpolymerized Immobilines. During this process the gel swells so that it is then necessary to reduce the gel, with the aid of a fan, to its original weight (17). This process is difficult to control and rather time-consuming. It is, therefore, preferable to use a rehydratable gel system for IPG gels (20,21). In this technique, gels, after polymerization, are washed, dried under a fan at room temperature, and stored in sealed plastic bags at –20°C for future use. This has the added advantage that batches of gels can be prepared, minimizing inter-gel variability.

Two strategies have been devised for rehydration of IPG gels. In the first method, weight is used to control the process (20,21). The alternative approach, which is less time-consuming and more reproducible, is to use volume to control rehydration (19,22). In the latter procedure, the gel is reassembled into a cassette of the same dimensions as that in which it was cast and reswollen in a controlled amount of the required solution.

Although narrow IPGs function very well, it has become apparent that problems are associated with the use of wide pH gradients. These problems include: (a) slow entry of sample proteins into the gel, (b) lateral band spreading, (c) prolonged focusing times, and (d) increased electroendosmosis. These difficulties are associated with the inherently low conductivity of the Immobiline system. It has been found that improved separations can be achieved by addition of low concentrations, typically 0.5% (w/v), of synthetic carrier ampholytes of the appropriate range to the solution used for gel rehydration (19,23,24). This technique has been termed "hybrid isoelectric focusing" by Altland (19) or "mixed carrier ampholyte-IPG IEF" by Righetti (25). Here we describe a hybrid IEF gel system for the separation of human serum proteins in a pH 4–7 gradient.

2. Materials

1. Gel Bond PAG sheets
2. Two 5 mm thick plastic plates (17 x 13 cm) for the gel-casting cassette. One plate has a 3 mm hole

drilled in one corner, 10 mm from the two sides, to form a gel solution entry port.

3. Silicon rubber gasket (0.5 mm thick, 7 mm wide) to fit around the edge of the glass plates.

4. PVC sheet (17 x 13 cm) with a 3 mm-diameter hole punched in register with the entry port in the plastic plate.

5. Rubber roller.

6. Acrylamide stock solution: 14.4 g acrylamide; 0.6 g Bisacrylamide, made up to 50 mL with distilled water.

   The solution is deionized with mixed-bed ion exchange resin (Duolite MB5113 mixed resin, BDH), 5 g/100 mL solution, for 1 h. After deionization, the solution is filtered. The stock solution is stored at 4°C for a maximum of 1 wk. This deionization step reduces ionic contamination (e.g., acrylic acid), which can adversely affect IEF.

7. Stock Immobiline solutions are prepared according to the method of Altland (*19*). This allows for polymerization of IPG gels under controlled conditions at room temperature. Acid Immobilines are titrated with 1*M* Tris to pH 6.8 and brought to a final concentration of 0.25*M* by adding 0.1*M* Tris-phosphate, pH 6.8. Basic Immobilines are titrated with 1*M* phosphoric acid to pH 6.8 and diluted to 0.25*M* in 0.1*M* Tris phosphate, pH 6.8. The volume required to obtain 0.25*M* immobiline solutions is calculated by the following formula:

$$V = M \times (1/C - 1/C_0 - V_0/M_0)$$

where V is the added total volume, M the amount of Immobiline in mmoles, M_0 the standard amount of Immobiline per bottle (5.11, 5.13, 5.14, 5.17, 5.19,

5.15, and 5.17 mmol for Immobilines of pK_a 3.6, 4.4, 4.6, 6.2, 7.0, 8.3, and 9.3, respectively), C the concentration of Immobiline to be adjusted for, C_0 the standard concentration of 0.2M in an Immobiline bottle after addition of the standard volume, V_0, of 25 mL. The titrated and diluted Immobilines should be aliquoted and stored at –20 to –70°C.

8. Recipe for stock Immobiline reagents required to prepare the pH 4–7 IPG IEF gel (total gel volume, 30 mL) (10):

Immobiline pK_a	Acidic solution, µL	Basic solution, µL
3.6	578	301
4.6	110	738
6.2	449	151
7.0	—	269
9.6	—	876
	1137	2235

Note that the recipes in ref. *10* give ratios of the Immobiline reagents required. The ratios must be multiplied by 5 to obtain a working concentration (0.2M) expressed as µL/mL of gel solution. Thus, for the 15 mL volumes of each solution used here (acidic, basic), the ratio value must be multiplied by a factor of 75.

9. The cassette will hold a total gel volume of 30 mL, so that 15 mL each of the acidic (light) and basic (dense) solutions are required to form the gradient. The final gel is T3C4 containing 4% (w/v) glycerol at the acidic end (top) and 15% (w/v) glycerol at the basic end (bottom) to stabilize the gradient until polymerization has occurred.

	Concentration	Acidic (light) solution	Basic (dense) solution
Acrylamide	T30C4	1.5 mL	1.5 mL
Glycerol	100%	0.6 mL	2.25 mL
Immobilines (acid)	0.2M	1137 µL	—
Immobilines (basic)	0.2M	—	2235 µL
TEMED	100%	7.5 µL	8.5 µL
Ammonium persulfate	10%	25 µL	57 µL
Distilled water	—	11.73 mL	8.85 mL
		15 mL	15 mL

10. Any gradient forming apparatus capable of delivering a linear gradient of the required volume can be used. The simplest device is a two-chamber gradient mixer (6,7), and an interesting modification has been published recently (26). We use an LKB Ultrograd electronic gradient former. Probably the best and most flexible approach is to use a computer-controlled, stepmotor-driven, high-precision buret system (14,15), which is available commercially from Desaga. The latter types of apparatus can be used to generate nonlinear IPGs of any desired shape that can be used to optimize protein resolution.

11. Rehydration solution: A solution of 8M urea containing 0.5% (w/v) ampholyte mixture (Servalyt pH 4–5, 5–6, 6–7 in a 1:1:1 mixture).

12. Any horizontal flat-bed IEF apparatus fitted with a cooling platten can be used. We currently use a Pharmacia FBE-3000 Flatbed system powered by a FBE-3000/150 powerpack fitted with a volt hour (Vh) integrator connected to a dual-channel chart recorder to give a continuous output of the voltage and current parameters.

13. Silicon fluid, Dow Corning 200/10 cs.
14. Electrode wicks (16 x 0.7 mm) cut from sheets of Whatman GF/B glass fiber paper. A stack of four wicks is used at each electrode position.
15. Electrolytes for the prerun: 0.5% (w/v) Servalyt 4–5, 5–6, 6–7 in a 1:1:1 mixture.
16. Silicon rubber sample application strip (available from Desaga), containing 27 slits for 20-µL samples or 54 holes for 10-µL samples.
17. Electrolytes for the focusing run.
    Catholyte—0.1M NaOH
    Anolyte—0.1M oxalic acid
18. Trichloroacetic acid (TCA) solution 20% (w/v) for gel fixation.
19. Staining solution: 0.5% (w/v) Coomassie Brilliant blue R250 in 45% methanol, 10% acetic acid.
20. Destaining solution: 45% methanol, 10% acetic acid.

3. Method

1. Cut a piece of GelBond PAG (17 x 13 cm). This material is specially treated so that the gel will adhere firmly to it. Only one surface (hydrophilic) is treated, the other surface being hydrophobic.
2. The IPG IEF gel is cast in a vertical cassette. The GelBond sheet is rolled onto a plastic plate with the hydrophobic surface of the GelBond in contact with the plate. A film of water is used between these surfaces to ensure good adherence, and air bubbles are removed with the roller. The U-shaped silicon rubber gasket is placed onto the GelBond/plastic plate assembly. This gasket provides a leak-proof seal without the use of grease. The PVC sheet is glued to the second plastic plate, ensuring the 3-

mm holes for entry of the gel solution are in register. Photographic "Spray Mount" adhesive is used to ensure good contact between the surfaces. The PVC sheet/plastic plate assembly is placed onto the gasket/GelBond/plastic plate assembly so that the gel solution entry port is at one of the lower corners of the cassette assembly. The cassette is held together along the two sides and the bottom edge with bulldog clips. The cassette is placed upright in a level, vertical position.

3. The gradient-forming device to be used is attached by tubing to the gel solution entry port at one lower corner of the cassette. The other side of the cassette is raised so that the bottom edge is at 45° to the horizontal. This forms a V shape with the solution input port at the lowest point. This ensures that the gel solution entering the cassette does so smoothly and is not subjected to excessive turbulence that would disturb the gradient. As the cassette fills, it is gently lowered to the horizontal position.

4. The gel is allowed to polymerize for 2 h at room temperature.

5. The cassette is dismantled, and the IPG gel on its GelBond support is carefully removed. The acidic (top) and the basic (bottom) ends of the gel are marked on the reverse side of the GelBond to ensure correct orientation of the gel during IEF.

6. The gel is washed twice, 30 min each, in 500 mL of 1% (w/v) glycerol. The presence of this low glycerol concentration in the gel prevents cracking and peeling of the gel during drying.

7. The gel is dried in a stream of air at room temperature. The drying time depends on the source of the air stream, but the gel must be completely dry before storage.

8. Gels can be stored indefinitely by sealing in plastic food-storage bags using a heat-sealer and storing in a dry container at −20°C.

9. Rehydration is performed using a cassette of the same dimensions as the one in which the gel was originally cast. The cassette is assembled as described previously using the dried gel on its support in place of the GelBond PAG sheet. A gasket of the same thickness is used. The solution used for rehydration can be introduced via the original input port or using a syringe in the top of the cassette. The gels are rehydrated with the rehydration solution containing 8*M* urea and 0.5% ampholytes for 2 h at room temperature. Rehydration is complete once all the solution in the cassette has been taken up by the gel.

10. The rehydrated gel is removed from the cassette and placed on the precooled (15°C) platten of the flat-bed IEF apparatus. A film of silicon fluid, which has excellent thermal conductivity and low viscosity, is used to provide good contact between the gel and the cooling plate. The electrode wicks wetted with prerunning electrolytes are applied to the ends of the gel.

11. The gel is prefocused at 3000 V, 5 mA, and 10 W limiting for 1 h. At this time the voltage will have risen to 3000 V whereas current will have dropped to below 5 mA. Refractive lines, running parallel to the electrode wicks, should be observed across the gel. This prerunning step removes any salts contained within the gel.

12. The electrode wicks are changed for others wetted with the running electrode solutions. The silicon rubber sample application strip containing sample application slits or holes is placed on the surface of

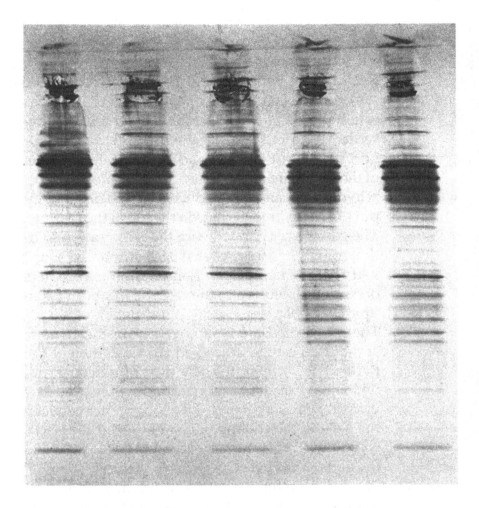

Fig. 1. Separation of human serum proteins using hybrid IEF in
a pH 4–7 IPG gel.

the gel near the cathode. Serum samples (6 μL) are
applied in the wells.

13. The gel is focused at 3000 V, 5 mA, 10 W limiting for
9000 Vh.

14. After IEF is complete, the gel is fixed with 20% (w/
v) TCA for 1 h.

15. The gel is washed for 30 min in water and stained for 30 min with Coomassie Brilliant blue R250. The gel is dried in air at 40°C. A typical separation of human serum proteins is shown in Fig. 1.

4. Notes

1. The catalyst levels used are chosen so that polymerization is complete 1 h after gel casting, allowing for maximum incorporation of Immobiline molecules into the polyacrylamide matrix without diffusion breaking down the gradient.
2. If nonionic (e.g., Triton X-100, NP-40) or zwitterionic (e.g., CHAPS) detergent is required in the gel to increase sample solubility, it should be added at the required concentration to the two solutions (acidic and basic) used to prepare the gel. [Levels of detergent between 0.5 and 2% (w/v) are compatible with the IEF gel system.] Detergent at the same concentration should also be included in the solutions used for washing the gel prior to drying. Detergent should not be added at the rehydration step, since it is already contained within the dried gel matrix. This method is used because it has been found to be difficult to rehydrate polyacrylamide gels containing *both* urea and detergent (*21,27,28*).

References

1. Bjellqvist, B., Ek, E., and Righetti, P. G. (1982) Isoelectric focusing in immobilized pH gradients: Principle, methodology and some applications. *J. Biochem. Biophys. Meth.* 6, 317–339.
2. Righetti, P. G. (1983) *Isoelectric Focusing: Theory, Methodology and Applications.* Elsevier, Amsterdam.

3. Righetti, P. G., Gianazza, E., and Bjellqvist, B. (1983) Modern aspects of isoelectric focusing: Two-dimensional maps and immobilized pH gradients. *J. Biochem. Biophys. Meth.* **8**, 89–108.

4. Righetti, P. G., Gianazza, E., and Gelfi, C. (1984) Immobilized pH Gradients: Analytical and Preparative Aspects, in *Electrophoresis 84* (Neuhoff, V., ed.) Verlag Chemie, Weinheim.

5. Righetti, P. G. (1986) Everything you wanted to know about Immobilines but were afraid to ask. *Sci. Tools* **33**, 1–4.

6. Righetti, P. G., Gelfi, C., and Gianzza, E. (1986) Conventional Isoelectric Focusing and Immobilized pH Gradients, in *Gel Electrophoresis of Proteins* (Dunn, M. J., ed.) Wright, Bristol.

7. LKB Application Note Number 324 (1984) LKB Produkter AB, Bromma, Sweden.

8. Dossi, G, Celentano, F., Gianazza, E., and Righetti, P. G. (1983) Isoelectric focusing in immobilized pH gradients: Generation of extended pH intervals. *J. Biochem. Biophys. Meth.* **7**, 123–142.

9. Gianazza, E., Dossi, G., Celentano, E., and Righetti, P. G. (1983) Isoelectric focusing in immobilized pH gradients: Generation and optimization of wide pH intervals with two-chamber mixers. *J. Biochem. Biophys. Meth.* **8**, 109–133.

10. Gianazza, E., Celentano, F., Dossi, G., Bjellqvist, B., and Righetti, P. G. (1984) Preparation of immobilized pH gradients spanning 2–6 pH units with two-chamber mixers: Evaluation of two experimental approaches. *Electrophoresis* **5**, 88–97.

11. Gianazza, E., Astrua-Testori, S., and Righetti, P. G. (1985) Some more formulations for immobilized pH gradients. *Electrophoresis* **6**, 113–117.

12. Gianazza, E., Giacon, P., Sahlin, B., and Righetti, P. G. (1985) Non-linear pH courses with immobilized pH gradients. *Electrophoresis* **6**, 53–56.

13. Altland, K., Hackler, R., Banzhoff, A., and von Eckardstein, A. (1985) Experimental evidence for flexible slope of immobilized pH gradients poured under computer control. *Electrophoresis* **6**, 140–142.

14. Altland, K. and Altland, A. (1984) Forming reproducible density and solute gradients by computer-controlled cooperation of stepmotor-driven burettes. *Electrophoresis* **5**, 143–147.

15. Altland, K. and Altland, A. (1984) Pouring reproducible gradients in gels under computer control: New devices for simultaneous delivery of two independent gradients, for more flexible slope and pH range of immobilized pH gradients. *Clin. Chem.* **30**, 2098–2103.

16. Righetti, P. G. (1984) Isoelectric focusing in immobilized pH gradients. *J. Chromatogr.* **300,** 165–223.

17. Righetti, P. G., Ek, K., and Bjellqvist, B. (1984) Polymerization kinetics of polyacrylamide gels containing immobilized pH gradients for isoelectric focusing. *J. Chromatogr.* **291,** 31–42.

18. Pietta, P., Pocaterra, E., Fiorino, A., Gianaza, E., and Righetti, P. G. (1985) Long-term storage of free and polyacrylamide gel-bound Immobiline chemicals. *Electrophoresis* **6,** 162–170.

19. Altland, K. and Rossmann, U. (1985) Hybrid isoelectric focusing in rehydrated immobilized pH gradients with added carrier ampholytes: Demonstration of human globins. *Electrophoresis* **6,** 314–325.

20. Gianazza, E., Artoni, G., and Righetti, P. G. (1983) Isoelectric focusing in immobilized pH gradients in presence of urea and neutral detergents. *Electrophoresis* **4,** 321–326.

21. Gelfi, C. and Righetti, P. G. (1984) Swelling kinetics of Immobiline gels for isoelectric focusing. *Electrophoresis* **5,** 257–262.

22. Altland, K., Banzhoff, A., Hackler, R., and Rossmann, U. (1984) Improved rehydration procedure for polyacrylamide gels in presence of urea: Demonstration of inherited human prealbumin variants by isoelectric focusing in an immobilized pH gradient. *Electrophoresis* **5,** 379–381.

23. Fawcett, J. S. and Chrambach, A. (1985) Electrofocusing in wide pH range Immobiline gels containing carrier ampholytes. *Protides Biol. Fluids* **33,** 439–442.

24. Fawcett, J. S. and Chrambach, A. (1986) The voltage across wide pH range immobilized pH gradient gels and its modulation through the addition of carrier ampholytes. *Electrophoresis* **7,** 266–272.

25. Righetti, P. G. and Gianazza, E. (1986) Two-Dimensional Maps with Immobilized pH Gradient in the First Dimension, in *Progres Recents en Electrophorese Bidimensionelle* M. -M. (Galteau, M. M. and Siest, G., eds.) Presses Universitaires, Nancy.

26. Fawcett, J. S. and Chrambach, A. (1986) Simplified procedure for the preparation of immobilized pH gradient gels. *Electrophoresis* **7,** 260–266.

27. Dunn, M. J., Burghes, A. H. M., and Patel, K. (1985) Two-dimensional electrophoretic analysis of genetic muscle disorders. *Protides Biol. Fluids* **33,** 479–482.

28. Dunn, M. J., Patel, K., and Burghes, A. H. M. (1986) New Developments for the First Dimension of 2D-PAGE, in *Progres Recents en Electrophorese Bidimensionelle* (Galteau, M. M. and Siest, G., eds.) Presses Universitaires, Nancy.

Chapter 16

Two-Dimensional Electrophoresis Using Immobilized pH Gradients in the First Dimension

Michael J. Dunn and Ketan Patel

1. Introduction

The high-resolution capacity of two-dimensional polyacrylamide gel electrophoresis (2-D PAGE) based on the method of O'Farrell (1) makes it the method of choice for the analysis of complex protein mixtures (2–4). In the standard procedure, cylindrical tube isoelectric focusing (IEF) gels are used. Unfortunately this system suffers from severe cathodic drift resulting in pH gradients that do not extend above pH 8, resulting in concomitant loss of basic proteins from 2-D maps. Gradients can be extended to pH 10 by special treatment of the glass IEF tubes (5) or by using a horizontal flat-bed IEF (5–7). In spite of the improvement in resolution of basic proteins that can be achieved with these procedures, difficulties are still associated with the first-dimension IEF gels as a consequence of the characteristic properties of the synthetic carrier ampholytes used to generate the pH gradients. These include problems of batch reproducibility of ampholytes, irreproducibility of separations, difficulty in control of pH gradient stability and shape (i.e., "pH gradient engineering"), and the possibility of artifacts caused by protein–ampholyte interactions.

Recent advances in IEF technology have provided alternative procedures that might overcome some of these problems. We have found one of these alternatives, buffer IEF, to be unsuitable or the first dimension of 2-D PAGE (8). The use of the Immobiline reagents (LKB), however, to generate immobilized pH gradients (IPGs) appears to be a much more promising approach for 2-D PAGE. The Immobilines are a series of seven substituted acrylamide derivatives with different pH values. IPG IEF gels are made by generating a gradient of the appropriate Immobiline solutions, so that during polymerization the buffering groups forming the pH gradient are covalently attached and immobilized via vinyl bonds to the polyacrylamide backbone. This immobilization results in the elimination of pH gradient drift (but *not* electroendosmosis) making pH gradients reproducible and infinitely stable—ideal properties for 2-D PAGE.

IPGs, reviewed in ref. 9 (and *see* chapter 15) were originally optimized for the generation of narrow and ultra-narrow pH gradients not generally suitable for 2-D applications. Recipes are now available, however for producing wide pH gradients (10), so that pH 4–10 gradients suited to 2-D PAGE can now be generated. Unfortunately, there are problems associated with the use of wide IPGs caused by the low conductivity of the

Immobiline system. These include slow entry of sample proteins, lateral band spreading, prolonged focusing times, and increased electroendosmosis. Improved separations can be obtained by the addition of low concentrations, typically 0.5% (w/v), of synthetic carrier ampholytes of the appropriate range to the IPG IEF gels (11–13). This technique has been termed "hybrid isoelectric focusing" by Altland or "mixed carrier ampholyte IPG IEF" by Righetti (*see* chapter 15).

IPG technology for 2-D PAGE is not fully developed, and it is true that the 2-D patterns obtained are generally inferior to those obtained using traditional methods. This appears to be caused not by problems with the IG IEF dimension *per se*, but by problems in elution and transfer of proteins from the first-dimension IPG gels to the second-dimension SDS-PAGE gls, resulting in straking of spots on 2-D maps (14–16). This phenomenon can probably be ascribed to the presence of fixed charges on the immobiline gel matrix, leading to increased electroendosmosis in the region of contact between the IPG IEF gel and the SDS-PAGE gel and resulting in disturbance of migration of proteins from the first to the second dimension gel.

Despite these problems, the ability of IPG technology to produce stable pH gradients whose shape can be engineered with ease makes this a method of great potential for increasing the reproducibility of 2-D maps. We will describe here a method of hybrid IEF using pH 4–10 IPG gels containing synthetic carrier ampholytes for the first dimension of 2-D PAGE.

2. Materials

1. GelBond PAG sheets.
2. Two 4-mm thick plastic plates (23 x 18 cm) for the

gel-casting cassette. One plate has a 3-mm hole drilled in one corner, 10 mm from each side, to form a gel solution entry port.

3. Silicon rubber gasket (0.5 mm thick, 7 mm wide) to fit around the edge of the plastic plates.

4. PVC sheet (23 x 18 cm) with a 3 mm-diameter hole punched in register with the entry port in the plastic plate.

5. Rubber roller.

6. Acrylamide stock solution: 14.4 g acrylamide; 0.6 g bisacrylamide made up to 50 mL with distilled water.

   The solution is deionized with mixed-bed ion exchange resin (Duolite MB5113 mixed resin, BDH), 5 g/100 mL solution, for 1 h. After deionization, the solution is filtered. The stock solution is stored at 4°C for a maximum of 1 wk. This deionization step reduces ionic contamination (e.g., acrylic acid) which can adversely affect IEF.

7. Stock Immobiline solutions are prepared according to the method of Altland (11). This allows for polymerization of IPG gels under controlled conditions at room temperature. Acid Immobilines are titrated with $1M$ Tris to pH 6.8. Basic Immobilines are titrated with $1M$ phosphoric acid to pH 6.8 and diluted to $0.25M$ in $0.1M$ LTris-phosphate, pH 6.8. The volume required to obtain $0.25M$ Immobiline solutions is calculated by the following formula:

$$V = M \times (1/C - 1/C_0 - V_0/M_0)$$

where V is the added total volume, M the amount of Immobiline in mmoles, M_0 the standard amount of Immobiline per bottle (5.11, 5.13, 5.14, 5.17, 5.19, 5.15, and 5.17 mmoles for Immobilines of pK_a 3.6,

4.4, 4.6, 6.2, 7.0, 8.3, and 9.3, respectively), C the concentration of Immobiline to be adjusted for, C_o the standard concentration of $0.2M$ in an Immobiline bottle after addition of the standard volume, V_o, of 25 mL. The titrated and diluted Immobilines should be aliquoted and stored at −20 to −70°C.

8. Recipe for stock Immobiline reagents required to prepare the pH 4–10 IPG IEF gel (total gel volume, 30 mL) (10):

Immobiline pK_a	Acidic solution, µL	Basic solution, µL
3.6	1102	—
4.6	—	113
6.2	458	53
7.0	90	488
8.5	338	158
9.6	—	360
	1988	1172

Note that the recipes in ref. *10* give ratios of the Immobiline reagents required. The ratios must be multiplied by 5 to obtain a working concentration ($0.2M$) expressed as L/mL of gel solution. Thus, for the 15-mL volumes of each solution (acidic, basic) used here, the ratio value must be multiplied by a factor of 75.

9. The cassette will hold a total gel volume of 30 mL, so that 15 mL each of the acidic (light) and basic (dense) solutions are required to form the gradient. The final gel is T3C4 containing 4% (w/v) glycerol at the acidic end (top) and 15% (w/v) glycerol at the basic end (bottom) to stabilize the gradient until polymerization has occurred.

	Concentration	Acidic (light)	Basic (dense)
Acrylamide	T30C4	1.5 mL	1.5 mL
Glycerol	100%	0.6 mL	2.25 mL
Immobilines (acid)	0.2M	1988 μL	—
Immobilines (Basic)	0.2M	—	1172 μL
TEMED	100%	7.5 μL	8.5 μL
Ammonium persulfate	10%	25 μL	57 μL
Distilled water		10.88 mL	10.01 mL
		15 mL	15 mL

10. Any apparatus capable of delivering a linear gradient of the required volume can be used. The simplest device is a two-chamber gradient mixer (9,17), and an interesting modification has been published recently (18). We use an LKB Ultrograd electronic gradient former. Probably the best and most flexible approach is to use a microcomputer-controlled, stepmotor-driven, high-precision buret system (19,20), which is available commercially from Desaga. The latter types of apparatus can be used to generate nonlinear IPGs of any desired shape that can be used to optimize protein resolution.

11. Rehydration solution: A solution of 8M urea containing 0.5% (w/v) ampholytes (Pharmalyte pH 3–10) and 15% (w/v) glycerol (16).

12. Any horizontal flat-bed IEF apparatus fitted with a cooling platten can be used. We currently use a Pharmacia FBE-3000 flat-bed system powered by a Pharmacia ECPS 3000/150 powerpack fitted with a VH-1 volt hour (Vh) integator connected to a dual-channel chart recorder to give a continous output of the voltage and current parameters.

13. Silicon fluid, Dow Corning 200/10 cs.

14. Electrode wicks (20 x 0.7 mm) cut from sheets of

Whatman GF/B glass fiber paper. A stack of four wicks is used at each electrode position.

15. Electrolytes for the prerun: 0.5% (w/v) Pharmalyte pH 3–10.

16. Silicon rubber (1 mm thick) sample application strip (available from Desaga) containing 27 slits for 20 μL samples. From this strip sample application pieces with individual slits are cut.

17. Electrolytes for the focusing run:
    Catholyte—0.1M NaOH
    Anolyte—0.1M oxalic acid

18. Equilibration buffer containing 8M urea, 1% (w/v) sodium dodecyl sulfate (SDS), 20 mM dithiothreitol (DTT), 15% (w/v) glycerol in 150 mM Tris buffer, pH 8.6.

3. Method

1. Cut a piece of GelBond PAG (23 x 18 cm). This material is specially treated so that the gel will adhere firmly to it. Only one surface (hydrophilic) is treated. The reverse (hydrophobic) surface of the GelBond sheet is marked with a permanent marker pen into lanes (1.2 cm wide, 14 cm long) that are numbered to aid identification of sample strips for the second dimension.

2. The IPG IEF gel is cast in a vertical cassette. The GelBond sheet is rolled onto a plastic plate with the hydrophobic surface of the GelBond in contact with the plate. A film of water is used between these surfaces to ensure good adherence, and air bubbles are removed with the roller. The silicon rubber gasket is placed onto the GelBond/plastic plate assembly. This gasket provides a leak-proof seal without the use of grease. The PVC sheet is glued

to the second plastic plate ensuring that the 3-mm holes for entry of the gel solution are in register. Photographic "Spray Mount" adhesive is used to ensure good contact between the surfaces. The PVC sheet/platic plate assembly is placed onto the gasket/Gel Bond/plastic plate assembly so that the gel solution entry port is at one of the lower corners of the cassette assembly. The gasket is held together along the two sides and the bottom edge with bulldog clips. The cassette is placed upright in a level, vertical position.

3. The gradient-forming device to be used is attached by tubing to the gel solution entry port at one lower corner of the cassette. The other side of the cassette is raised so that the botttom edge is at 45° to the horizontal. This forms a V shape with the solution input at the lowest point. This ensures that the gel solution entering the cassette does so smoothly and is not subjected to excessive turbulence that would degrade the gradient. As the cassette fills, it is gently lowered to the horizontal position.

4. The gel is allowed to polymerize for 2 h at room temperature.

5. The cassette is dismantled, and the IPG gel on its GelBond support is carefully removed. The acidic (top) and the basic (bottom) ends of the gel are marked on the reverse side of the GelBond to ensure correct orientation of the gel during IEF.

6. IPG gels cannot be used directly for IEF, but must first be washed to remove salts, catalysts, and unpolymerized Immobilines. The gel swells during this process so that it must be reduced to its original weight prior to IEF. This is best achieved using a rehydratable gel system. In this technique the gel is washed twice, 30 min each, in 500 mL 1%

(w/v) glycerol. The presence of this low glycerol concentration in the gel prevents cracking and peeling of the gel during drying.

7. The gel is dried in a stream of air at room temperature. The drying time depends on the source of the air stream, but the gel must be completely dry before storage.

8. Gels can be stored indefinitely by sealing in plastic food-storage bags using a heat-sealer and storing in a dry container at –20°C.

9. Rehydration is performed using a cassette of the same dimensions as the one in which the gel was originally cast. The cassette is assembled as described previously using the dried gel on its support in place of the GelBond PAG sheet. A gasket of the same thickness is used. The solution used for rehydration can be introduced via the original input port or using a syringe into the top of the cassette. The gels are rehydrated with solution containing 8M urea and 0.5% ampholytes for 2h at room temperature. Rehydration is complete once all the solution in the cassette has been taken up by the gel.

10. The rehydrated gel is removed from the cassette and placed on the precooled (15°C) platten of the flat-bed IEF apparatus. The silicon fluid, which has excellent thermal conductivity and low viscosity, is used to provide good contact between the gel and the cooling plate. The electrode wicks wetted with prerunning electrolytes are applied to the ends of the gel.

11. The gel is prefocused at 3000 V, 5 mA, and 10 W limiting for 1 h. At this time the voltage will have risen to 3000 V, whereas the current will have dropped to below 5 mA. Refractive lines, running

dropped to below 5 mA. Refractive lines, running parallel to the electrode wicks, should be observed across the gel. This prerunning step removes salts contained within the gel.

12. The electrode wicks are changed for others containing the running electrode solutions. The individual silicon sample rubber application wells are applied to the surface of the gel in each lane near the cathode. Samples (10 µL) are applied in the wells. Inclusion of Bromophenol blue in the samples helps in checking that the wells do not leak, since any mixing of adjacent samples is disastrous for 2-D PAGE.

13. The gel is focused at 3000 V, 5 mA, 10 W limiting for 15,000 Vh.

14. The gel is wrapped in plastic food-wrapping film and stored at –70°C.

15. For the second dimension, the gel is thawed and the appropriate sample lanes cut off with scissors using the lane markings on the back of the GelBond PAG sheet as a guide.

16. Strips are incubated at room temperature for 2 x 15 min in 7 mL of equilibration buffer.

17. Preparation of the second dimension SDS-PAGE gels will not be described, since they are of the normal Laemmli (21) type, as used in the standard 2-D PAGE procedure. The gels we use are linear 8–20% gradient SDS-PAGE gels (1 mm thick, 16 cm high, and 18 cm wide with a 2 cm high stacking gel) (*see* vol. 1 of this series).

18. The IPG IEF gel strips are applied to the SDS gels by filling the space in the cassettes above the stacking gels with equilibration buffer and gently sliding the strips into place. The strips should be in close contact with the top of the stacking gels. There is no

need to cement the strips in place with agarose.

19. The second-dimension SDS gels can be run in any suitable vertical slab gel apparatus. The gels are run at 12 mA/gel overnight until the Bromophenol blue tracking dye reaches the bottom of the gels.

20. After 2-D electrophoresis is complete, the gels can be subjected to any desired procedure to detect the separated components (e.g., electroblotting, fixation and staining, autoradiography and so on). An example of [^{35}S]-methionine-labeled human skin fibroblast proteins separated by this procedure is shown in Fig. 1.

4. Notes

1. The catalyst levels used are chosen so that polymerization is complete 1 h after gel casting, allowing for maximum incorporation of Immobiline molecules into the polyacrylamide matrix without diffusion resulting in breakdown of the gradient.

2. If nonionic (e.g., Triton X-100, NP-40) or zwitterionic (e.g., CHAPS) detergent is required in the gel to increase sample solubility, it should be added at the required concentration to the two solutions (acidic and basic) used to prepare the gel. The gel system is compatible with levels of detergent between 0.5% and 2% (w/v). Detergent at the same concentration should also be included in the solutions used for washing the gel prior to drying. Detergent should not be added at the rehydration step since it is already contained within the dried gel matrix. This method is used because it has been found difficult to rehydrate polyacrylamide gels containing *both* urea and detergent (*14,22,23*).

Fig. 1. Autoradiograph of a 2-D PAGE separation of human skin fibroblast proteins labeled with [³⁵S]-methionine using a pH of 4–10 IPG IEF gel in the first dimension.

References

1. O'Farrell, P. H. (1975) High resolution two-dimensional electrophoresis of proteins. *J. Biol. Chem.* **250,** 4007–4021.
2. Dunn, M. J. and Burghes, A. H. M. (1983) High resolution two-dimensional polyacrylamide gel electrophoresis. I. Methodological procedures. *Electrophoresis* **4,** 97–116.
3. Dunn, M. J. and Burghes, A. H. M. (1983) High resolution two-dimensional polyacrylamide gel electrophoresis. II. Analysis and applications. *Electrophoresis* **4,** 173–189.
4. Dunn, M. J. and Burghes, A. H. M. (1986) High Resolution Two-Dimensional Polyacrylamide Gel Electrophoresis, in *Gel Electrophoresis of Proteins* (Dunn, M. J., ed.). Wright, Bristol.
5. Burghes, A. H. M., Dunn, M. J., and Dubowitz, V. (1982) Enhancement of resolution in two-dimensional gel electrophoresis and simultaneous resolution of acidic and basic proteins. *Electrophoresis* **3,** 354–363.
6. Dunn, M. J., Burghes, A. H. M., Witkowski, J. A., and Dubowitz, V. (1985) Analysis of genetic muscle disorders by two-dimensional electrophoresis. *Protides Biol. Fluids* **32,** 973–976.
7. Dunn, M. J. and Patel, K. (1987) Two-Dimensional Polyacrylamide Gel Electrophoresis Using Flatbed Isoelectric Focusing in the First Dimension, in *Methods in Molecular Biology*, vol. 3 (Walker, J. M., ed.). Humana, Clifton, New Jersey.
8. Burghes, A. H. M., Patel, K., and Dunn, M. J. (1985) Comparison of focusing in buffers and synthetic carrier ampholytes for use in the first dimension of two-dimensional polyacrylamide gel electrophoresis. *Electrophoresis* **6,** 453–461.
9. Righetti, P. G., Gelfi, C., and Gianazza, E. (1986) Conventional Isoelectric Focusing and Immobilized pH Gradients, in *Gel Electrophoresis of Proteins* (Dunn, M. J., ed.). Wright, Bristol.
10. Gianazza, E., Astrua-Testori, S., and Righetti, P. G. (1985) some more formulations for immobilized pH gradients. *Electrophoresis* **6,** 103–146.
11. Altland, K. and Rossmann, U. (1985) Hybrid isoelectric focusing in rehydrated immobilized pH gradients with added carrier ampholytes: Demonstration of human globins. *Electrophoresis* **6,** 314–325.
12. Fawcett, J. S. and Chrambach, A. (1985) Electrofocusing in wide pH range Immobiline gels containing carrier ampholytes. *Protides Biol. Fluids* **33,** 439–442.

13. Fawcett, J. S. and Chrambach, A. (1986) The voltage across wide pH range immobilized pH gradient gels and its modualtion through the addition of carrier ampholytes. *Electrophoresis* **7**, 266–272.

14. Dunn, M. J., Burghes, A. H. M., and Patel, K. (1985) Two-dimensional electrophoretic analysis of genetic muscle disorders. *Protides Biol. Fluid* **33**, 479–482.

15. Dunn, M. J. and Patel, K. (1986) Two dimensional polyacrylamide gel electrophoresis using immobilized pH gradients in the first dimension. *Protides Biol. Fluids* **34**, 695–699.

16. Gorg, A., Postel, W., and Weser, J. (1985) IEF and 2D-electrophoresis with narrow and ultranarrow immobilized pH gradients for the analysis of genetic variants. *Proteides Biol. Fluids* **33**, 467–470.

17. LKB Application Note Number 324 (1984) LKB Produkter AB, Bromma, Sweden.

18. Fawcett, J. S. and Chrambach, A. (1986) Simplified procedure for the prepation of immobilized pH gradient gels. *Electrophoresis* **7**, 260–266.

19. Altland, K. and Altland, A. (1984) Forming reproducible density and solute gradients by computer-controlled cooperation of stepmotor-driven burettes. *Electrophoresis* **5**, 143–147.

20. Altland, K. and Altland, A. (1984) Pouring reproducible gradients in gels under computer control: New devices for simultaneous delivery of two independent gradients, for more flexible slope and pH range of immobilized pH gradients. *Clin. Chem.* **30**, 2098–2103.

21. Laemmli, U. K. (1970) Cleavage of structural proteins during the assembly of the head of bacteriophage T4. *Nature* **227**, 680–685.

22. Dunn, M. J., Patel, K., and Burghes, A. H. M. (1986) New Developments for the First Dimension of 2D-PAGE, in *Progres Recents en Electrophorese Bidimensionelle* (Galteau, M. M. and Siest, G. eds.) Presses Universitaires, Nancy.

23. Gelfi, C. and Righetti, P. G. (1984) Swelling kinetics of Immobiline gels for isoelectric focusing. *Electrophoresis* **5**, 257–262.

Chapter 17

Two-Dimensional Polyacrylamide Gel Electrophoresis Using Flat-Bed Isoelectric Focusing in the First Dimension

Michael J. Dunn and Ketan Patel

1. Introduction

Electrophoretic techniques are, perhaps, the most important group of procedures for the separation and analysis of complex protein mixtures because of their high resolution capacity. This resolving power is maximized in high-resolution, two-dimensional polyacrylamide gel electrophoresis (2-D PAGE) based on the technique developed by O'Farrell (1). This methodology involves a combination of a first-dimension separation in cylindrical isoelectric focusing (IEF) gels containing 9M urea and 2% (w/v) of the detergent NP-40 with

discontinuous gradient polyacrylamide gel electrophoresis in the presence of sodium dodecyl sulfate (SDS-
PAGE) in the second dimension. Detailed reviews of the
theoretical basis, methodological variations, and applications of this technology can be found in refs. 2–5.

The standard 2-D PAGE procedure based on the
O'Farrell (1) method using rod IEF gels in the first dimension produces high-resolution protein separations
(*see* vol. 1 in this series), but certain problems are associated with its use. The most important of these problems
is severe cathodic drift during IEF, resulting in pH gradient instability. Thus, pH gradients rarely extend to
above pH 8, leading to the absence of basic proteins from
2-D maps. Methylcellulose treatment of the walls of the
glass tubes reduces electroendosmosis during IEF by
masking the charges present on the glass. In this way,
pH gradients can be extended to pH 10, but we have
found that the basic proteins in the resultant 2-D maps
are present as streaks rather than as discrete spots (6).

To overcome the problem of protein resolution at
basic pH, a special system, termed nonequilibrium pH
gradient electrophoresis (NEPHGE), using rod gels was
developed (7). In this technique, proteins are loaded at
the anodic end of the gel and electrophoresed for a
relatively short time. The proteins are thus separated
electrophoretically according to their mobilities in the
presence of a rapidly migrating pH gradient. Their
separation is dependent on the time of electrophoresis,
and their final positions on the gel do not correspond to
their isoelectric points. The non-steady-state nature of
this procedure compared with conventional IEF makes
it inherently rather irreproducible and very sensitive to
small changes in experimental conditions.

Gradient stability can also be improved by using
modified electrolytes (8,9). It should also be remem-

bered that polyacrylamide gels can contain agents that will contribute to cathodic drift: (a) trace impurities of acrylic acid, (b) covalent incorporation of polymerization catalysts as terminal groups in polyacrylamide chains, and (c) hydrolysis of amide groups to acrylic acid above pH 10. It is, therefore, essential to prepare freshly and deionize acrylamide solutions prior to IEF gel preparation (*10*).

In contrast to tube IEF gels usually used in 2-D PAGE systems, horizontal flat-bed gels are almost universally used for one-dimensionsl IEF separations. A significant advantage is that flat-bed apparatus is specifically designed for IEF applications, in contrast to the modified conventional electrophoresis apparatus normally used for tube-gel IEF. Thus, small electrolyte volumes can be used, and the electrodes can be positioned close to the ends of the gels in order to minimize pH gradient drift and loss of proteins from the gels (*11*). Perhaps most importantly, flat-bed IEF gels are capable of resolving basic proteins into discretely separated components rather than poorly resolved streaks.

Initial attempts to adapt flat-bed IEF technology to 2-D separations had limited success, probably because of the thick gels that were used (*12–14*). More successful procedures have been developed, however, using thin (*6*) and ultrathin (*15,16*) IEF gels.

In the method we have developed (*6,17,18*), 0.5 mm thick first-dimension IEF gel slabs are cast on a plastic support in a simple glass cassette with a silicon rubber gasket. The success of flat-bed IEF for 2-D separations depends on reliable binding of IEF gels to the plastic sheet, since they must be able to withstand the handling procedures involved in equilibration and transfer to the second-dimension SDS-PAGE gels. Originally (*6*) we treated polyester sheets with Dow Corning Prime Coat

1200, followed by reaction with silane A174. This method was not entirely satisfactory, but the commercial gel supports available at that time were not reliable for IEF gels containing both urea and nonionic detergent. By increasing catalyst concentrations two- to threefold, however, we are now able to reliably bind polyacrylamide IEF gels to GelBond PAG sheets in the presence of 8M urea and 0.5% (w/v) NP-40 or CHAPS detergents (17).

After polymerization is complete, the IEF gel is placed on a suitable flat-bed apparatus. The upper surface of the gel is covered with a plastic sheet, containing punched-out holes for sample application, to protect the gel from desiccation and the effects of the atmosphere. Small volumes of electrolyte can be used to advantage in flat-bed IEF, but they must be somewhat stronger than those used for rod-gel IEF. For the electrode wicks containing the electrolytes, we use strips of glass fiber paper (19). The gels should be run at a controlled temperature of 15°C, since at lower temperatures there is risk of crystallization of urea within the gel matrix. One of the major advantages of the flat-bed system is that there is a great flexibility in the choice of sample application position, thus avoiding problems associated with sample application under extreme acid or alkaline conditions that are unavoidable in the tube-gel IEF system.

After IEF is complete, the cover-sheet is removed and the gel is covered with plastic food-wrapping film. For nonradioactive samples, individual sample tracks are cut using a template of lane markings made during gel preparation. The sample strips are frozen rapidly and stored at –70°C. For appropriately radiolabeled samples, the wrapped gel can be placed in contact with a sheet of X-ray film and exposed overnight (or longer) at –70°C. The film is then developed and the resulting autoradiograph used as a guide to excise the sample tracks.

Gel strips are equilibrated for 5–10 min in buffer-containing SDS prior to application to the second-dimension SDS-PAGE gels. Since the gels are firmly attached to a solid support, they are relatively easy to handle, and the stretching that often occurs with cylindrical IEF gels, resulting in distorted 2-D maps, can be avoided. The top of the SDS-PAGE gel cassette is filled with equilibration buffer, and the IEF strip is slid into place, avoiding entrapment of air bubbles. It is usually not necessary to cement the strips in place, but this can be done, if required, using buffered agarose. The second, SDS-PAGE dimension then proceeds in the normal manner as in the standard 2-D PAGE technique, and after 2-D electrophoresis is complete the resultant 2-D protein maps can be visualized using any appropriate detection procedure.

We will describe how this technique is used to separate human skin fibroblast whole-cell proteins that have been radiolabeled with [^{35}S]-methionine.

2. Materials

1. GelBond PAG sheets (17.5 x 23 cm).
2. Glass plates, 17.5 x 23 cm, for cassettes.
3. Silicon rubber gasket, 0.5 mm thick x 7 mm wide, to fit around the edge of the glass plates.
4. PVC sheets (17.5 x 23 cm).
5. Rubber roller.
6. Solution A: Acrylamide stock solution (T30/C4): 14.4 g of acrylamide; 0.6 g *of N,N'*-methylene bis-acrylamide (Bis) made up to 50 mL with distilled water.

The solution is deionized with mixed-bed ion exchange resin (Duolite MB5113 mixed-bed resin, BDH), 5 g/100 mL solution, with stirring for 1 h. After degassing using a vacuum pump, the solution is filtered using a sintered glass filter under vacuum. The solution can be stored at 4°C for a maximum of 1 wk. The deionization step reduces ionic contaminants, which can increase cathodic drift and pH gradient instability.

7. Urea (BRL enzyme grade) should be stored dry at –20°C to reduce the rate of breakdown of urea with the formation of cyanate ions, which can react with protein amino groups to form stable carbamylated derivatives of altered charge.

8. Solution B: Acrylamide–urea solution: 5.32 mL acrylamide stock (solution A), 19.40 g urea, 14.48 mL distilled water.

   After the urea is completely dissolved, the mixed-bed ion exchange resin (5 g/100 mL) is added, and the solution is deionized with stirring for 1 h. This removes charged contaminants from the urea, most importantly any cyanate ions that are present. The solution is filtered under vacuum.

9. Any synthetic carrier ampholyte preparations can be used, but we use a mixture (*see* section 3, step 3) of the following:
   Pharmalyte 3–10 (Pharmacia)
   Pharmalyte 5–8 (Pharmacia)
   Servalyt 2–11 (Serva)
   Servalyt 4-6 (Serva)
   Ampholine 3.5–10 (LKB)
   The appropriate ampholyte mixture for a particular type of sample must be established (*see* section 4, Note 5).

10. Solution C: 10% (w/v) 3-[(cholamidopropyl)-dimethylammonio]-1-propane sulfonate (CHAPS).

The product from CalBiochem is salt free, but rather expensive. The Sigma product is cheaper, but must be deionized using mixed-bed ion exchange resin (5 g/100 mL) with stirring overnight at 4°C.

11. N,N,N',N'-Tetramethylethylenediamine (TEMED).

12. Solution D: Ammonium persulfate, 10% (w/v), made fresh daily.

13. Silicon fluid, Dow Corning 200/10 cs.

14. Electrode wicks (20 x 0.7 cm) cut from sheets of Whatman GF/B glass fiber paper.

15. Solution E: Catholyte, 1M NaOH.

16. Solution F: Anolyte, 1M oxalic acid.

17. Solution G: Bromophenol blue, 0.1% (w/v) in distilled water (BDH).

18. Solution H; (Sample lysis buffer): 9M urea, 3% (w/v) CHAPS, 20 mM dithiothreitol (DTT).

19. Solution I: Equilibration buffer–10% (w/v) SDS (final concentration, 2.3%), 11.5 mL–1M Tris buffer, pH 6.8 (final concentration, 124 mM), 6.2 mL DTT (final concentration, 20 mM), 0.154 g. Sodium β-mercaptoacetate (final concentration, 5 mM) 0.114 g, made up to 50 mL with distilled water.

3. Method

1. Cut a piece of GelBond PAG measuring 17.5 cm x 23 cm. The gel will adhere only to the hydrophilic side of the sheet. The reverse side (hydrophobic surface) is marked using a permanent marker pen into individual lanes (13.5 x 1.2 cm). Fifteen lanes can be accomodated, and these are flanked on each side by a further 1-cm wide lane. The sample lanes are numbered (1–15) to aid in identification of the sample strips at the completion of the IEF run.

2. The IEF gel is cast in a vertical cassette fitted with a U-shaped gasket. The GelBond PAG sheet is rolled onto a glass plate (17.5 cm x 23 cm) with the marked, hydrophobic side in contact with the glass plate. A small film of water is used between the glass and plastic sheets to ensure good adherence. Small air bubbles must be removed using a roller. On a second glass plate of the same dimensions, a sheet of PVC (17.5 x 23 cm) is rolled. The U-shaped silicon rubber gasket, which provides a leak-proof seal without the use of grease, is placed onto the Gel-Bond PAG/glass plate assembly. The PVC/glass plate is then placed onto the gasket/GelBond PAG/glass plate to form the complete cassette, which is held together with bulldog clips on the two sides and along the bottom edge.

3. Samples for 2-D PAGE are solubilized and denatured using high levels of urea and detergent (*see* solution H) so that the proteins can be separated as individual polypeptide chains. Thus the IEF gels also contain urea and zwitterionic detergent (CHAPS). To 17.5 mL of the acrylamide-urea mixture (solution B) are added the ampholytes (0.46 mL Pharmalyte 3–10, 0.46 mL Servalyt 2–11, 0.46 mL Ampholine 3.5–10, 0.12 mL Pharmalyte 5–8, 0.06 mL Servalyt 4–6). TEMED (15 µL) is then added, and the solution is degassed under vacuum. After degassing, 1 mL of CHAPS (solution C) is added, followed by 60 µL of ammonium persulfate (solution D). The solution is mixed and poured, avoiding the formation of air bubbles, into the cassette using a syringe barrel with a needle fitted with plastic tubing. The gel is allowed to polymerize at room temperature for 1 h. The resulting gel contains 8M urea, 0.5% (w/v) CHAPS, and 3.12%

(w/v) ampholytes. The gel cannot be stored and must be used for IEF immediately.

4. The bulldog clips are removed, and the PVC/glass plate is gently removed. The gel on its GelBond PAG support can now be removed from the cassette.

5. Any flat-bed IEF apparatus can be used. We are currently using a Pharmacia FBE-3000 flat-bed system powered by a Pharmacia ECPS 3000/150 powerpack fitted with a VH-1 volt hour (Vh) integrator. The Vh integrator has been fitted with outputs to monitor constantly V and mA output using a dual-channel chart recorder. The flat-bed apparatus is cooled to 15°C using a thermostatically controlled circulator.

6. A thin film of silicon fluid is applied to the cooling plate. The gel on its plastic support is then placed onto the cooling plate. The film of silicon fluid, which has excellent thermal conductivity properties and a low viscosity, allows for good contact between the gel support and the cooling plate.

7. A cover sheet (20 x 13.5 cm), having the same lane markings as the GelBond sheet and with slits (3 x 10 mm) punched for sample application, is applied to the surface of the gel. The lane markings should be superimposable, and care must be taken to eliminate air bubbles between the surface of the gel and the cover sheet, which is weighted with glass plates.

8. A stack of four wicks is used at the anode and cathode. The wicks are moistened (not too wet) with the electrode solutions (solutions E and F). The wicks are then applied at either end of the cover sheet with the cathode arranged to be at the sample-application side of the gel. A 1 mm gap should

remain between the wicks and the cover sheet to prevent electrical arcing. Bromophenol blue (solution G) (2 µL) is applied at each sample application site, and the electrode assembly is placed on the gel.

9. The gel is prefocused at 600 V, 30 mA, 15 W limiting for 600 Vh. After prefocusing the Bromophenol blue should have migrated about half way to the anode, and all lanes should have migrated with equal velocity.

10. The samples, solubilized in sample lysis buffer (solution H), are then applied in the holes in the cover sheet. We load 10^6 dpm of [^{35}S]-methionine-labeled proteins; it is best to limit the sample volume to 5 µL. Only the central lanes (3–12) should be used for samples, with the remainder being loaded with 2 µL of Bromophenol blue (solution G). The gel is focused at 700 V, 30 mA, 15 W limiting for a total of 14,000 Vh. The voltage is then increased to 1000 V (30 mA, 15 W limiting) to sharpen the bands to give a total of 15,000 Vh.

11. After completion of IEF, the gel is removed from the apparatus. The wicks and cover sheet are carefully removed. A lane to which a sample was not applied is cut off for pH gradient determination. The remainder of the gel is covered with food-wrapping film and stored between two glass plates at –70°C. Autoradiography can be used at this stage to identify sample lanes and to monitor the quality of the separation (*see* section 1).

12. The pH gradient is determined with a surface pH electrode (Ingold, Desaga) at 1-cm intervals along the length of the gel strip.

13. If an autoradiograph has been made, sample tracks are excised using the X-ray film as a template. Otherwise, sample lanes must be cut off using the lane markings as a guide.

14. Strips are equilibrated for 7 min in 5 mL of equilibration buffer (solution I).

15. The preparation of second-dimensions SDS-PAGE gels will not be described here, since they are of the normal Laemmli (20) type, as used in the standard 2-D PAGE technique (*see* vol 1 of this series). These gels can be either of a suitable constant percentage concentration of polyacrylamide or of a linear or nonlinear polyacrylamide concentration gradient (*see* vol. 1 of this series). We currently use modified nonlinear (8–20%) gradient SDS-PAGEg gels (21). The gels (0.75 mm thick) are 16 cm wide and 18 cm long. A short (2 cm) stacking gel is used.

16. The IEF strips are appplied to the SDS gels by filling the space in the cassettes above the stacking gels with equilibration buffer (solution I) and gently sliding the strips into place. Good contact between the IEF strips and the tops of the stacking gels must be achieved. There is no need to cement the strips in place with agarose.

17. The second-dimension SDS-PAGE gels can be run in any suitable vertical electrophoresis apparatus. The gels are run at 12 mA/gel overnight until the Bromophenol blue reaches the bottom of the separating gels.

18. The gels can be subjected to any suitable procedure to detect the separated proteins (e. g., electroblotting, autoradiography, protein staining, and so on). A typical separation of [^{35}S]-methionine-labeled human skin fibroblast proteins using this technique is shown in Fig. 1.

4. Notes

1. Avoid touching the hydrophilic surface of GelBond PAG, since this reduces gel adherence.

Fig. 1.   Autoradiograph of a two-dimensional separation of human skin fibroblast proteins labeled with [^{35}S]-methionine using flat-bed IEF in the first dimension.

2.   Different sizes of IEF gel can be readily used in this technique within the constraints of the size of the cooling plate in the IEF apparatus being used.

3.   The cassette must be standing upright and level during gel preparation.

4. We use the zwitterionic detergents CHAPS in the IEF gel, but it is also possible to use nonionic detergents such as Triton X-100 or NP-40 with equally good results.

5. We use mixtures of different commercial synthetic carrier ampholytes, since the greater diversity of ampholyte species increases resolution. The mixture has also been optimized for the separation of human skin fibroblast proteins. Such pH gradient engineering using synthetic carrier ampholytes is a rather empirical process, however. For any particular type of sample to be separated by 2-D PAGE, it is good practice to establish which ampholyte or mixture of ampholytes produces the best separation for that sample.

6. The IEF gel can be run without a cover-sheet, but the gel should then be exposed to a humid N_2 atmosphere, or an NaOH solution should be present within the apparatus to absorb CO_2.

7. If no cover sheet is used, samples should be applied using silicon rubber sample application strips (Desaga) or similar devices. It is not recomended to apply samples as droplets to the gel surface (they tend to spread) or to use pieces of filter paper on which the sample has been absorbed (some proteins bind irreversibly).

References

1. O'Farrell, P.H. (1975) High resolution two-dimensional electrophoresis of proteins. *J. Biol. Chem.* **250**, 4007–4021.
2. Dunn, M.J. and Burghes, A.H.M. (1983) High resolution two-dimensional polyacrylamide gel electrophoresis. I. Methodological procedures. *Electrophoresis* **4**, 97–116.

3. Dunn, M.J. and Burghes, A.H.M. (1983) High resolution two-dimensional polyacrylamide gel electrophoresis. II. Analysis and applications. *Electrophoresis* 4, 173–189.

4. Dunn, M.J. and Burghes, A.H.M. (1986) High Resolution Two-Dimensional Polyacrylamide Gel Electrophoresis, in *Gel Electrophoresis of Proteins* (Dunn, M.J., ed.) Wright, Bristol.

5. Celis, J.E. and Bravo, R., eds. (1984) *Two-Dimensional Gel Electrophoresis of Proteins* Academic, New York.

6. Burghes, A.H.M., Dunn, M.J., and Dubowitz, V. (1982) Enhancement of resolution in two-dimensional gel electrophoresis and simultaneous resolution of acidic and basic proteins. *Electrophoresis* 3, 354–363.

7. O'Farrell, P., Goodman, M.M., and O'Farrell, P.H. (1977) High resolution two-dimensional electrophoresis of basic as well as acidic proteins. *Cell* 12, 1133–1142.

8. Duncan, R. and Hershey, J.W.B. (1984) Evaluation of isoelectric focusing running conditions during two-dimensional isoelectric focusing/sodium dodecyl sulfate-polyacrylamide gel electrophoresis: Variation of gel patterns with changeing conditions and optimized isoelectric focusing conditions. *Anal. Biochem.* 138, 144–155.

9. Tracy, R.P., Currrie, R.M., Kyle, R.A., and Young, D.S. (1982) Two-dimensional gel elctrophoresis of serum specimens from patients with monoclonal gammopathies. *Clin. Chem.* 28, 900–907.

10. Burghes, A.H.M., Dunn, M.J., Statham, H.E., and Dubowitz, V. (1982) Analysis of skin fibroblast proteins in Duchenne muscular dystrophy. I. Sodium dodecyl sulphate polyacrylamide gel electrophoresis. *Electrophoresis* 3, 177–185.

11. Righetti, P.G. (1983) *Isoelectric Focusing: Theory, Methodology and Applications* Elsevier, Amsterdam.

12. Goldsmith, M.R., Rattner, E.C., Macy, M., Koehler, D., Balikov, S.R., and Bock, S.C. (1979) Two-dimensional electrophoresis of small molecular weight proteins. *Anal. Biochem.* 99, 33–40.

13. Iborra, G. and Buhler, J.M. (1976) Protein subunit mapping. A sensitive high resolution method. *Anal. Biochem.* 74, 503–511.

14. Rangel-Aldao, R., Kupiec, J.W., and Rosen, O.M. (1979) Resolution of the phosphorylated and dephosphorylated cAMP-binding proteins of bovine cardiac muscle by affinity labelling and two-dimensional elctrophoresis. *J. Biol. Chem.* 254, 2499–2508.

15. Gorg, A., Postel, W., Westermeier, R., Gianazza, E., and Rightetti, P.G. (1980) Gel gradient electrophoresis, isoelectric focusing and two-dimensional techniques in horizontal, ultrathin polyacrylamide layers. *J. Biochem. Biophys. Meth.* **3**, 273–284.

16. Gorg, A., Postel, W., and Westermeier, R. (1981) SDS-Gel Gradient Electrophoresis, Isoelectric Focusing and High-Resolution Two-Dimensional Electrophoresis in horizontal, Ultrathin-Layer Polyacrylamide Gels, in *Electrophoresis '81* (Allen, R.C. and Aranaud, P., ed.) DeGruyter, Berlin.

17. Dunn, M.J., Burghes, A.H.M., Patel, K., Witkowski, J.A., and Dubowitz, V. (1984) Analysis of Genetic Neuromuscular Diseases by 2D-PAGE: A System for Comparisons, in *Electrophoresis '84* (Neuhoff, V., ed.) Verlag Chemie, Weinheim.

18. Dunn, M.J., Burghes, A.H.M., Witkowski, J.A., and Dubowitz, V. (1985) Analysis of genetic muscle diseases by two-dimensional electrophoresis. *Protides Biol. Fluids* **32**, 973–976.

19. Cuono, C. B. and Chapo, G. A. (1982) Gel electrofocusing in a natural pH gradient of pH 3–10 generated by a 47-component buffer mixture. *Electrophoresis* **3**, 65–75.

20. Laemmli, U.K. (1970) Cleavage of structural proteins during the assembly of the head of bacteriophage T4. *Nature* **227**, 680–685.

21. Burghes, A.H.M., Patel, K., and Dunn, M.J. (1985) Comparison of buffer focusing in buffers and synthetic carrier ampholytes for use in the first dimension of two-dimensional polyacrylamide gel electrophoresis. *Electrophoresis* **6**, 453–461.

Chapter 18

Preparative Aspects
of Immobilized pH Gradients

Pier Giorgio Righetti and Cecilia Gelfi

1. Introduction

We will give here guidelines on the preparative aspects of immobilized pH gradients (IPGs), the latest and most powereful fractionation technique in electrophoresis. The load ability of IPG gels has been demonstrated to be at least 10 times higher than in conventional isoelectric focusing (IEF), thus approaching or even passing the load limit of isotachophoresis (1). The following aspects will be illustrated: (a) explorative runs aimed at defining the load capacity; (b) optimization of experimental parameters, such as ionic strength (I), pH gradient width, and gel thickness; (c) general considerations on the load ability and ionic strength of the milieu; (d) protein load as a function of %T in the matrix.

At the methodological level, we will present the following procedures: (a) casting of soft and thick Immobiline gels; (b) recovery of protein zones from an Immobiline gel by electrophoretic elution into hydroxyapatite beads; (c) recovery by electrophoretic transfer into ion-exchange resins.

2. Theory

Although in the past, in preparative electrophoretic or IEF runs, most separations have been performed by trial and error, with no real knowledge of the capacity of the system, it is now possible in IPGs to predict theoretically and from a few analytical runs the load capacity of the system compatible with a given degree of resolution. We will discuss here the various parameters needed for optimizing a preparative run.

2.1. Theoretical Prediction of Acceptable Protein Loads in IPGs

For practical preparative work, an equation has been derived correlating the maximum protein load in a single zone to the p*I* (isoelectric point) distance (Δp*I*) with the nearest contaminant, to the gel cross-sectional area and to the slope of the pH gradient. The equation

$$M=[[(\Delta \text{p}I)/([d(\text{pH})]/(dx)-L]2C_{M}A \qquad [1]$$

where M = protein load in a single zone (major component) in mg; Δp*I* = p*I* difference between major component and nearest contaminant (in pH units); $d(\text{pH})/dx$ = slope of the pH gradient along the separation track (pH units/cm); L = protein free space, between the major band and the impurity, that is needed to cut the gel without loss of protein or without carrying over the impurity (in general 1 mm is an acceptable distance); C_M = average concentration in the focused zone of the major component (mg/mL); A = cross sectional area of the gel perpendicular to the focusing direction (in cm²) (2).

It can be seen that protein load can be maximized by increasing A (the liquid volume available to the focused zone) and by decreasing the slope of the pH gradient (i.e., by focusing in narrow or ultranarrow pH gradients). As a practical guideline, a graph has beeen constructed correlating these three basic experimental parameters: protein load in a single zone, ΔpI between the band of interest and nearest contaminant and slope of the pH gradient along the separation axis [$d(\text{pH})/dx$]. This graph is essentially a plot of Eq. [1], taking as a concentration limit (C_M) a common upper limit experimentally found of 40 mg/mL (in reality, as shown below, this "solubility" limit depends on the %T; in a 2.5 %T matrix, the C_M value can be as high as 90 mg protein/mL gel phase). Figure 1 shows how the graph is laid out: the abscissa reports the ΔpI value (in pH units), and the ordinate the protein load (mg/cm²) for a given A value. The ΔpI vs. protein load plane is cut by lines of different slopes representing pH gradients of different widths along the IPG gel length. It is seen that ultranarrow pH gradients (e. g., 0.02 pH units/cm) allow for extremely high protein loads (up to 80 mg/cm²), while still retaining a resolution better than $\Delta pI = 0.01$. At the opposite extreme, broad pH gradients (e.g., 0.2 pH units/cm) would allow a resolution of only $\Delta pI = 0.1$ with a protein load of less than 40 mg/cm².

2.2. Optimization of Environmental Parameters

We have also performed a thorough study on the optimization of environmental parameters [I (ionic

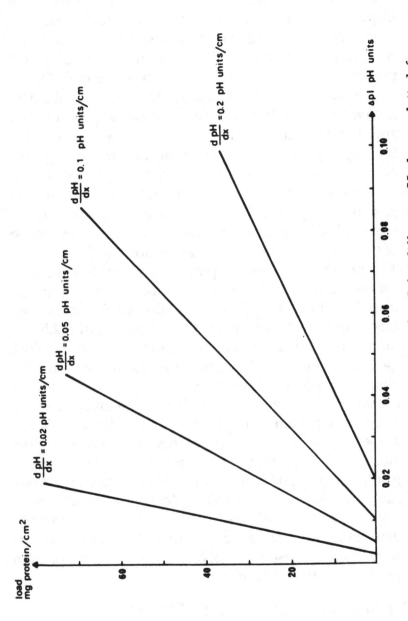

Fig. 1.   Acceptable protein load as a function of ΔpI for different pH slopes plotted for a mean concentration of 45 mg/mL in the major protein zone. This is a graphical representation of Eq. [1] (ref 2).

strength), gel thickness, pH gradient width] for maximizing protein loads in Immobiline matrices. These aspects are summarized in Fig. 2. By increasing the ionic strength of the gel from 1.25 to 7.5 mEq/L, a fourfold increment in load capacity is obtained; above this level, a plateau is abruptly reached around 10–12 mEq/L. By increasing the gel thickness from 1 to 5 mm, a proportional fivefold increment in protein load ability is achieved. The system does not level off. A 5-mm thickness seems to be optimal, however, since thicker gels begin to develop thermal gradients in their transverse section, generating skewed zones. Finally, by progressively decreasing the width of the pH interval, there is a linear increase in protein load capability. Here too the system does not reach a plateau. Because of the very long focusing times required by narrow pH gradients, however, aggravated by the high viscosity of protein zones at high loads, it is suggested not to attempt to fractionate large proteing amounts in pH ranges narrower than 0.5 pH units (3).

2.3. General Considerations on the Load Ability and the Ionic Strength of the Milieu

We have seen above that IPGs have a load ability at least 10 times higher than conventional IEF. We believe this is mostly because of strong difference in ionic strength characteristic of the two systems. At the very low I values typical of IEF (approximately 1 mEq/L) (4), macromolecules will have a very low solubility minimum and will tend to aggregate and flocculate. An increase of I (approximately 7–10 mEq/L) as characteristic of IPGs, is thus beneficial since according to the Debye-huckel equation:

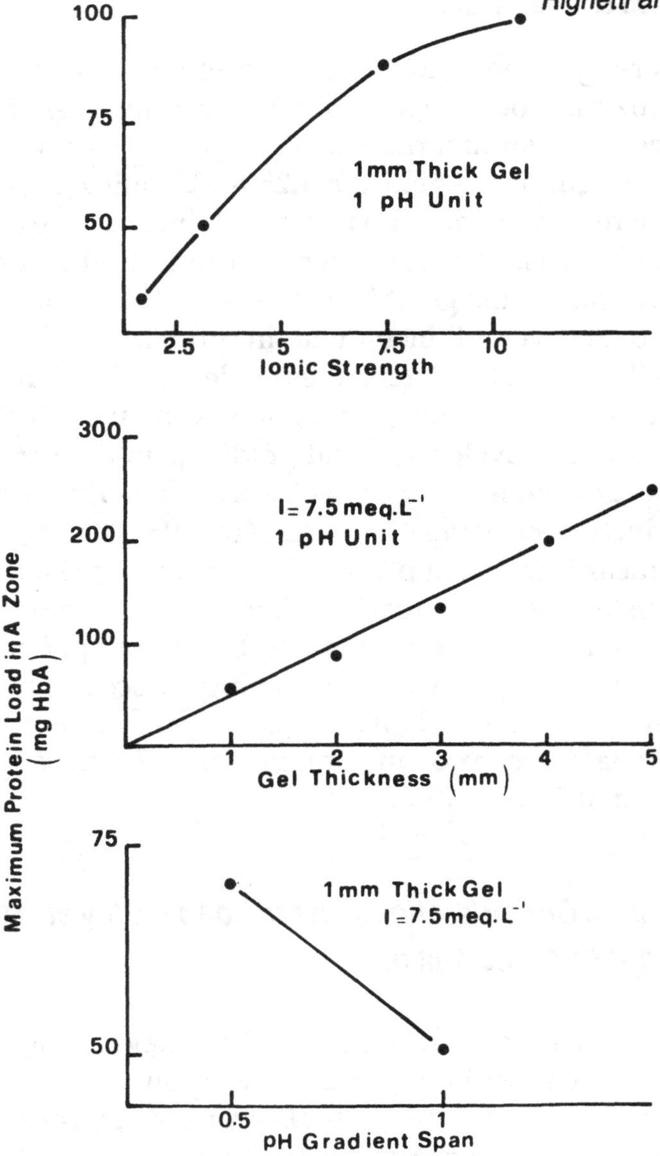

Fig. 2. Loading capacity of IPGs. The maximum load in a single protein zone is plotted: (top) as a function of ionic strength at constant gel thickness in a 1-pH unit interval; (middle) as a function of gel thickness at constant ionic strength in a 1-pH unit interval; (bottom) as a funtion of pH gradient width at constant ionic strength and constant gel thickness (ref. 3).

$$-\log \gamma = \log (S/S_0) = 0.51\ Z^2 \sqrt{I} \qquad [2]$$

where γ is the activity coefficient of an ion of charge Z, S and S_0 are the solubilities of a protein at the pI at a given ionic strength and as extrapolated to zero ionic strength, respectively. Thus, as the environmental I increases, and the γ values of the ions (both in solution and in the protein) decrease, the protein solubility increases: This is the well known "salting in" effect described already in 1936 by Cohn (5). We are tempted to equate IPG gels to "salting-in" media and IEF gels to "salting out" milieus. It might be argued that, as long as the proteins precipitate at their pI, and this material is confined in the isoelectric zone, this should not affect the load capacity in gel matrices, since the precipitated zone is gravitationally stable. The point of the matter is that in real cases this does not quite happen. As demonstrated by Gronwall (6), the solubility of an isoionic protein, plotted against pH near the isoionic point, is a parabola, with a fairly narrow minimum at relatively high I, but with progressively wider minima, on the pH axis, at decreasing I values.

This phenomenon can be appreciated in Fig. 3, taken from a 1942 classic. This graph has more than one degree of complexity, worth discussing. On the one hand, by increasing the environmental ionic strength from 1 to 20 mM NaCl, the solubility of β-lactoglobulin, at a given pH (approximately 5.3, very close to the pI value), increases by a factor of eight. As a second phenomenon, it can be seen that, at moderately high ionic strength (20 mEq/L), all the protein molecules tend to be isoelectric at a precise point on the pH axis, whereas at low I values (1 mEq/L) they are smeared over as much as 0.4 pH unit. For easier reading, we have replotted these physical phenomena in the two follow-

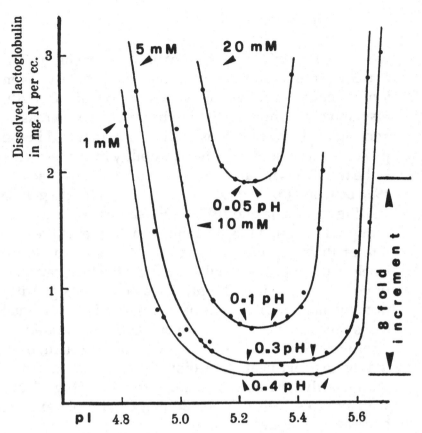

Fig. 3. Solubility of β-lactoglobulin in the pH range of 4.8–5.6 as a function of the ionic strength of the environment, from 1 to 20 m*M* NaCl. The amount of protein dissolved has been determined as Kjeldahl nitrogen. The pH values on the parabolas give the pH span of the solubility minimum, centered on a theoretical p*I* of 5.3 (modified from ref. 6).

ing graphs: the dependence of the solubility of the protein at p*I* from the environmental ionic strength is displayed in Fig. 4, and the width of the apparent p*I* zone on the pH axis, as a function of the *I* of the milieu, is plotted in Fig. 5, and is seen to vary in a funnel-shaped fashion. This last graph deserves further comments: it

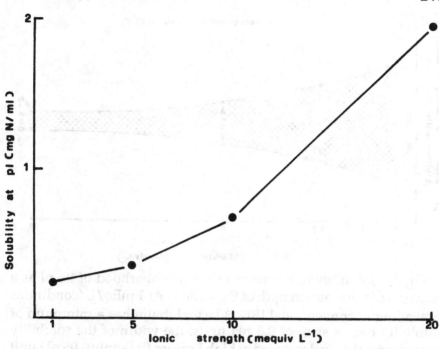

Fig. 4. Plot of solubility at p*I* vs. ionic strength of the milieu for β-lactoglobulin (data from Fig. 3). The solubility increases by a factor of 8 for a 20-fold increase in ionic strength.

can be seen that, at the prevailing *I* values typical of conventional IEF (1 mEq/L), the solubility minimum of β-lactoglobulin spans a 0.4 pH unit interval, whereas in IPGs (average *I* = 10 mEq charges, but easily adaptable to any other *I* value), the pH width of solubility minima is strongly decreased in a funnel-shaped fashion down to only 5/100 of a pH unit. In other words, what is detrimental in conventional, preparative IEF runs is not isoelectric precipitation, but near-isoelectric precipitation. The precipitate is not confined at the p*I* position, but is usually smeared over as much as 1/2 pH unit interval, thus being completely detrimental to the resolution of adjacent species (7).

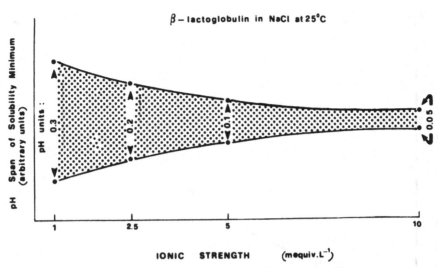

Fig. 5. Solubility of a protein in the neighborhood of its p*I* as a function of the ionic strength of the milieu. At 1 mEq/L (conditions prevailing in conventional IEF), β-lactoglobulin has a minimum of solubility over a span of 0.3 pH units; the width of the solubility funnel is markedly decreased at high I values (it is only 0.05 pH unit at 10 mEq/L, values typical of an IPG milieu) (from ref. 3).

2.4. Protein Load as a Function of %T in the Matrix

We have seen, so far, three parameters influencing a gel load ability (ionic strength, gel thickness, pH gradient width). In reality, there is a fourth, hidden parameter, perhaps even more important than the others, that we might have grasped immediately if we had paid more attention to a theoretical article by Bode (8) on the structure of hydrophilic matrices. We have seen above that, no matter how the experimental conditions are optimized and the ionic strength is increased, a common upper load limit for all proteins investigated has been found, (approximately 40 mg protein/mL gel solution). We have been intrigued by this barrier and have tried to find an explanation for it. The key to this apparent "solu-

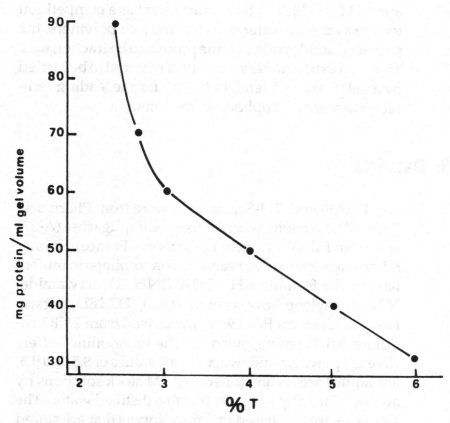

Fig. 6. Relation between loading capacity (in terms of mg protein/mL gel volume) and %T (T = grams of acrylamide and crosslinker per 100 mL gel volume) value of the gel matrix. Notice that, although in the range of 3–6%T, the protein load decreases linearly, in softer gels (3% T) it increases exponentially (from ref. 9).

bility limit" is to be found in Fig. 6: The amount of protein accepted by a gel matrix is directly related to its composition (%T). The limit of 40 mg protein/mL gel is only valid for a 5% T polyacrylamide matrix: as the number of fibers in the gel is decreased, progressively more protein can be loaded in the system, so that in a 2.5% T gel as much as 90 mg proteins/mL gel can be applied (9). This has been interpreted as a competition for the available water between the two polymers, the polyacrylamide coils, and the protein to be fractionated. This is an extraordinary amount of material to be carried by a gel phase and renders IPG by far the leading technique in any electrophoretic fractionation.

3. Materials

DEAE- and CM-Sephadexes were from Pharmacia (Uppsala, Sweden), whereas low-gelling agarose (A-37) was from IBF (Villeneuve La Garenne, France). Immobiline chemicals (LKB brand of nonamphoteric buffers having the formula $CH_2=CH–CONH–R$), acrylamide, N,N'-methylene bisacrylamide (Bis), TEMED, persulfate, and GelBond PAG were purchased from LKB Produkter AB, Bromma, Sweden. The Immobiline buffers (five are powders and two, with pK values of 8.5 and 9.3, are liquid) are reconstituted to 0.2M stock solutions by adding 25 mL (by weight) of double distilled water. The solutions are dispensed in 5-mL aliquots that are stored frozen (only one 5-mL aliquot should be kept in the refrigerator at any given time as a working solution; stability at 4°C is not guaranteed after 3 wk, whereas in the frozen state, they are stable for at least 6 m. Acrylamide and Bis are mixed as a stock 30% *T*, 4%C solution that can be kept at 4°C in a brown bottle for up to a month (by slow hydrolysis, acrylic acid will form with time. *Use with care. Acrylamide, Bis, and Immobilines are neurotoxins; do not inhale powders, and handle liquids with gloves.* After polymerization, they are inert. A stock, 40% solution of persulfate is made fresh every week and stored at 4°C in a dark bottle (0.5–1 mL will be plenty as a week's supply). We routinely use a Multiphor II or Ultrophor chamber, coupled to a Macrodive V power supply and to a Multitemp thermostat (LKB Produkter AB). In addition to our published recipes for 2–6 pH unit wide IPG gradients (10), formulations for 58 IPG intervals spanning 1 pH unit, at increments of 0.1 pH unit, are given in Table 1. Only these recipes are provided, since it is quite likely that in preparative runs only narrow IPG gradients will be utilized.

4. Methods

We will give here an example for the purification of normal human adult hemoglobin (HbA) over a 1-pH unit IPG range. The nearest contaminant is its glycosylated derivative (Hb A1c) with a ΔpI of only 0.04 pH unit. A preparative experiment in general requires a 2-d period, as follows: (a) day one: gel casting, washing, weight reconstitution, sample application, overnight run; (b) day two: excision of the major protein zone, casting of the agarose plate, preparation of a trench with either hydroxyapatite or DEAE (diethyl aminoethyl)-Sephadex, electrophoretic transfer and protein elution from the exchanger beads. For the materials check list, see section 3 above. In addition, we will use agarose M or, even better, the low-gelling type (Agarose A-37) and, for protein recovery, two different methodologies: adsorption on a weak [hydroxyapatite (HA)] or a stronger (DEAE-Sephadex) ion exchanger. For the preparative run, we will use a 2-mm thick gel, one half the size of the cooling block of the Ultrophor (11 x 11 cm), over a 1-pH unit interval (pH 6.8–7.8), of low matrix content (3% T), thus being able to fractionate at least 150 mg of total sample. The operative steps are described below. The gel composition of the IPG run is given in Table 2,

Table 1: Narrow pH Gradients[a]

Control pH at 20°C	Volume (µL) 0.2M Immobiline pK acidic dense solution							pH range	Mid-point	Control pH at 20 C	Volume (µL) 0.2M Immobiline pK basic light solution						
	3.6	4.4	4.6	6.2	7.0	8.5	9.3				3.6	4.4	4.6	6.2	7.0	8.5	9.3
3.84 0.03	—	750	—	—	—	—	159	3.8–4.8	4.3	4.95 0.06	—	750	—	—	—	—	591
3.94 0.03	—	710	—	—	—	—	180	3.9–4.9	4.4	5.04 0.07	—	810	—	—	—	—	667
4.03 0.03	—	—	755	—	—	—	157	4.0–5.0	4.5	5.14 0.06	—	—	745	—	—	—	584
4.13 0.03	—	—	713	—	—	—	177	4.1–5.1	4.6	5.23 0.07	—	—	803	—	—	—	659
4.22 0.03	—	—	689	—	—	—	203	4.2–5.2	4.7	5.33 0.08	—	—	884	—	—	—	753
4.32 0.03	—	—	682	—	—	—	235	4.3–5.3	4.8	5.42 0.10	—	—	992	—	—	—	871
4.42 0.03	—	—	691	—	—	—	275	4.4–5.4	4.9	5.52 0.12	—	—	1133	—	—	—	1021
4.51 0.04	—	—	716	—	—	—	325	4.5–5.5	5.0	5.16 0.14	—	—	1314	—	—	—	1208
4.61 0.05	562	—	600	863	—	—	—	4.6–5.6	5.1	5.69 0.04	—	—	863	—	—	—	105
4.75 0.05	458	—	675	863	—	—	—	4.7–5.7	5.2	5.79 0.04	—	—	863	863	—	—	150
4.86 0.04	352	—	750	863	—	—	—	4.8–5.8	5.3	5.90 0.04	—	—	863	863	—	—	202
4.96 0.03	218	—	863	863	—	—	—	4.9–5.9	5.4	5.99 0.03	—	—	863	863	—	—	248
5.07 0.03	158	—	863	863	—	—	—	5.0–6.0	5.5	6.09 0.04	—	—	863	803	—	—	338
5.17 0.04	113	—	863	863	—	—	—	5.1–6.1	5.6	6.20 0.04	—	—	863	713	—	—	443
5.24 0.18	1251	—	—	1355	—	—	—	5.2–6.2	5.7	6.34 0.04	337	—	—	724	—	—	—
5.33 0.12	1055	—	—	1165	—	—	—	5.3–6.3	5.8	6.43 0.03	284	—	—	694	—	—	—
5.43 0.12	899	—	—	1017	—	—	—	5.4–6.4	5.9	6.53 0.03	242	—	—	682	—	—	—
5.52 0.09	775	—	—	903	—	—	—	5.5–6.5	6.0	6.63 0.03	209	—	—	686	—	—	—
5.62 0.07	676	—	—	817	—	—	—	5.6–6.6	6.1	6.73 0.03	182	—	—	707	—	—	—
5.71 0.06	598	—	—	755	—	—	—	5.7–6.7	6.2	6.82 0.03	161	—	—	745	—	—	—
5.81 0.06	536	—	—	713	—	—	—	5.8–6.8	6.3	6.92 0.03	144	—	—	803	—	—	—
5.91 0.05	486	—	—	689	—	—	—	5.9–6.9	6.4	7.02 0.03	131	—	—	884	—	—	—
6.01 0.05	447	—	—	682	—	—	—	6.0–7.0	6.5	7.12 0.03	120	—	—	992	—	—	—
6.10 0.04	416	—	—	691	—	—	—	6.1–7.1	6.6	7.22 0.03	112	—	—	1133	—	—	—
6.11 0.11	972	—	—	—	1086	—	—	6.2–7.2	6.7	7.21 0.03	262	—	—	—	686	—	—
6.21 0.09	833	—	—	—	956	—	—	6.3–7.3	6.8	7.31 0.03	224	—	—	—	682	—	—
6.30 0.08	722	—	—	—	857	—	—	6.4–7.4	6.9	7.41 0.03	195	—	—	—	694	—	—
6.40 0.07	635	—	—	—	783	—	—	6.5–7.5	7.0	7.50 0.03	171	—	—	—	724	—	—
6.4? 0.0?	5?5	—	—	—	7?2	—	—	6.6–7.6	7.1	7.6? 0.03	15?	—	—	—	771	—	—

control pH						pH range	control pH							pH
6.59 ± 0.05	509	—	—	699	—	6.7–7.7	7.70 ± 0.03	137	—	—	840	—	—	7.2
6.69 ± 0.05	465	—	—	683	—	6.8–7.8	7.80 ± 0.03	125	—	—	934	—	—	7.3
6.78 ± 0.04	430	—	—	684	—	6.9–7.9	7.90 ± 0.03	116	—	—	1058	—	—	7.4
6.88 ± 0.04	403	—	—	701	—	7.0–8.0	8.00 ± 0.03	108	—	—	1217	—	—	7.5
6.98 ± 0.04	381	—	—	736	—	7.1–8.1	8.09 ± 0.03	103	—	—	1422	—	—	7.6
7.21 ± 0.06	1028	—	—	750	—	7.2–8.2	8.36 ± 0.05	548	—	—	750	750	—	7.7
7.31 ± 0.06	983	—	—	750	—	7.3–8.3	8.46 ± 0.05	503	—	—	750	750	—	7.8
7.41 ± 0.05	983	—	—	750	—	7.4–8.4	8.56 ± 0.05	458	—	—	750	750	—	7.9
7.66 ± 0.15	1230	—	—	1334	—	7.5–8.5	8.76 ± 0.04	331	—	—	—	720	—	8.0
7.75 ± 0.12	1037	—	—	1149	—	7.6–8.6	8.85 ± 0.03	279	—	—	—	692	—	8.1
7.85 ± 0.10	885	—	—	1004	—	7.7–8.7	8.95 ± 0.03	238	—	—	—	682	—	8.2
7.94 ± 0.08	764	—	—	893	—	7.8–8.8	9.05 ± 0.06	206	—	—	—	687	—	8.3
8.04 ± 0.07	667	—	—	810	—	7.9–8.9	9.14 ± 0.06	180	—	—	—	710	—	8.4
8.13 ± 0.06	591	—	—	750	—	8.0–9.0	9.24 ± 0.06	159	—	—	—	750	—	8.5
8.23 ± 0.06	530	—	—	710	—	8.1–9.1	9.34 ± 0.06	143	—	—	—	810	—	8.6
8.33 ± 0.05	482	—	—	687	—	8.2–9.2	9.44 ± 0.06	130	—	—	—	893	—	8.7
8.43 ± 0.04	443	—	—	682	—	8.3–9.3	9.54 ± 0.06	119	—	—	1004	—	—	8.8
8.52 ± 0.04	413	—	—	692	—	8.4–9.4	9.64 ± 0.06	111	—	—	1149	—	—	8.9
8.62 ± 0.04	389	—	—	720	—	8.5–9.5	9.74 ± 0.06	105	—	—	1334	—	—	9.0
8.40 ± 0.14	1208	—	—	—	1314	8.6–9.6	9.50 ± 0.06	325	—	—	—	—	716	9.1
8.49 ± 0.12	1021	—	—	—	1133	8.7–9.7	9.56 ± 0.06	275	—	—	—	—	691	9.2
8.59 ± 0.10	871	—	—	—	992	8.8–9.8	9.69 ± 0.06	235	—	—	—	—	682	9.3
8.68 ± 0.08	753	—	—	—	884	8.9–9.9	9.79 ± 0.06	203	—	—	—	—	689	9.4
8.78 ± 0.07	659	—	—	—	803	9.0–10.0	9.88 ± 0.06	177	—	—	—	—	713	9.5
8.87 ± 0.06	584	—	—	—	745	9.1–10.1	9.98 ± 0.06	157	—	—	—	—	755	9.6
8.97 ± 0.05	525	—	—	—	707	9.2–10.2	10.08 ± 0.06	141	—	—	—	—	817	9.7
9.07 ± 0.04	478	—	—	—	686	9.3–10.3	10.18 ± 0.06	129	—	—	—	—	903	9.8
9.16 ± 0.07	440	—	—	—	682	9.4–10.4	10.28 ± 0.06	119	—	—	—	—	1017	9.9
9.26 ± 0.07	410	—	—	—	694	9.5–10.5	10.38 ± 0.06	111	—	—	—	—	1163	10.0

*Volumes of Immobiline for 15 mL of each starting solution (two gels). From LKB Application Note No. 324. The pH range (given in the middle two lanes) is the one existing in the gel during the run at 10°C. For controlling the pH of the starting solutions, the values (control pH) are given at 20°C (LKB Produkter AB, Bromma, Sweden).

Table 2
Formulation for a Preparative pH 6.8–7.8 IPG Run[a]

Chemicals	Acidic (heavy) sol.	Basic (light) sol.
Immobiline pK 3.6[b]	387.5 µL	103 µL
Immobiline pK 7.0[b]	569 µL	781 µL
Acrylamide stock (30 %T, 4 %C) for (T% = 3)	1.25 mL	1.25 mL
Glycerol (87%)	2.50 mL	
Water	7.79 mL	10.36 mL
Total volume	12.5 mL	12.5 mL
TEMED	9 µL	9 µL
Persulfate (40% sol.)	12 µL	12 µL

[a]Gel thickness, 2 mm; size, 11 x 11 cm.
[b]From reconstituted, stock 0.2*M* solutions.

whereas two alternative formulations for the electrophoretic elution are listed in Table 3. The hemoglobin (Hb) lysate is prepared as follows: The red cells are washed three times in normal saline and lysed in 2 vol of distilled water (final concentration approximately 8% Hb). The stroma is removed by shaking the lysed cells with 1 vol of carbon tetrachloride and centrifuging at full speed in a bench centrifuge (approximately 3000*g*) for 20 min. We prefer not to add KCN, since it often disturbs the run. It is best to gas the sample with CO for 30 s and then stored in 35% ethylene glycol at –20°C. The stability is guaranteed for at least 1 yr.

4.1. First Day

1. Assemble the cassette (15 min).
2. Mix the IPG solutions (Table 2), and check the pH (30 min).

Table 3
Eluting Systems for Proteins from an Immobiline Matrix

	HA-Ultrogel	DEAE-Sephadex
Agarose	M, 0.8 %	A-37, 1%
	800 mg	1 g
Buffer	100 mM Tris-Gly[a] pH 9.1; 12.1 g Tris + 4.1 g Gly in 1 L final volume (20°C)	100 mM Tris-acetate[b] pH 8.5; 12.1 g Tris + 35 mL 1M CH_3COOH in 1L final volume (20°C)
Ion exchanger	HA-Ultrogel	DEAE-Sephadex
	15 mL Equilibrated in 150 mL Tris-Gly buffer pH 9.1 to fill a 12.5 × 1 × 0.5 cm trench	500 mg swollen and equilibrated in 100 mL of Tris acetate buffer, pH 8.5, to fill a 12.5 × 1 × 0.5 cm trench
Elution buffer	0.2 M Phosphate buffer, pH 6.8; 2.84 g Na_2HPO_4 + 1.75 g NaH_2PO_4 in 100 mL final volume (20°C)	0.2M Tris-Gly pH 9.5 in 0.2M NaCl; 2.42 g Tris + 0.21 g Gly + 1.168 g NaCl in 100 mL final volume (20°C)

[a]Electrode buffer: double molarity, same pH.
[b]Electrode buffer: same molarity and pH.

3. Pour the gradient, carefully rinse the apparatus, and dry the outlet with ethanol (20 min).
4. Polymerize the gel (1 h in a forced ventilation oven at 50°C or in a water bath at the same temperature) (*11*).
5. Weigh the gel and mark the weight on the plastic backing.
6. Wash the gel (2 x 1 h) in 1 L of distilled water each time.
7. Blot the gel suface with soft tissue (to prevent local gel swelling caused by droplets on the surface) and reconstitute to its original weight (1 h). (During washing the gel swells as it is a weak ion-exchanger. Eliminate adsorbed excess water by blowing warm air with a fan at approximately 1 m distance from the vertically standing gel; check periodically with a balance till weight reconstitution) (*12*).
8. Apply the sample in the trench precast in the gel with the aid of a Pasteur or a Gilson pipet (150 mg in 0.9 mL), and start the experiment (15 min). Run at low voltage (400 V maximum) until all the sample has entered the gel. The run will continue overnight at 2000 V/10 cm.
9. Reswell the ion-exchanger overnight (Table 3) (if DEAE-Sephadex is chosen for elution) (*13*).

4.2. Second Day

For all materials in this section, refer to Table 3.

1. Wash the ion-exchanger (30 min).
2. Boil the agarose in buffer (15 min).
3. Cut the IPG gel strip with the protein band of interest (the matrix stays bound to the plastic foil).
4. Pour the agarose into the tray and allow to cool at 42°C.

5. Embed the IPG gel strip into the melted agarose A-37 plate, with the plastic backing against the glass bottom of the tray (this and following operations are shown in Fig. 7).

6. Cut a trench (2 cm wide, 24 cm long) anodic to the IPG strip (*see* Fig. 7).

7. Fill this trench with a slurry of ion-exchanger (DEAE-Sephadex or with HA-Ultrogel, according to the elution system chosen);

8. Start the experiment (1 h at 30 W, approximately 300 V, 5°C).

9. Remove the ion exchanger and transfer to a syringe (5 min).

10. Wash with elution buffer (20 min; in case of DEAE-Sephadex, refer to Fig. 8).

11. For testing protein recovery, read the eluate at 546 nm (for Hb).

5. Notes

1. It is best to start the preparative run at low voltage (maximum of 300–400 V) until all the protein has entered the trench (1 h or more). The reason: If the sample contains salts, the conductivity in the sample zone will be very high, so that, if the experiment is started immediately at the maximum voltage, the protein will be heat-denatured. In addition, in moving from the free phase (the trench) into a restrictive medium, the protein concentrates against the wall of the trench, and there is a risk of aggregation and precipitation. We recommend this procedure also for analytical runs.

2. The advantage of the soft gels will be apparent during the run: The viscoelastic forces of the matrix are weakened, the osmotic force in the protein zone

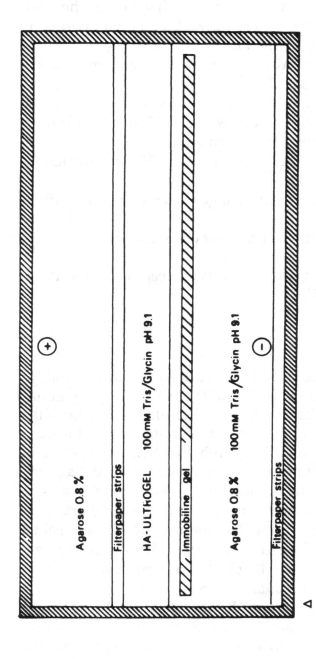

Fig. 7. Assembly for electrophoretic protein recovery from an immobiline strip. The glass tray (LKB 90.00.0157) is the same as that used for preparative IEF runs in granulated gel layers. It is filled with 100 mL of melted 1% agarose in appropriate buffer. Before gelation, the IPG strip is embedded in the agarose phase, to ensure optimum contact between the two matrices upon gel formation. Five millimeters in front of the IPG strip a trench is dug, to be filled either with HA-Ultrogel or with the DEAE-Sephadex, according to the chosen protocol. Electrical contact is ensured with filter paper strips stretching from the agarose gel surface to the electrolyte reservoirs of the Multiphor chamber (from ref. 2).

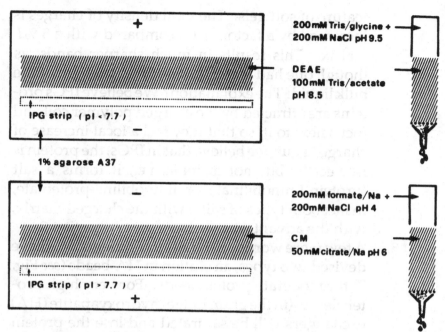

Fig. 8. Preparative set-up for elution into ion-exchange resins. The assembly is the same as in Fig. 7, except that the trench is filled with either DEAE-Sephadex (for acidic to neutral proteins, up to p*I* values of approximately 7.7) or CM-Sephadex (for decidedly alkaline proteins, p*I* > 7.7) (*see* the left part of the figure; notice that in the case of CM, the polarity is reversed). Upon electrophoretic transfer, the beads are transferred to a syringe or small column and the proteins eluted by titrating the resins with the two types of buffers indicated in the right part of the figure (upper formulations refer to the elution system; lower formulations to the buffers used during the electrophoretic recovery; these latter formulations are indicated by bidirectional arrows) (from ref. 15).

takes over so that, proportionally to the amount loaded in a single zone, the gel will stretch locally forming a peak in the focused protein band, the peak height representing an equilibrium between the two opposite forces. There is another important feature of soft gels: The local density of charges is increased by a factor of 1.6 compared with a 5 %T matrix. This results in much sharper bands, as though one had increased the ionic strength of the bulk liquid. The explanation? We believe IPGs proteins are attracted by the charged polymer coil and focus next to it, so that they feel a local increase of charge. Thus, we believe that in IPGs, the protein is isoelectric, but not isoionic, i.e., it forms a salt (protein-Immobilinate or Immobiline-proteinate, in fact both types of salts) with the charged matrix, with the advantages we have seen in section 2.

3. A few more words on the elution systems. We have devised two types, an elution in HA-beads (*see* Fig. 7) for moderate protein loads. For very high protein levels (in the gram range) hydroxyapatite (HA) exchangers will be saturated and lose the protein during the electrophoretic step at the anode. Therefore, we have devised an elution in DEAE- (for proteins up to pI 7.7) or in CM-Sephadex (for alkaline proteins, pI > 7.7), as illustrated in Fig. 8. From these last systems, the proteins are eluted by titrating (i. e., deprotonating the DEAEs or protonating the CMs) the exchanger, with recoveries better than 90% protein in all cases tested (from HA beads, the classical eluant is 0.2M phosphate buffer, pH 6.8). As an alternative, for very small protein loads, one could resort to the canal-Immobiline system devised by Bartels and Bock (14). Finally, we have described a fourth elution system that, unlike the previous three, allows for direct recovery in a free liquid phase in a dialysis bag (16).

Acknowledgments

For the development of preparative IPG techniques, we have been supported by a 3-year grant from CNR (Roma) P. F. Chimica Fine e Secondaria, Tematica Metodologie di Supporto, and by Ministeri Della Pubblica Istruzione E Sanitá.

References

1. Righetti, P.G. (1984) Isoelectric focusing in immobilized pH gradients. *J. Chromatogr.* **300**, 165–223.
2. Ek, K., Bjellquist, B., and Righetti, P.G. (1983) Preparative isoelectric focusing in immobilized pH gradients. I. General principles and methodology. *J. Biochem. Biophys. Meth.* **8**, 135–155.
3. Gelfi, C. and Righetti, P.G. (1983) Preparative isoelectric focusing in immobilized pH gradients. II. A case report. *J. Biochem. Biophys. Meth.* **8**, 157–172.
4. Righetti, P.G. (1980) On the molarity and ionic strength of carrier ampholytes in isoelectric focusing. *J. Chromatogr.* **190**, 275–282.
5. Cohn, J.E. (1936) Salting-in and salting-out of proteins. *Chem. Rev.* **19**, 241–255.
6. Gronwall, A. (1942) Solubility of β-lactoglobulin at the pI as a function of ionic strength. *Comp. Rend. Lab. Carsberg, Ser. Chem.* **24**, 185–195.
7. Righetti, P.G. (1983) *Isoelectric Focusing: Theory, Methodology and Applications* Elsevier, Amsterdam.
8. Bode, H. (1980) Partitioning and Electrophoresis in Flexiblle Polymer Networks, in *Electrophoresis '79* (Radola, B.J., eds.) de Gruyter, Berlin.
9. Righetti, P.G. and Gelfi, C. (1984) Immobilized pH gradients for isoelectric focusing. III. Preparative separations in highly diluted gels. *J. Biochem. Biophys. Meth.* **9**, 103–119.
10. Gianazza, E., Astrua-Testori, S., and Righetti, P.G. (1985) Some more formulations for immobilized pH gradients. *Electrophoresis* **6**, 113–117.

11. Righetti, P. G., Ek,K., and Bjellqvist, B. (1984) Polymerization kinetics of polyacrylamide gels containing immobilized pH gradients for isoelectric focusing. *J. Chromatogr.* **291,** 31–42.

12. Gelfi, C. and Righetti, P.G. (1984) Swelling kinetics of Immobiline gels for isoelectric focusing. *Electrophoresis* **5,** 257–262.

13. Casero, P., Gelfi, C., and Righetti, P.G. (1985) Preparative isoelectric focusing in immobilized pH gradients. IV. Recovery of protein zones from Immobiline matrices into ion-exchange resins. *Electrophoresis* **6,** 59–69.

14. Bartels, R. and Bock, L. (1984) Immobiline-Canal Isoelectric Focusing: A Simple Preparative Technique for Protein Recovery from Narrow pH Gradient Gels, in *Electrophoresis '84* (Neuhoff, V., ed.) Verlag Chemis, Weinheim.

15. Righetti, P.G. and Gianazza, E. (1987) Isoelectric Focusing in Immobilized pH Gradients: Theory and Methodology, in *Methods of Biochemical Analysis* vol. 32 Glick, D., ed.) Wiley, New York.

16. Righetti, P.G., Gelfi, C., Morelli, A., and Westermeier, R. (1986) Direct recovery of proteins into a free-liquid phase after preparative isoelectric focusing in immobilized pH gradients. *J. Biochem. Biophys. Meth.* **13,** 151–160.

Chapter 19

Isoelectric Focusing Under Denaturing Conditions

Christopher F. Thurston
and Lucy F. Henley

1. Introduction

The introduction of methods for the electrophoretic analysis of proteins under denaturing conditions (1,2) is something of a landmark in the development of methodologies for the analysis of proteins. Not only was analysis giving good correlation to molecular weight possible with small samples of complex mixtures of proteins, but in addition, analysis of normally insoluble proteins became possible in a relatively simple manner. These methods were based on solubilizing protein in sodium dodecyl sulfate (SDS) such that the intrinsic charge differences between proteins were overwhelmed by the relatively strong charge of the dodecyl sulfate ion, which resulted in very good correlation (for most, but not all, proteins) between electrophoretic mobility and molecular weight (*see* Chapter 6 in Vol. 1 of this series).

It is precisely the property of proteins masked by dodecyl sulfate that is exploited in isoelectric focusing (IEF) analysis, and consequently if denatured protein is to be analyzed by IEF, an alternative method of solubilization is required. We describe here the IEF analysis of protein samples solubilized by the use of $8M$ urea, a powerful denaturant that is uncharged, based on the method of O'Farrell (3).

Complete denaturation of many proteins involves, in addition, the reduction of inter- and/or intramolecular disulfide bonds that is accomplished by inclusion of a thiol reagent such as 2-mercaptoethanol or dithiothreitol. This is simple to do for electrophoretic analysis because the dissociated/unfolded polypeptides are stabilized in this state by SDS, but here again the requirements of IEF are different. The redox state of a protein is affected by thiol reagents and so consequently is their behavior during IEF analysis. We therefore show an example of such effects and give conditions for maintaining protein in the fully reduced state during the focusing process (*see* Figs. 1 and 2).

2. Materials

1. Silane A174 (γ-methacryloxypropyltrimethoxysilane).
2. Glacial acetic acid, analytical grade.
3. Urea. Use the highest purity available. "Ultrapure" from Schwartz-Mann is suitable. Although urea solutions can be deionized with a mixed-bed ion-exchange resin, the procedure requires solid urea at

DITHIOTHREITOL (M)

2ˑ 0ˑ 0 0·25 0.5 I I·5 2 2.5

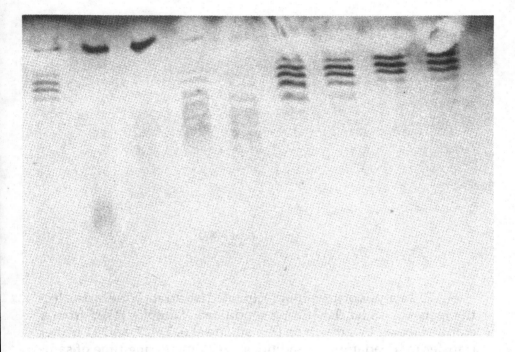

Fig. 1. Isoelectric focusing gel stained with Coomassie blue. All tracks were loaded with 10 µg pure isocitrate lyase. The concentration of dithiothreitol in the samples is shown above each track. Asterisks indicate tracks loaded with immunoprecipitated isocitrate lyase (*see* Chapter 12). The anti-isocitrate lyase immunoglobulins present in these samples did not focus as distinct bands. The heavily stained bands near the top of the photograph in tracks containing no dithiothreitol are artifactual; in the absence of reducing agent, much of the protein commonly precipitates in the position of the applicator strip or along one end of the strip. The region of the gel shown is from about pH 6 (top) to about pH 4 (bottom) (from ref. 5, with permission).

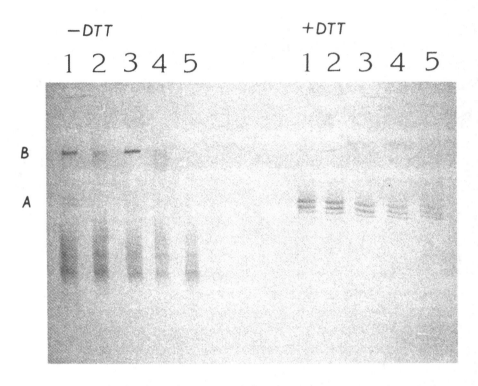

Fig. 2. Analysis of immunoprecipitated isocitrate lyase by isoelectric focusing under denaturing conditions. Samples taken from a culture labeled by growth on [^{35}S]-sulfate at 0, 2, 4, 7, and 11 h after transfer to "deadapting" conditions. With increasing time of sampling, the amount of radioactivity precipitable with anti-isocitrate lyase decreases from 11,000 cpm at 0 h (track 1) to 5000 cpm at 11 h (track 5). Fully reduced isocitrate lyase focuses as two main bands and one or two minor bands, average p*I*, 5.6, as seen when the samples were focused in the presence of 2*M* dithiothreitol (+DTT). In the absence of dithiothreitol (-DTT), only the initial sample contained detectable, fully reduced isocitrate lyase (marked A), and in successive samples the enzyme becomes progressively more oxidized focusing at progressively more acidic p*I* values. Bands marked B are precipitated protein at the position of the applicator strips. The pH range shown in the fluorograph is approximately from pH 8 (top) to pH 3.5 (bottom) (from ref. 5, with permission).

least for sample preparation, so it must be recrystallized if purification is attempted.

4. Nonidet NP40: 10% (v/v) solution in distilled water.

5. 30% Acrylamide: 28.38% (w/v) acrylamide and 1.62% (w/v) *N,N'*-methylene bisacrylamide, dissolved in distilled water, filtered through a 0.22-μm pore-size membrane filter and stored in the dark at 4°C. This reagent can also be contaminated with charged impurities (acrylic acid), which can be removed by recrystallization or treatment with ion-exchange resin, but there are now highly purified grades available from most suppliers.

6. Ampholytes: Use premixed preparations to give the desired pH range, such as LKB "Ampholines" or Pharmacia "Pharmalyte."

7. *N,N,N',N'*-Tetramethylethylenediamine (TEMED).

8. Ammonium persulfate. Use analytical grade and store the solid at 4°C. The solid can decompose if heated, without change in appearance, so do not rely on a bottle of uncertain provenance.

9. 10 m*M* Phosphoric acid.

10. 20 m*M* NaOH.

11. Fixing solution: 0.7*M* trichloroacetic acid, 0.16*M* sulfosalicylic acid (TCA/SSA).

12. Destain solution: glacial acetic acid:ethanol:water; 1:2.5:6.5 by volume.

13. Staining solution: 2.5 g Coomassie blue R250, 100 mL glacial acetic acid, 500 mL ethanol, 400 mL distilled water. Filter through Whatman No. 1 paper on the day of use. (Dissolve Coomassie in ethanol component first.)

14. 1*M* Sodium salicylate or a commercial water-miscible fluorographic solution such as "Amplify"

from Amersham-Searle or "Enlightning" from New England Nuclear.

15. Flatbed isoelectric focusing apparatus with provision for cooling the gel-carrying plate.

3. Method

3.1. Casting the Gel

1. Adjust the pH of 1 L of distilled water to 3.5 with glacial acetic acid and mix in 4 mL of Silane A174 by stirring for 15 min. Soak a glass gel-plate in the solution for 1 h, rinse with distilled water, and dry in air (*see* Note 2 in section 4).

2. Use a few drops of ethanol to adhere a clean plastic sheet to a second glass plate (*see* Note 2 in section 4).

3. Mount the silanized plate and the plate-backed plastic sheet separated by a 1-mm casting gasket, with the open corner of the gasket uppermost.

4. Prepare 20 mL of gel-forming solution in a sidearm flask as follows: 9.6 g urea, 4 mL distilled water, 4 mL 10% nonidet, 2.66 mL acrylamide solution, and 0.5 mL ampholyte mixture are gently swirled to dissolve the urea (*see* note 1 in section 4). Place an oversize rubber bung over the top of the flask and apply a vacuum from a water pump via the sidearm for 10 min with occasional further swirling to thoroughly degas the mixture. Allow air to re-enter the flask slowly (*see* Note 3 in section 4).

5. Add 22 µL of TEMED and 16 µL of freshly prepared 10% (w/v) ammonium persulfate, swirl briefly to mix, take up into a disposable syringe, and deliver immediately into the gel-casting box from step 3 above. Completely fill the system and cover the opening in the casting gasket with parafilm. The

mixture should polymerize within about 2 h, but may conveniently be left overnight at this stage (*see* Note 4 in section 4).

3.2. Prefocusing the Gel

1. When the gel has polymerized, unclamp the plates, separate the plastic sheet from its backing plate, then very slowly peel off the plastic sheet from the acrylamide gel layer. Lift away the casting gasket and remove any liquid from the edge of the gel with clean tissue.
2. Mount the gel-plate on the cooling plate of the isoelectric focusing apparatus. Circulate water at 10°C through the system.
3. Apply electrode wicks soaked in 10 mM phosphoric acid (anode) and 20 mM NaOH (cathode) to the edges of the gel and connect the electrodes.
4. Using a dc power source, set to give constant voltage, apply 200 V for 15 min, 300 V for 30 min, and 400 V for 30 min, to establish the pH gradient within the gel.

3.3. Sample Preparation

To samples in solution, add solid urea to give a concentration of 8M then mix with an equal vol of 8M urea, 2% (w/v) Nonidet NP4D, 2% (w/v) ampholytes. Protein precipitates may be dissolved directly in this mixture. The sample mixture must contain a relatively high concentration of protein since the maximum volume that can be loaded is 35 µL. The sample mixture may in addition contain 2M dithiothreitol (*see* Note 5 in section 4).

3.4. Focusing

Disconnect the power supply from the prefocused gel and place paper sample applicators near the cathode end of the gel (*see* Note 6 in section 4). Deliver the sample (preferably no more than 10 µL) onto each applicator and focus for 30 min at 500 V. Disconnect the power supply and remove the paper sample applicators from the gel with flat-ended forceps. Focus for a further 90 min at 800 V.

3.5. Measuring the pH Gradient

When focusing is finished, remove the electrode wicks and cut a track of 5-mm square sections from the position of the anode to the position of the cathode. Soak each section overnight in 0.25 mL distilled water and measure the pH with a microelectrode without removing the slice of gel.

3.6. Staining

Fix for 1 h in TCA/SSA (this step also removes ampholytes that would otherwise stain with Coomassie blue), and wash for 30 min in destain solution. Stain with Coomassie blue with constant agitation for 2–16 h. Destain with repeated changes of destain solution until the background of the gel is cleared of stain.

3.7. Fluorography

For detection of radioactive bands on the gel, it is necessary to detach the gel from the silanized glass plate. By the time the gel is fully destained, it has usually started to detach at the edges. Hold the plate upside down over a tray containing destain solution, and separate the gel from the plate with a sharp scalpel. Soak the gel in 1*M* sodium salicylate or a commercial fluor for 2 h, and dry down onto Whatman 3-mm paper and expose to X-ray film (*see* Chapters 16 and 17 in vol. 1 of this series).

4. Notes

1. The quantities given are for a gel 230 × 115 mm. Plates of this size, plastic sheets, and casting gaskets are obtained from Pharmacia. The Pharmacia manual *Isoelectric Focusing—Principles and Methods* is an extremely useful reference work for both the background and practicalities of isoelectric focusing.

2. The thickness of the gel (1 mm) is optimal for focusing, but makes handling difficult. The polymerized gel is covalently linked to the glass plate by the silane in order to prevent creepage and/or buckling during focusing and is polymerized against a plastic sheet rather than a second glass plate, since it is easier to detach from the polymerized gel after casting.

3. For a usable focusing pH range of 4–9, use 0.4 mL pH 3–10 ampholytes and 0.1 mL of pH 4–6 ampholytes. Note that the electrode solutions described are for this pH range and are not applicable to all ranges.

4. The polymerization of acrylamide mixtures is inhibited by oxygen and impurities. Shaking up the solution to mix it must be avoided, particularly after degassing. If the dispensing syringe is large enough to hold all of the mixture, the process of filling the syringe is probably sufficient to mix the final additions. If the gel does not set, try degassing

longer or try a different bottle of ammonium persulfate, and if still unsuccessful, double the amount of TEMED included. If this does not work, the reagents are not pure enough. If the gel sets unevenly, the glass plate was not clean.

5. The small sample volume allowed makes precipitated samples desirable. Chloroform/methanol precipitation (4) is a good way of doing this, and has the additional advantage of removing lipid. The description given here is for small volumes, such as might be processed for electrophoretic analysis, but the volumes may be scaled up in proportion 10-fold, using 15 mL Corex tubes. All steps are performed at room temperature. Take 150 µL of protein solution in a 1.5-mL microfuge tube, add 600 µL methanol, mix by three or four inversions of the capped tube, and centrifuge for 10 s. Add 150 µL chloroform, blend briefly on a vortex mixer, and centrifuge for 10 s. Add 450 µL of distilled water and blend thoroughly on a vortex mixer. Centrifuge for 10 s to separate the two liquid phases that formed on addition of the water. Carefully remove the upper phase and discard. Do not attempt to remove it all, but leave a residual layer 1–3 mm deep, because most of the protein is at the interface. Add 450 µL methanol, vortex mix briefly, and centrifuge for 4 min. After centrifugation, a white protein pellet should be visible. Aspirate off and discard (as chlorinated solvent) the bulk of the supernatant. Remove remaining supernatant and dry the pellet in a vacuum desiccator or under a stream of dry N_2. Pellets dissolve readily and without heating in sample buffers used for electrophoresis or IEF and may be stored indefinitely at –20°C.

If the sample contains proteins that are susceptible to oxidation, clear resolution may only be
obtained if the samples are made $2M$ in dithiothreitol (see Figs. 1 and 2). There is a need for this very
high concentration to prevent migration of pro-
tein(s) and thiol reagent in different directions
during focusing from depleting the protein(s) entirely of reducing agent.

6. The best position for sample loading strips has to be
determined by trial and error. The samples we
have run all contain proteins that precipitate at
acidic pH, so it has been preferable to place the
applicators in the pH 8 region of the (prefocused)
gel.

References

1. Weber, K. and Osborne, M. (1969) The reliability of molecular
weight determinations by dodecyl sulfate-polyacrylamide
gel electrophoresis. *J. Biol. Chem.* **244,** 4406–4412.
2. Laemmli, U.K. (1970) Cleavage of structural proteins during
the assembly of the head of bacteriophage T4. *Nature* **227,**
680–685.
3. O'Farrell, P.H. (1975) High resolution two-dimensional electrophoresis of proteins. *J. Biol. Chem.* **250,** 4007–4021.
4. Wessel, D. and Flugge, U.I. (1984) A method for the quantitative recovery of protein in dilute solution in the presence of
detergents and lipids. *Anal. Biochem.* **138,** 141–143.
5. Henley, L.F. and Thurston, C.F. (1986) The disappearance of
isocitrate lyase from the green alga *Clorella fusca* studied by
immunoprecipitation. *Arch. Microbiol.* **145,** 266–271.

Chapter 20

Computer Analysis of 2-D Electrophoresis Gels

Image Analysis, Data Base, and Graphic Aids

H. H-S. Ip

1. Introduction

Although two-dimensional (2-D) gel electrophoresis has been used for the past ten years, a large amount of potentially useful information is still generally unused, since accurate and rapid methods for extracting, managing, and analyzing the vast quantity of data that can be derived from the gels are not widely available. The difficulty is partly caused by the large number of spots as well as artifacts present in a 2-D gel and the non-uniformity in the electrophoresis process that causes distorted spot patterns and makes comparison among gels difficult. With the advent of data base management systems (DBMS) and computer graphics, 2-D gel data can be organized and manipulated with relative ease.

This chapter describes work that has been done toward automating the process of extracting useful quantitative information from 2-D gels and comparing spot data derived from separate gels. The design and implementation of a data base and a data analysis system for 2-D gels are also outlined.

An approach to the problem is to present a computer with an image of a 2-D gel and automate the process of locating and analyzing protein spots seen on the gel image. Several systems have evolved along this line (for examples, *see* refs. 1–4).

The computational problems one faces in this approach are as follows:

1. Elimination of background level and "noise" from the gel image.
2. Delineation of areas in the image corresponding to protein spots on the gel.
3. Measurements of protein spot attribute values, e.g., the position of the spot on the gel, the size of the spot, and so on.
4. Pairing of corresponding spots seen on separate gel images.

Each of the above subtasks involves the processing of a large amount of data. Our work takes advantage of the processing power of a 2-D array computer, CLIP4S (5), developed at University College London. An array computer that processes large quantity of data "in parallel" reduces the time and cost of performing data-intensive tasks compared with a conventional computer, which carries out computations serially.

We have also investigated the applicability of a commercially available data base management system (DBMS) to storing and managing data relevant to 2-D gel analysis. The data base is linked to the program modules, which provides various utilities for visualizing, manipulating, and analyzing protein spot patterns by means of computer graphics. This includes procedures for reconstructing the protein spot pattern as seen on the original gel on a computer terminal or on paper, procedures for matching spots across gels, and graphics aids for manipulating and editing spot data.

2. System Overview

The computing hardware for abstracting protein spot data from a 2-D gel image comprise the CLIP4S array computer and a LSI11-23™ minicomputer. Apart from controlling the operations and data movements in and out of the array computer, the minicomputer also provides the tools and environment for software developments under the UNIX™ operating system. Mass data storage is provided by magnetic disks.

Briefly a 2-D gel is back-illuminated from a lightbox with a Vidicon television camera mounted directly above it. The television signals of the 2-D gel image are converted into a digitized image of 512 x 512 picture elements (pixels): each pixel can have one of 256 (8 bits) gray values or intensity (0 = black, 255 = white).

The CLIP4S array computer consists of a 512 x 4 strip of processing cells. Each processing cell is a simple computer in its own right, and each cell is connected to its eight nearest neighbors, as shown in Fig. 1, so that information can be passed between them. The array computer operates on successive strips of 512 x 4 pixels in the gel image, starting from the top of the image and working downward. This scanning process is completely automatic. The processing power of the array computer comes from the design; all available 2048 processing cells operate on their assigned pixels simultaneously. For example, 2048 additions can be completed in the

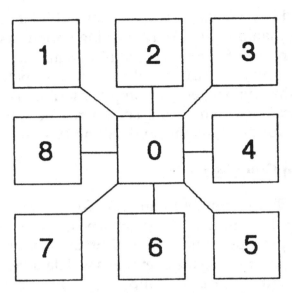

Fig. 1. Interconnections of the CLIP4S processing cells.

time it takes one processing cell to complete one addition.

Protein spot measurements derived from the gel image by the array computer are transferred via magnetic tapes to the DEC2060 mainframe computer at the Imperial Cancer Research Fund. This Mainframe computer holds the 2-D gel data base and the relevant data analysis programs. The user interacts with 2-D gel data using a SIGMEX 6130 graphics wokstation and a Summargraphic data tablet. The former is connected to the mainframe computer. The system configuration is summarized in Fig. 2.

3. Processing of 2-D Gel Electrophoresis Images

3.1. "Noise" Reduction

To reduce digitization (electrical noise in the gel image), the camera scans the gel eight times, producing

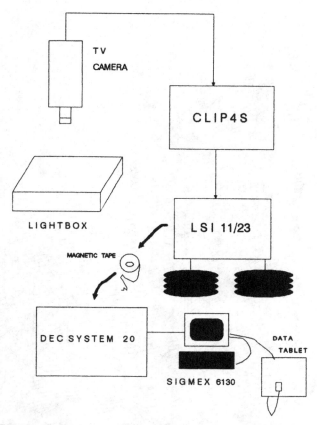

Fig. 2. System configuration and overview.

eight digitized images of the gel. These images are added, and the resultant image is divided by eight to obtain an averaged image. Each pixel gray value in the averaged image is, therefore, an average of the corresponding pixel gray values in the eight input images.

Data in the averaged image can be "smoothed" in a statistical sense by fitting them with a high-ordered polynomial function. Figure 3 shows the result of fitting the image data with a cubic B-spline function. The advantage of using B-spline functions over other datasmoothing functions, such as weighted local average or

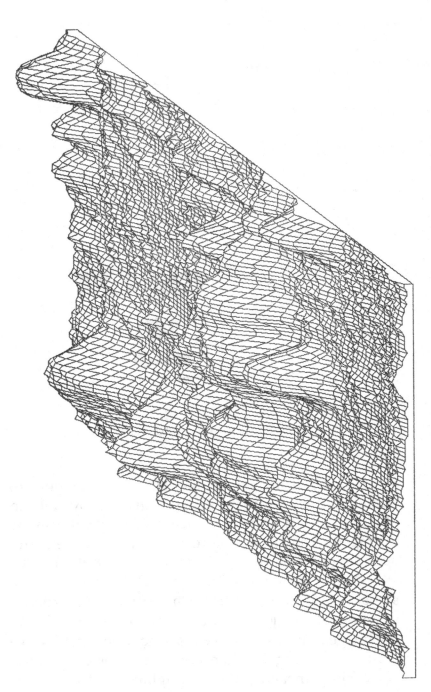

Fig. 3. Noise reduction. (a) 3-D intensity plot of the original gel data.

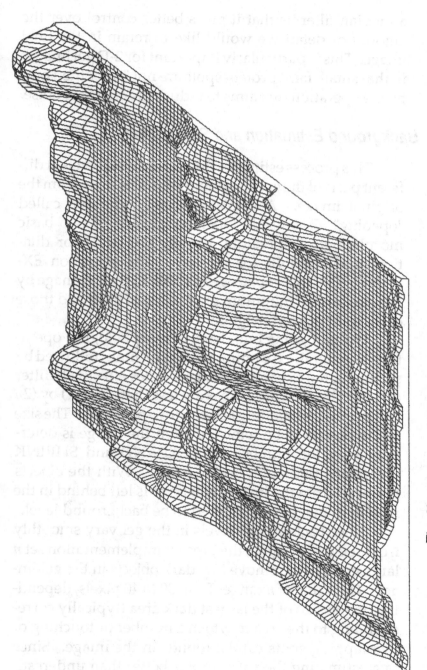

Fig. 3b. Results of fitting a cubic B-spline function to (a).

a median filter, is that it gives better control over the amount of detail we would like to retain in the final image. This is particularly important for 2-D gel images in that small, faint protein spots are not "smoothed out" by the operation that aims to reduce the effects of noise.

3.2. Background Estimation and Subtraction

This process estimates the background levels at different parts of the gel image and subtracts them from the original image. A spatial filtering technique called "opening" (6) is used. Opening consists of two basic morphological operators, namely EXPAND (or dilation) and SHRINK (or erosion). The operation EXPAND/SHRINK replaces each pixel in the gel image by the maximum/minimum gray value of itself and those of its eight nearest neighbors.

Without going into detail of morphological operators, in general, expanding an image n times followed by shrinking the expanded image n times effectively filter out any objects whose areas are less than $(2n + 1)$ by $(2n + 1)$ pixels, i.e., opening acts as a low-pass filter. The size of the objects to be removed from the image is determined by the number n of EXPAND and SHRINK operations that apply to the image. With the objects removed from the image, whatever is left behind in the resultant image is considered to be background level.

Since background levels in the gel vary smoothly from region to region, the present implementation set n large enough to remove any dark objects in the gel image. The value of n ranges from 20 to 40 pixels, depending on the size of the largest dark area (typically corresponding to the area in which a number or touching or overlapping spots can be found) in the image. Since overestimating the value of n is better than underestimating it (the latter may result in the elimination or reduction of genuine spot signals in the image), the value of *n* is usually set to around 40 pixels for most gel images. Figure 4 illustrates this background subtraction process.

3.3. Spot Delineation

There are two basic components in the spot-delineation process: the detection of the spot and the determination of spot boundaries. The former aims to detect the location of a protein spot in the image with emphasis on resolving overlapping spots; the latter estimates the boundary of the detected spots. Ideally, the boundary should include most of the protein loading of the detected spot and, at the same time, take into account the effects of overlapping spots.

The spot-detection process models the intensity profile of a protein spot on a Gaussian function and locates the "central core" of the spot using a method described by Lemkin and Lipkin (2). Our method, however, differs from that described in ref. 2 in the forms of the Laplacian or second derivative masks that are used to locate the core of protein spots and the way spot boundaries are determined. Instead of determining the spot boundaries by propagating out from the central cores according to a set of heuristic rules that are based on observations on the shape of the Gaussian function and its derivatives, the boundaries of the spots are computed by simply EXPANDing the central cores a number of times. Possible merging of separate central cores during this EXPANDing phase is prevented by the "exoskeletons" which will be described in detail later.

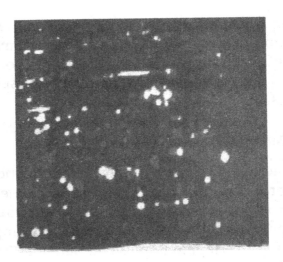

Fig. 4a. Original gel image.

Fig. 4.b. Background.

3.3.1. *Central Core Detection*

The central core detection algorithm can be summarized as follows. The image gray values are "convolved" using the set of one-dimensional masks:

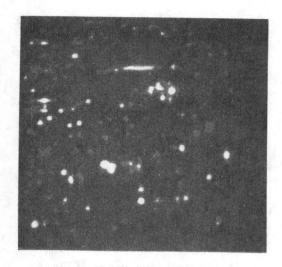

Fig. 4c. Background subtracted gel image.

[1 0 0 0 - 2 0 0 0 1]   [1 0 0 0 - 2 0 0 0 1]ᵀ

The convolution operation forms a weighted sum, for each pixel in the image, of pixel values with weighting factors and neighborhood defined by the above masks. Points on the image that give rise to negative values for both masks are defined as the central cores of the spots. Isolated points are considered as noise and are removed. Each group of eight-connected points remaining is taken as the central core of a spot. Figure 5 shows the central core image as computed by this process.

The two masks are variations to the second derivative masks used in Lemkin's method. It was observed that there are fluctuations in the intensity values within a protein spot that cause local departure of the protein spot intensity distribution from the Gaussian function that is being modeled. Consequently more than one central core may be generated from a single spot when using the local 3 by 3 difference formula of Lemkin. The

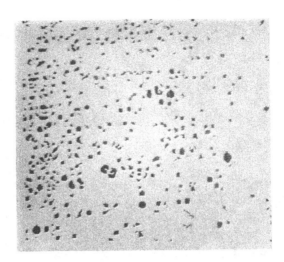

Fig. 5. Central core image of Fig. 4c.

effects of this local fluctuation can be reduced by using the modified second derivative masks that effectively "step over" such local fluctuations.

This fluctuation of intensity values within a spot may be attributed to local variations of polypeptide loading within the spot as a result of the nonuniformity of the electrophoresis process. Other possible contributing factors are the digitization and staining processes.

The size of the modified masks, however, determines the resolution at which overlapping spots can be resolved. The trade off between using different size masks is that, if the mask is relatively too small (e.g., 3 by 3), a single large spot may result in several distinct central cores; if the mask is large (e.g., 15 by 15), overlapping spots may give rise to a single central core. These observations suggest the use of multi-resolution masks for different size spots; the research problem in this case would be to find a suitable control strategy for the application of the different-sized masks and a way to interpret the resulting image.

Central cores arising from artifacts in the gel can be removed by deleting any central cores smaller than a given size, e.g., isolated points, and those with intensities below a given threshold value.

Although streaks in the gel image can be filtered out using directional filter, e.g., EXPAND followed by SHRINK in the direction perpendicular to the direction of the streaks, they are sometimes retained in order to account for the total protein loads of the spots. In the presence of a streak, the above delineation method would produce a series of central cores along the length of the streak.

3.3.2. Determination of Spot Boundaries

Central cores generated from the above process typically cover only the central portion of the spots. Some way of propagating the central core outward in order to account for all the protein load attributes to the spot is needed.

This can be done by computing the "exoskeleton" of the central cores in the image. Exoskeleton is a planar graph that marks the midlines of the space between the set of objects (central cores) in an image. Each arc in the exoskeleton forms an equidistance boundary between a pair of nearest objects. The exoskeleton of a set of central cores can be computed by finding the medial axes of the background in the image. The following set of thinning masks (7) are used:

00X	X00	X1X	X1X
011	110	110	011
X1X	X1X	X00	00X

000	1X0	X11	0XX
X1X	110	X1X	011
11X	XX0	000	0X1

where 1 = pixel, which must have value 1 (white)
    0 = pixel, which must have value 0 (black)
    X = don't care

This set of masks or bit patterns is matched successively to all the points in the "inverse" of the central core image. When a match is found, the central pixel is deleted (i.e., switched from a value of 1 to 0). The process is repeated by cycling through these masks until the output image stabilizes. Free ends in the resulting skeletons are "chewed" off using the following thinning masks:

X00	X1X	00X	000
110	010	011	010
X00	000	00X	X1X
000	XX0	00X	000
X10	010	01X	010
X00	000	000	0XX

To determine the spot boundaries, the central cores are EXPANDed K times. After each EXPAND, the expanded central cores image is logically "AND" [i.e., corresponding pixels having values of 1 (white) in both operand images remain white, whereas all other pixels are switched to values of 0 (black) in the resultant image] with the inverse of the exoskeleton image. The latter step has the effect of preventing the expanding spot areas from merging. The value of K is currently set to 5, which is an estimate of the extent that the polypeptide fragment load of a spot may "spread" in the gel. This value also reflects the focusing capability of the run. Figure 6 illustrates the spot boundaries determination process. When the process is completed, two output images for each scanned gel are stored, namely the

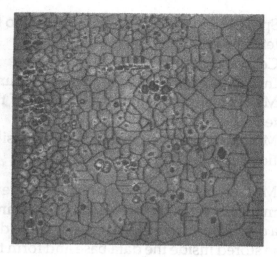

Fig. 6a. Exoskeleton of Fig. 4.

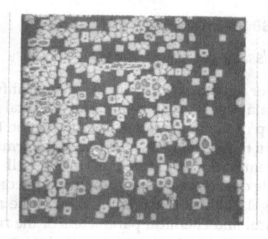

Fig. 6b. Estimated spot boundaries.

background subtracted image and the central core image.

3.4. Measurement of Spot Attribute Values

The set of attribute values to be measured for a protein spot includes:

1. Spot center coordinates with respect to the bottom left corner of the image.
2. Central core area.
3. Integrated density within the spot boundary.
4. Maximum gray value (peak value) of the spot intensity distribution.
5. Minimum gray value of the spot intensity distribution.

Measurements of these attributes are computed automatically from the processed images and grouped into a data record, one for each spot. These data records can be stored inside the data base and form the input to the data-analysis programs in the system.

4. Data Base for 2-D Gels

4.1. Design Issues

For a given experiment, the relevant information in our data base is organized in a hierarchical fashion. At the top of the hierarchy is information on the cell line used in the experiment, followed by information on the different samples derived from that cell line. Each sample data record contains information on the associated gel, such as the amount used in the run and the physical and chemical parameters of the run. Successive levels in the hierarchy contain information on the scanning parameters and the numerous spot data records derived from each scan of the gel. The organization of the data base is shown in Fig. 7.

We have investigated the construction of a data base for 2-D gel data using a commercially available data base management system (DBMS). In this way, we can take advantage of the tools provided by the DBMS, such as facility to construct queries, mechanism for

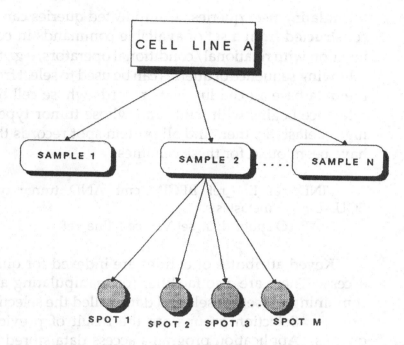

Fig. 7. Organization of a 2-D gel data base.

searching and updating relevant data records, and so on. The initial design was concentrated on the partition of all relevant information into separate entities and the relationships between entities. A specific query is answered by combining information on the relevant entities through relational operators. A clear understanding of the set of most likely queries would be advantageous in the sense that the access paths to the relevant data can be optimized during the early stages of the design.

The system is a 1022 Data Base Management System (8), in which different entities are stored in separate data sets, and information stored in different data sets can be combined using various operators, e.g., MAP, JOIN. The system also provides a simple syntax for formulating user queries. Complicated queries can be constructed from a set of available commands in conjunction with relational/conditional operators, e.g., the following sequence of queries can be used to select from the data base all cell line data records whose cell line reference begins with "cmt" and whose tumor type is high metastatic, then find all protein spot records that have been found for these cell lines:

FIND cell_line_ref BEGIN cmt AND tumor_type EQUAL high_metastasis.
MAP TO spots_data_set VIA cell_line_ref

Keyed attributes of entities are indexed for quick access. There are also facilities for manipulating and combining subsets of selected data, called the selection groups. Selection groups are the result of previous queries. Application programs access data stored inside the data base through a set of FORTRAN interface subroutines.

4.2. Implementation Issues

There are a number of limitations in our version of the 1022 DBMS, which makes the implementation of the 2-D gel data base structure entirely within the framework of 1022 impractical. One of them is the fixed length storage structure of 1022, which makes it necessary for the implementor of the data base to predefine the storage space that will be used to store each attribute of the entity. Consequently, more storage space is assigned than is necessary, because each data record takes up the amount of space that will be needed for the worst-case entry, even though most of the assigned space is left empty for a typical record. Another limitation is that deleted record space cannot be reclaimed until the entire data base is rebuilt (DUMPed). This means for a data base whose entries will be frequently edited (deleted and added), a large amount of storage space will be locked up and frequent "DUMPing" is needed in order to retrieve this space.

In the context of 2-D gel data base data on cell lines, samples are relatively static in the sense that once it is entered, the data record is not likely to be deleted or updated. On the other hand, data on protein spots and across gels matches are highly dynamic. Incorrect matches may be deleted (in the case of no match) or updated (in the case of mismatch); spot data records may also need to be deleted or updated because of possible errors in computer delineation. More importantly, since each gel may produce thousands of spots, using a fixed amount of storage for each protein spot data record is not suitable. This is because slight overestimation of the storage space required for a spot data record will result in significant wastage of storage space when multiplied by thousands of spot records. Under the circumstances, it is more appropriate to use only the minimum amount of storage space required by each spot record. This means using a variable-length record structure. Moreover, the storage overhead of each spot record resulting from entry in the key table for fast data base searches is also significant when multiplied by several thousands.

As a result of the above considerations, it was decided that data on cell lines, samples, and runs will be stored using 1022 data record structures, and data on protein spots and matches will normally be held in flat ASCII files that can be loaded into 1022 data record structures when needed. Application programs are designed to access spot data from these flat files, and conversion routines are provided to convert spot data from 1022 record format to flat file format and vice versa.

Details of the implementation will not be covered here, except to give a brief description of what information is stored in the data sets. The 1022 2-D gel data base comprises four separate data sets, namely cell lines, samples, scans, and spots. The information stored in each data set is shown in Table 1. Using the 1022 MAP function, attribute values of associated data records stored in separate data sets can be extracted.

The data base on cell lines, samples, and scans is maintained and updated by the biologists as and when data become available. The spot data base contains data records of spots that are needed for data base queries, e.g., associating a subset of protein spots with data on the cell lines or samples from which these spots are derived.

There is no reason, however, to prevent these data from being incorporated into a more flexible DBMS that supports variable size records. Combinations of information stored inside these two types of file can be used in multiple gel analysis. For example, corresponding spots found across a series of gels can be compiled by "chaining" through the corresponding match files and having their respective attribute values collated, e.g., the variation of densities of corresponding spots across a series of exposures of the same autorad gel.

5. Multiple-Gels Analysis

The computation of individual spot data has already been covered above: here we describe the process of matching protein spots across gels.

Table 1

Attributes for the Four Data Sets in the 2-D Gel Data Base

Data set	Attribute name	Example
Cell lines	Cell line reference	CMT167/1
	Hybrid	Yes
	Hybrid partner	CMT170
	Metastasis	High
	Tumor	Yes
	Comments	None
Samples	Cell line reference	CMT167/1
	Sample	20/E
	Sample date	25/6/85
	Run date	10/2/86
	To use (microliter)	19.6
	Optical density	0.12
	Photo	Yes
	Comments	Exposure at 1,2,3 min
Scan	Cell line reference	CMT167/1
	Sample	20/E
	Scan date	5/5/86
	Scan file name	CMT167F21
	Comments	None
Spot	Scan file name	CMT167F21
	Spot id	148
	Centroid X-coordinate	204
	Centroid Y-coordinate	392
	Area	148
	Integrated intensity	784
	Maximum intensity	9
	Minimum intensity	2

5.1. Spot-Pairing Algorithm

The spot-pairing algorithm works on the assumptions that (a) the two sets of spots are roughly aligned and (b) the distortion of the position of each spot can be reduced using a linear transformation applied globally to each spot on one of the matching gels. Assumption (a) can be satisfied by manually aligning the gel with an image of another gel that has already been processed. The camera system of CLIP allows an image from the video camera to be superimposed with another stored image. Alternatively, some form of "flickering" system may be used to roughly align the images. For the purpose of multiple gel analysis, it is better to roughly align the set of gels in an experiment against a standard or reference gel. This gel is usually the one considered by the experimenter as the cleanest gel of all those used in the experiment in terms of spot clarity and occurrence.

During gel matching, the set of spots delineated from the gel images is compared, and the pairing algorithm proceeds as follows:

1. Compute an area threshold above which $k\%$ of the spots in gel 1 lies.
2. Extract the subset of spots from gel 1 whose area lies above the area threshold.
3. Repeat steps (1) and (2) from gel 2.
4. Pair spots in the two subsets using a nearest-neighbor scheme.
5. Compute the linear transformation parameters from the matched pairs.
6. Apply the transformation to all the spots in one of the gels.
7. Pair the rest of the spots using the same nearest-neighbor scheme.

The rationale behind the algorithm is that, given that two gels are derived from samples that are in some way related, e.g., low and high metastasis states of the same cell line, their gene products should be similar in terms of variety and abundancy. Consequently, the set of proteins that is synthesized in abundance should be similar for the two samples. This implies that the pattern of the set of dominant spots (large/dark spots) should also be similar for the two gels. Matching these two patterns initially has the following advantages: (a) it provides a means of estimating the transformation parameters reliably because artifacts that can potentially distort the correction transformation are not likely to show up as major spots on the two gels, and (b) the distribution of major spots is normally less clustered than that of all the spots seen in a gel: consequently, pairing based on nearest-neighbor criterion can be carried out more reliably.

The value of k in step (a) above is predefined in the program and is chosen such that a reasonable number of major spots are selected for the transformation parameters estimation. The percentile method is used in order to account for the variation of overall spot size in different gels. It effectively chooses a suitable area threshold for each gel depending on the overall statistics of spot sizes in the gel.

The set of matched major spots plays the role of the "landmark set" in subsequent pairing. This landmark set is usually defined manually in other interactive pairing algorithms (e.g., 3,9). For widely distorted gels, instead of using the relatively computationally cheap nearest-neighbor criterion, the set of major spots can be matched using more expensive and sophisticated matching algorithms (e.g., 10,11). It is expected that, for the above reasons, instead of applying an expensive matching algorithm to all the spots in the gels, matching the set of major spots only using the expensive algorithm would result in highly accurate matches that can then be used to guide the match for the rest of the spots using a less sophisticated, but computationally cheaper, algorithm.

The estimated transformation is applied to the rest of the spots in one of the gels, and the set of transformed spots is then matched against the other set of untransformed spots. When the distortion in the two gels is small, nearest-neighbor match produces reliable results. Confusion usually occurs in areas in which the local distribution of spots is highly clustered.

For highly distorted gels, it is possible to use the system to partition the spots into separate subsets. Corresponding subsets of spots in the two gels are then matched separately.

The results of the matches are output onto another file that contains pairs of the matched spot indices. These values enable the associated spot data to be retrieved from the spot data files. Mismatches may be corrected by means of various graphics aids, which are described below.

6. Graphics Aids

The top-level menu provides access to the various modules in the 2-D gel anlysis system, e.g., data base, data manipulation, hard copy service, and so on. These modules are accessed by moving a cross-wire to the appropriate slot in the displayed menu. Once inside the selected module, different utilities, e.g., gel display, can be invoked similarly via submenus. The set of utilities available for studying spot data is shown in Table 2. During gel display, which reconstructs the spot pattern in a gel, each of the spots detected in the gel is drawn as

Table 2
Menu for Analyzing Spot Data

READ FILE	Read in a set of spots from a file and make it "current"
WRITE FILE	Store current group in a file
LIST	List all sets of spots currently held in the system
GETPAIR	Pair two sets of spots
AND	Logical AND of two sets of spots
NEWNAME	Give a new name to a set of spots
EXOR	Logical exclusive OR two sets of spots
DISPLAY	Graphics display of current group submenu: Display data of a spot Define a calibration set Cut out part of current group
STORE	Store current group in the working memory
GSEARCH	Pick out spots satisfy given constraints from current group
TOP	Return to TOP level, e.g., to exit
PRINT	Print numerical data of current group
SETCURRENT	Define a new current group

an "equivalent circle" (circle having the same area as the original spot) on the terminal screen. The equivalent circles can be color-coded according to the intensity of the associated spots in the gel.

To display numerical data such as attribute values of a spot, the user simply moves a cross-wire using a "mouse" to the spot on the display. This facility can also be used to select and highlight a subset of spots that is of particular interest to the researcher. These user-selected spots can be stored and analyzed separately from the rest of the spots in the same gel.

Graphics facilities for multiple gel analysis include superimposition of a pair of gel displays. If a match file exists for the pair of gels, vectors linking corresponding spots are also drawn. Numerical data on a pair of matched spots can be displayed similarly by moving a cross-wire to the spots. This provides a simple way of inspecting and comparing quantitative data on any pair of spots. Matches can also be edited (deleted or altered) using the "mouse," in which case the corresponding match vector will disappear when the match is cancelled or point to another spot in the case of a match correction. The updated match information is written out onto another file on exit.

Hard copies of the displays are drawn using a Hewlett-Packard plotter (Fig. 8). An option is available to plot spots on a gel that is common with other gels in one color and the rest of the spots in another color. When this form of display is plotted on a transparency (acetic paper) and is scaled such that the equivalent circles superimposed exactly with the spots on the gel, gel comparison results as suggested by the computer can be confirmed or rejected by the investigator by overlaying the plot on the gels concerned.

Experience has shown that it is not always easy to directly relate a given equivalent circle plotted on a terminal screen to the actual spot on the original gel, mainly because of the lack of intensity information (shadings) in such a display.

The graphics facilities in the system as it stands provide a means by which the investigator can easily index spot data, to combine or subtract subsets of spots from different gels and to edit and study matches in multiple gels analysis. The various graphics display utilities serve to highlight results of the computer analysis that are to be studied in conjunction with the original gels.

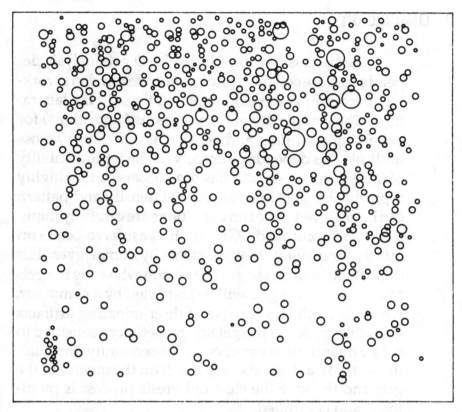

Fig. 8. A computer-reconstructed spot pattern.

7. Applications

The system is currently used in a number of separate feasibility studies. They are (a) a study of differences in the spot patterns associated with gene products produced by cells in high and low metastatic states, (b) a study of protein rates of synthesis, and (c) a study of relationships between radioactivities and spot intensities in autoradiographs. The system is also being assessed for the automatic detection of qualitative and quantitative differences in the analysis of a cDNA recombinant library by differential screening (*12*).

8. Discussion

The success of computerized 2-D gel analysis depends on close collaboration between researchers working on computer development and those in the laboratory. The complexity of the programs (algorithms) for extracting spot data and consistent matchings across multiple gels depends ultimately on the reproducibility of the 2-D gel process. It has been observed that highly reproducible gels in terms of spot density and pattern can be achieved if the run conditions are carefully monitored and controlled. Gels in the past have been run mostly for qualitative assessments by human eyes. It is important, in this case, for the person who runs the gels to realize that the gels will be inspected by a computer, which is much less capable of discriminating artifacts and ambiguities in the gel image, yet more sensitive to subtle quantitative variations. Consequently, the quality of the data depends very much on the quality of the gels and the way the electrophoresis process is monitored and controlled.

In quantitative gel analysis, as observed by other workers in the field, to account for systematic variations in and across the gels, some form of data normalization is carried out on spot data prior to data base searches or statistical analysis. The choice of data-normalization method varies depending on the kind of analysis being performed. More importantly, different results may be obtained for different types of data normalization and the subset of spot data used in the normalization process. Different forms of data normalization (prior to data base searches) have been discussed by Lemkin and Lipkin (13). In this system, the matched spots data can be output in such a way that they can be readily analyzed using standard statistical packages, such as

GLIM™ and MINITAB™, to take advantage of the vast number of statistical tools available in these packages.

The system is organized so that new program modules (e.g., an implementation of an improved spot-pairing algorithm) can be added to the system at a latter date and with few side effects on other unrelated program modules.

Acknowledgments

The technical assistance of Miss Shaila Pandya is gratefully acknowledged. I would also like to thank Mr. Pavel Vitek, Dr. Leonard M. Franks, and Prof. Michael J. Duff for helpful discussions.

References

1.   Anderson, N.L., Taylor, J., Coulter, B.P., Scandora, A.E. Jr., and Anderson, N.G. (1981) The TYCHO system for computer analysis of two-dimensional gel electrophoresis patterns. *Clin. Chem.* **27**, 1807–1820.

2.   Lemkin, P. and Lipkin, L. (1981) A computer system for 2D gel electrophoresis analysis. I. Segmentation and system preliminaries. *Comput. Biomed. Res.* **14**, 272–297.

3.   Garrels, J.I., Farrar, J.T., and Burwell, IV, C.B. (1984) The QUEST System for Computer-Analyzed Two-Dimensional Electrophoresis of Proteins, in *Two-Dimensional Gel Electrophoresis of Proteins* (Celis, J.E. and Bravo, R., eds.) Academic, London.

4.   Potter, D.J. (1985) Computer-assisted analysis of two-dimensional electrophoresis images using an array processor. *Comput. Biomed. Res.* **18**, 347–362.

5.   Fountain, T.J. (1983) A Study of Advanced Array Processor Architectures and Their Optimisation for Image Processing Tasks. PhD Thesis, University of London.

6. Serra, J. (1982) *Image Analysis and Mathematical Morphology* Academic, London.

7. Archelli, C., Cordella, L., and Levialdi, S. (1975) Parallel thinning of binary pictures. *Electronic Lett.* **11**.

8. User's Reference Manual (1986) System 1022 Data Base Management System. Software House, Cambridge, Massachusetts.

9. Lemkin, P. F. and Lipkin, L.E. (1981) GELLAB; A computer system for 2D gel electophoresis analysis. II. Pairing spots. *Comput. Biomed. Res.* **14**, 355–380.

10. Miller, M.J., Olson, A.D., and Thorgeirsson, S.S. (1984) Computer analysis of two-dimensional gels: Automatic matching. *Electrophoresis* **5**, 297–303.

11. Skolnick, M.M. (1985) Automatic comparison of 2-D electrophoretic gels. *IEEE Proc. Computer Vision Pattern Recog.* 48–54.

12. Ip, H. H-S. and Van-Wile, M. (in preparation) Computer-assisted differential screening of recombinant libraries.

13. Lemkin, P.F. and Lipkin, L.E. (1983) 2-D electrophoresis gel data base analysis: Aspects of data structures and search strategies in GELLAB. *Electrophoresis* **4**, 71–81.

Chapter 21

Two-Dimensional (Crossed) Immunoelectrophoresis

John M. Walker

1. Introduction

Two-dimensional (2-D) immunoelectrophoresis, also known as crossed immunoelectrophoresis, is a particularly useful technique for the quantitation of mixtures of proteins and the analysis of the composition of protein mixtures. The method consists of two sequential electrophoretic steps:

1. The first dimension, during which the protein mixture to be analyzed is separated by electrophoresis in an agarose gel.
2. The second dimension, during which the separated proteins are electrophoresed at right angles into a freshly applied layer of agarose containing a predetermined amount of antibody.

During the second electrophoretic step, the proteins migrate into the gel until they are completely precipitated by the antiserum, each protein forming a separate precipitate peak. This technique exploits the fact that at pH 8.4 gamma globulins migrate toward the cathode. Antibody and antigen therefore migrate toward each other and form precipitation peaks of antigen–antibody complexes. The area under any peak is directly proportional to the concentration of that protein in the sample and inversely proportional to the concentration of specific antibody in the gel.

This technique was originally introduced as a method for quantitatively measuring changes in serum protein patterns in human disease and has proved particularly successful in this respect (1–6). As many as 40 different serum proteins may be quantitated in a single experiment. As a demonstration of this technique, therefore, this chapter will describe the 2-D immunoelectrophoresis of a serum sample. This method can be applied, however, to a range of problems in the field of molecular biology. It is important to remember that the choice of protein(s) to be quantitated is controlled by the antiserum used in the second dimension. For example, isoenzymes will normally separate during electrophoresis in the first dimension. Using the appropriate antiserum (which will recognize all isozymes) in the second dimension will result in overlapping peaks corresponding to each isozyme, the area under each peak being proportional to the concentration of each isozyme. Similarly, the relative proportions of different proteins (e.g., lens crystallin proteins) at different stages of tissue development can be quantitated with this technique. Other examples of where this technique has been used include the measurement of protein complex formation (7), the differentiation of morphologically similar species (8), the determination of the antigenic composition of cell-wall extracts (9), the measurement of protein expression in different tissue types (10), the detection of variant forms of proteins (11), the measurement of enzyme–inhibitor complexes (12), screening for the presence of, and quantitation of, specific proteins in different species (13), the binding of radiolabeled ligands to carrier proteins (14), the study of the structure and morphogenesis of viruses (15), and the quantitation of protein synthesis during larval development (16).

2. Materials

1. Barbitone buffer (0.06M, pH 8.4) is prepared from sodium barbitone and barbitone. Dilute this buffer 1:1 with distilled water for use in the electrophoresis tanks.
2. Agarose, 2% w/v (aqueous).
3. Protein stain: 0.1% Coomassie Brilliant Blue in 50% methanol/10% acetic acid. (Dissolve the stain in the methanol component first, and then add the appropriate volumes of acetic acid and water.) Destain: 10% methanol/7% acetic acid.
4. Glass plates, 5 x 5 cm, 1.0–2.0 mm thick.
5. Bovine serum.
6. Reference serum: bovine serum containing 1 drop of concentrated bromophenol blue solution.
7. Anti-bovine serum.
8. Flat-bed electrophoresis apparatus with cooling plate.

3. Method

1. Dissolve the 2% agarose by heating to 100°C, then place the bottle in a 52°C water bath. Stand a 5-mL

pipet in this solution and allow time for the agarose to cool to 52°C. At the same time the pipet will be warming.

2. Place a test tube in the 52°C water bath and add 2.8 mL of 0.06*M* barbitone buffer to the test tube.

3. Leave for 3 min to warm, then add 2.8 mL of agarose solution. This transfer should be made as quickly as possible to avoid the agarose setting in the pipet. Briefly Whirlymix the contents and return to the water bath.

4. Thoroughly clean a 5 x 5 cm glass plate with methylated spirit. When dry, put the glass plate on a leveling plate or level surface.

5. Remove the test tube from the water bath and pour the contents of the tube onto the glass plate. Keep the neck of the tube close to the plate and pour slowly. Surface tension will keep the liquid on the surface of the plate. Alternatively, tape can be used to form an edging to the plate.

6. Allow the gel to set for 5 min. While the gel is setting, pour 0.03*M* barbitone buffer into each reservoir of the electrophoresis tank and completely wet the wicks. The wicks are prepared from six layers of Whatman No. 1 filter paper. They should be cut to exactly the width of the gel to be run, and should be well wetted with barbitone buffer.

7. Make two holes in the gel 0.8 cm in from one edge of the plate and approximately 1.5 cm in from the side of the plate. This is most easily done by placing the gel plate over a predrawn (dark ink) template (such as the one shown in Fig. 1), when well positions can easily be seen through the gel. The wells can be made using a Pasteur pipet or a piece of metal tubing attached to a weak vacuum source (e.g., a water pump).

First dimension electrophoresis

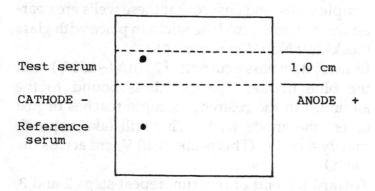

After electrophoresis the gel is cut where indicated by the dotted line.

Second dimension electrophoresis

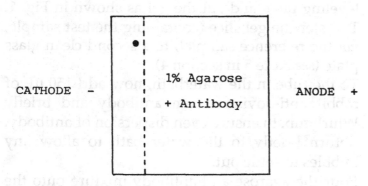

Strip cut from first dimension

Fig. 1. Diagram showing sample loading and gel-slicing positions.

8. Place 0.5 µL of serum (bovine) in the test well and 0.5 µL of reference serum in the other well. "Top up" the wells with electrode buffer (~ 1 µL).

9. Place the gel plate on the cooling plate of an electrophoresis tank, and place the electrode wicks over

the edges of the gel. Take care not to overlap the sample wells, and ensure that these wells are nearest the cathode. Hold the wicks in place with glass blocks (*see* Note 4 in section 4).

10. Immediately pass a current of 20 mA (~250 V) until the blue marker dye (which is bound to the albumin) in the reference sample reaches or just enters the anode wick. This will take approximately 40 min. (This is about 10 V/cm across the plate.)

11. Toward the end of this run, repeat steps 2 and 3, using a new test tube, but this time use only 2.2 mL each of buffer and agarose. (You will be covering a smaller surface of glass plate with this second dimension solution.)

12. When the plate has run, return the plate to the leveling table and cut the gel as shown in Fig. 1. Transfer the gel slice (containing the test sample, *not* the reference sample!) to a second clean glass plate (*see* Note 5 in section 4).

13. To the tube in the waterbath, now add 150 µL of rabbit anti-bovine serum antibody and briefly Whirlymix to ensure even dispersion of antibody. Return briefly to the water bath to allow any bubbles to settle out.

14. Pour the agarose and antibody mixture onto the second plate adjacent to the strip of gel and allow the gel to set for 5 min.

15. Place the plate in the electrophoresis apparatus as before and run at 3.0 mA/plate overnight (1 V/cm), making sure that the sample strip is at the cathode end (i.e., protein separated in the first dimension now move toward the anode).

16. At the end of the overnight run, precipitation peaks will be seen in the gel. These are not always easy to

visualize and are best observed using oblique illumination of the gel over a dark background (e.g., using a dark background illuminator).

17. A more clear result can be obtained by staining these precipitation peaks with a protein stain. First the antiserum in the gel, which would otherwise stain strongly for protein, must be removed. One way of doing this is to wash the gel with numerous changes of saline (0.14M) over a period of 2–3 d. The washed gel may then be stained directly or dried and then stained, to give a permanent record. To dry the washed gel, place it on a clean sheet of glass or gel bond, wrap a piece of wet filter paper around the gel, and place it in a stream of warm air for about 1 h. Wet the paper again with distilled water and remove it to reveal the dried gel fixed to the glass plate. Any drops of water on the gel can be finally removed in a stream of warm air.

18. A more convenient and quicker method of preparation is to blot the freshly run gel dry. With the gel on a sheet of clean glass or gel bond, place eight sheets of filter paper (Whatman No. 1) on top of the gel and apply a heavy weight (lead brick or beaker of water) for 1 h. After this time, carefully remove the now wet filter papers to reveal a flattened and nearly dry gel. The last sheet of filter paper may be stuck to the gel and should be thoroughly wetted with distilled water before removing. Complete the drying process by heating the gel in a stream of hot air for about 1 min. When completely dry, the gel will appear as a glassy film.

19. Whichever method of gel preparation is used, the gel can now be stained by placing in stain (with shaking) for 10 min, followed by washing in destain. Blue peaks will be seen on a clear back-

ground. Additional peaks that were not apparent
when viewed as immunoprecipitates should now
be evident.

4. Notes

1. When possible, the gel plates used should be as thin
    as possible to maximize the effect of the cooling
    plate. This is particularly important in the first
    dimension run when considerable heat is evolved.
2. The amount of antiserum to be used in the gel de-
    pends of course on the antibody titer and the
    amount of protein run in the first dimension, and is
    best determined by trial and error. If peaks go off
    the top of the gel, then either the antiserum concen-
    tration must be increased in the second dimension
    or the amount of protein run in the first dimension
    reduced. The reverse applies if the peaks produced
    are too small for accurate measurement. The
    amount of antiserum quoted in the section 3 (150
    μL) is the volume of commercially available anti-
    bovine serum used in our laboratory when produc-
    ing 2-D immunoelectrophoresis gels from bovine
    serum. This is an ideal test system for someone
    setting up the technique for the first time.
3. The working temperature for the agarose gel (52°C)
    is just above the setting temperature of agarose, and
    great care must be taken to ensure that agarose sol-
    utions are not allowed to drop below this tempera-
    ture prior to pouring the gel. This temperature is
    chosen because we wish to minimize the change of
    denaturing the antiserum, which would be consid-
    erably less stable at higher temperatures, but of
    course the agarose must stay in solution. In any
    case, once the antiserum has been added to the

agarose, the gel should be poured immediately to prevent the antibody from being exposed unnecessarily to high temperatures.

4. Because of the thickness of the electrophoresis wicks used, they are quite heavy when wet and have a tendency to "slip off" the gel during the course of the run. This can be prevented by placing thick glass blocks on top of the wicks, where they join the gel, or a heavy glass sheet across both wicks. This, by also covering the gel, has the added advantage of reducing evaporation from the gel.

5. When cutting the gel strip from the first dimension, make sure that sufficient gel remains adjacent to the sample track to allow for the placing on a wick on this gel in the second dimension, i.e., cut a gel slice with the sample to one side of this slice (*see* Fig. 1). It is important that the wicks do not cover the sample when running the second dimension.

6. A uniform field strength over the entire gel is critical in rocket immunoelectrophoresis. For this reason the wick should be exactly the width of the gel.

7. When setting up the electrophoresis wicks, it is important that they be kept well away from the electrodes. Since the buffer used is of low ionic strength, if electrode products are allowed to diffuse into the gel, they can ruin the gel run. Most commercially available apparatus is therefore designed such that the platinum electrodes are masked from the wicks and also separated by a baffle. To minimize these effects, the use of large tank buffer volumes (800 mL per reservoir) is also encouraged.

8. The time quoted for electrophoresis is suitable for most samples. For proteins with low electrophoretic mobility, however, the times given for the elec-

trophoretic runs may be too short. In the protocol described, the bromophenol blue spot that is visualized in the first dimension is not "free" bromophenol blue, but the dye bound to albumin. Since albumin is one of the fastest-running proteins in the first dimension (only a few minor prealbumin proteins run faster), when the blue dye reaches the anode wick we know that the majority of the serum proteins are now spread along the full length of the gel. When studying other protein mixtures, however, the appropriate time for the first dimension run should be determined in initial trial experiments. Free bromophenol blue can be used as a marker dye to indicate the staus of an electrophoresis run at any given time. Similarly, in the second dimension, the time of electrophoresis given may not be long enough to allow for complete development of precipitation peaks for proteins with low electrophoretic mobility. Do note that it is not possible to "overrun" the second dimension electrophoresis. Once the precipitation peaks are formed, they are quite stable to further electrophoresis.

9. If no results are obtained, check that the proteins being analyzed run to the anode under the conditions used. This assumption is implicit in the method described here.

10. Some workers include 3% w/v polyethylene glycol 6000 in the gel, which is said to enhance precipitation of antigen–antibody complexes and thus increase sensitivity.

11. For extreme sensitivity, a silver stain method for detecting immunoprecipitates in agarose gels has been described (17).

12. An excellent review, with appropriate photographs, of problems that can arise with immuno-

electrophoretic techniques is given in Axelsen (1983) (*see* Further Reading).

Further Reading

Axelsen, N.H., Kroll, J., and Week, B., eds. (1973) *A Manual of Quantitative Immunoelectrophoresis—Methods and Applications* Universitatsforgleget, Oslo.

Axelsen, N.H., ed. (1983) *Handbook of Immunoprecipitation-in-Gel Techniques* Blackwell Scientific, Oxford, London.

References

1. Laurell, C-B. (1965) Antigen-antibody crossed electrophoresis. *Anal. Biochem.* **10**, 358.
2. Peeters, H., ed. (1967) A Quantitative Immunoelectrophoresis Method, in *Protides of the Biological Fluids* Elsevier, Amsterdam.
3. Clarke, H. and Freemann, T. (1968) Quantitative immunoelectrophoresis of human serum proteins. *Clin. Sci.* **35**, 403–413.
4. Weeke, B. (1970) The serum proteins identified by means of the Laurell crossed electrophoresis. *Scand. J. Clin. Lab. Invest.* **25**, 269–275.
5. Davies, D., Spurr, E., and Versey, J. (1971)Modifications to the technique of two-dimensional immunoelectrophoresis. *Clin. Sci.* **40**, 411–417.
6. DAKO Information Sheet No. 4. Provided by DAKO-immunoglobulins a/s, Copenhagen, Denmark (or local supplier).
7. Podack, E.R., Dahlback, B., and Griffin, J.G. (1986) Interaction of S-protein of complement with thrombin and antithrombin III during coagulation protection of thrombin by S protein from antithrombin III inactivation. *J. Biol. Chem.* **261**, 7387–7392.
8. Klima, F. (1985) Immunochemical investigation to the systematics of larvae of the genus *Hydropsyche trichoptera hydropsychidae*. *Entomol. Nachr. Ber.* **29**, 161–169.
9. Cox, R.A.C. and Britt., L.A. (1985) Antigenic heterogeneity of an alkali-soluble water-soluble cell wall extract of coccidioides-immitis. *Infect. Immunol.* **50**, 365–369.

10. Martin, B., Vranckx, R., Denoulet, P., and Nunez, E.a. (1985) Alpha fetoprotein expression in intraembryonic and extra-embryonic fluids of developing chick embryo. *Dev. Biol.* **111**, 352–358.

11. Howarth, D.J., Samson, D., Stirling, Y., and Seghatchian, M.J. (1985) Antithrobin III Northwick Park: A variant antithrombin with normal affinity for heparin but reduced heparin cofactor activity. *Thromb. Haemostasis* **53**, 314–319.

12. Stockley, R.A. and Afford, S.C. (1984) Effect of leuocyte elastase on the immunoelectrophoretic behavior of Alpha-1 antitrypsin. *Clin. Sci.* (Lond.) **66**, 217–224.

13. Townsend, R. and Archer, D.B. (1983) A fibril protein antigen specific to spiroplasma. *J. Gen. Microbiol.* **129**, 199–206.

14. Scott, B.J., Bateman, J.E., and Bradwell, A.R. (1982) Detection of tritium labeled ligands and their carrier proteins using a multi wire proportional counter. *Anal. Biochem.* **123**, 1–10.

15. Milleville, M., Cousin, C., and Boulanger, P. (1979) Contribution of 2-dimensional immunoelectrophoresis techniques to the study of the structure and morphogenesis of human adenovirus. *CR Seances Soc. Biol. Fil.* **133**, 288–302.

16. Chalaye, D. (1979) Immunochemical study of the hemolymph and oocyte proteins of *Rhodnius-prolixus. Can. J. Zool.* **57**, 329–336.

17. Porro, M., Viti, S., Antoni, G., and Saletti, M. (1982) Ultrasensitive silver-stain method for the detection of proteins in polyacrylamide gels and immunoprecipitates on agarose gels. *Anal. Biochem.* **127**, 316–321.

Chapter 22

Peptide Synthesis

Brian Austen

1. Introduction

Peptides are required for many different aspects of biotechnology. Since DNA sequences are generally functional only in the protein for which they code, the explosion in sequences that are known has led to an increase in the requirement for synthetic peptides. Quite often, the peptide will be haptenized to a protein carrier and then used to raise antibodies that will then recognize the parent protein. In other instances, synthetic peptides with biological activity will be required in high purity for testing in functional assays. The synthesis of peptide analogs that differ from the naturally occurring sequence in key residues are also often in demand. The methods described here are suitable for making peptides up to about 20 residues in length and do not require sophisticated equipment or highly toxic chemicals such as hydrogen fluoride.

A rapid method of synthesis of peptides on a solid-phase support will be described. The essential feature of this method is the use of a polymer that functions to protect the C-terminal carboxyl and at the same time renders the growing peptide completely insoluble. The polymer is permeable to the various reagents and solvents used so that excess reagents may be removed by filtration and washing. The method was first developed by Merrifield (1). A number of different variations of methods of solid phase synthesis are in use in many laboratories throughout the world. These methods have a common strategy in that the peptide is assembled from the C-terminal end, and the C-terminal amino acid with its α-amino group protected is first attached through an ester linkage to the solid-phase polymer. The protecting group is then chemically cleaved off, and the next amino acid, again with its α-amino group protected, is coupled via peptide linkage to the α-amino group of the insolubilized C-terminal residue. This cycle of deprotection and coupling an additional amino acid residue is repeated until the peptide is fully assembled on solid phase. Since the side chains of many amino acids would react with the reagents used in making the peptide bonds, they are first protected with chemical groups that are stable to the conditions used to deprotect the α-amino group, but may be removed at the termination of the synthesis.

In many laboratories, a strategy initially developed by Meienhofer and coworkers (2), and also by Sheppard and colleagues at the MRC laboratories in Cambridge (3), is widely used. In this, temporary protection of the α-amino group is achieved by the base-labile 9-fluorenylmethoxycarbonyl (Fmoc) group, first described by Carpino and Han (4). Side chain protecting groups are chosen so that they may be removed by mildly acidic conditions at the end of the synthesis. The growing peptide is attached to the solid-phase support by a variety of different types of linkers, one of which is cleaved under the same acid conditions as side chain protectants to yield C-terminal acids, and another which is cleaved with ammonia to yield a C-terminal amide or hydrazine to yield C-terminal hydrazides.

The Cambridge group has developed the use of functionalized polyamide gel as a solid-phase support. This support has the advantage that it is fairly hydrophilic and swells dramatically in dimethylformamide (DMF), a solvent favored for peptide synthesis. Kent (5) has shown, however, that polystyrene is equally good as a solid-phase support providing the correct choice of solvent (DMF) is made. Polyamide has the disadvantage that it is more difficult to handle, sticks to glass, tends to be difficult to wash, and has a tendency to become granular as the synthesis proceeds. Moreover, only polystyrene is suitable for the T-bag method (*see* next chapter). In consequence we use polystyrene with an acid-sensitive linking group such as that present in the *p*-alkoxybenzyl alcohol resin of Wang (6), in which the peptide-resin bond is labilized by the electron-donating ability of the para-ether oxygen. Peptide amides or hydrazides may be synthesized on 4-(hydroxymethyl)-benzoylmethyl resin in which the peptide-to-resin bond is labilized to nucleophiles by the *para*-ester group. Norleucine, included by Sheppard as an internal linker to aid analysis of the degree of substitution, may be dispensed with since it is easy to weigh small samples of polystyrene prior to amino acid analysis.

2. Materials

2.1. Chemicals

1. 4-Alkoxybenzyl-resin. (4-(Hydroxymethyl) phen–oxymethyl-copolystyrene-1% divinylbenzene)

(200–400 mesh) resin is suitable for preparation of peptide acids (0.7 mEq/g reactive groups). This resin may be obtained from Nova Biochem, Peninsula Labs, Bachem, or Pierce.

This resin may also be more cheaply prepared from Merrifield resin (chloromethylated copolystyrene-1% divinylbenzene, about 0.9 mEq/g, 200–400 mesh), also obtained from these suppliers (6). Merrifield resin (5.1 g; 4.6 mmol) in 25 mL of dimethylacetamide is allowed to react with 4-hydroxybenzyl alcohol (0.74 g; 5.9 mmol) and sodium methoxide (0.32 g; 6.1 mmol) at 80°C for 24 h. The resin is collected and washed with dimethylformamide, dioxane, dichloromethane, and methanol.

2. 4-(Hydroxymethyl)-benzoyloxymethyl-copoly (styrene-1% divinyl benzene) resin is suitable for preparation of peptide amides or hydrazides (0.7 mmol/g). It may be obtained from Nova Biochem.

3. Fmoc amino acids may be obtained from Cambridge Research Biochemicals (CRB), Bachem, or Peninsula. It is possible to purchase preactivated derivatives (pentafluorophenyl esters) that are now available from CRB. Reaction with these esters are, however, reportedly less rapid than using carbodiimide, but are necessary for coupling Fmoc-Gln, Fmoc-Asn, and Fmoc-Lys (Boc). As more chemical companies produce Fmoc amino acid derivatives, it is hoped that prices will fall. Alternatively, they may be readily synthesized from free or side chain-protected amino acids by the method of Paquet (7), using 9-fluorenylmethyl succinimidyl carbonate and an organic extraction at low pH to isolate the product. The purity of the products may be established by reverse-phase HPLC on a C8 column (e.g., Brownlee RP-300) in 0.1% TFA with

an acetonitrile gradient; they are eluted at quite high concentrations (about 40%) of acetonitrile. The following side chain-protected amino acids are recommended:

Fmoc-Arg(Mtr) (mesitylene trisulfonyl)

Fmoc-Asp(OtBu) (*O-t*-butyl ester)

Fmoc-Glu(OtBu) (*O-t*-butyl ester)

Fmoc-Cys(Trt) (Trityl). This protecting group comes off with trifluoracetic acid to give the free sulfhydryl, which may then be used to couple directly to protein carriers via the bifunctional linker *m*-maleimidobenzylsuccinimide (*8*).

Fmoc-Cys(Acm) (Acetamidomethyl). This group is stable to the trifluoracetic acid used to cleave the peptide from the resin and is useful if extensive purification of the peptide is required. After purification the Acm group may be removed by treatment with mercuric acetate (see below).

Fmoc-Lys(Boc)-Pfp ester. The pentafluoro-phenyl ester of ε-*N-t*-butoxycarbonyl-Lys). Obtained from CRB.

Fmoc-Asn or Gln-Pfp ester. Obtained from CRB.

Fmoc-Ser(tBu) (*O-t*-butyl ether)

Fmoc-Thr(tBu) (*O-t*-butyl ether)

Fmoc-Tyr(tBu) (*O-t*-butyl ether)

Fmoc-His(Bum)[$N(\pi)$-*t*-butoxymethyl] (Nova Biochem)

All other Fmoc amino acids are available from CRB or Bachem. In our experience both companies produce high-purity products. These derivatives are very slightly cheaper from Peninsula, but our experience with this supplier is very limited.

4. α-*N*-Boc-amino acids. These are added as the N-terminal amino acid only. They are cheaper than

Fmoc-amino acids, and may be obtained from Fluka, Bachem, or Sigma.

5. Dimethylformamide (DMF). This solvent must be free of dimethylamine, an impurity that is often present. The quality of DMF may be assessed by a colorimetric reaction with dinitrofluorobenzene (FDNB). Mix equal volumes of DMF with a fresh FDNB solution (1 mg/mL) in 95% ethanol, let stand for 30 min, and read the absorbance at 381 nm. Blank FDNB (0.5 mg/mL) usually reads 0.2. Good quality DMF should have an OD of no more than 0.15 higher than the blank. Reagent grade DMF may be purified by distillation from ninhydrin under vacuum from a water pump, using a nitrogen bleed, and stored over molecular sieve 4A (BDH Chemicals). Alternatively, purchase the Puriss grade from Fluka, add about 1 in of molecular sieve 4A to the bottom of a 21 bottle, flush with nitrogen, and store sealed in the cold room for at least 4 d before use.

6. Dichloromethane (DCM). Obtained from Fisons. Reagent grade may be used without purification.

7. Piperidine. Obtained from Fluka. Distill this reagent from potassium hydroxide in a good fume-cupboard.

8. Trifluoroacetic acid (TFA). This should be free of aldehydes; distillation should be avoided because of the danger of forming anhydrides. An acceptable grade can be purchased from Kali-Chemie, West Germany, or Peboc Ltd., Llangefni, Anglesey.

9. Diisopropylcarbodiimide (Fluka).

10. 1-Hydroxybenzotriazole (BDH).

11. Dimethylaminopyridine (BDH).

12. N-Methyl-morpholine (Fluka). Distill from ninhydrin.

13. Methanol (UV-grade) (Fisons).

14. Siliconizer (BDH).
15. Pyridine (BDH). Distill from ninhydrin in a fume cupboard.
16. Benzoyl chloride (Fluka).
17. Diisopropylethylamine (Fluka). Distill from ninhydrin.
18. *t*-Amyl alcohol (BDH).
19. *N*-Acetylimidazole (BDH).

2.2. Apparatus

1. Synthesis glass vessel fitted with No. 3 glass sintered frit and No. 14/20 Quickfit ground-glass joint (*see* Fig. 1). The dimensions suitable for 2 g of resin would be about 40 mm across and 75 mm in length.
2. Motorized shaker capable of rotating the shaker through about a 105° arc. This can be made by any competent workshop using a 24 rpm motor attached by a crank to a disc that is itself attached to a laboratory clamp. Commercial shakers are apparently available from Sequemat, 109 School St., Watertown, Massachusetts, 02172.
3. General apparatus, i.e., oil pump, water pump, rotary evaporator, fraction collector, centrifuge, spectrophotometer.
4. Access to an amino acid analyzer and HPLC.

2.3. Chromatographic Materials

1. Ion-exchange cellulose (CM52 or DE52) from Whatman.
2. Sephadex G-15 and SP-Sephadex (C-25) from Pharmacia.
3. Reverse-phase column (4.6 x 250 mm). Brownlee RP-300 (Anachem).
4. Amberlite MB-3 mixed ion-exchange resin from BDH.

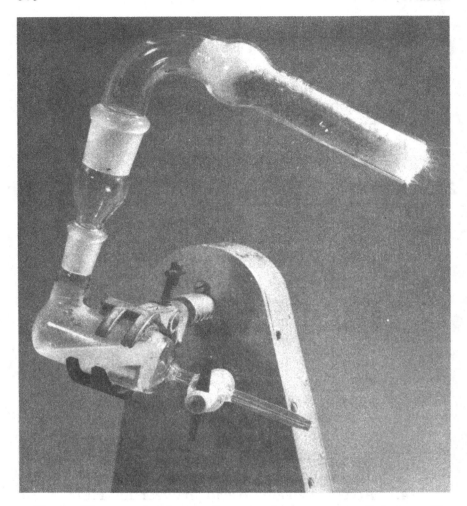

Fig. 1. Glass synthesis vessel attached to a motorized shaker. The vessel contains polystyrene resin and is fitted with a silica gel drying tube.

3. Ninhydrin Monitoring

The extent of coupling may be monitored by the color developed by the resin with a ninhydrin reagent, which is indicative of the amount of free amino group present. The conditions are described by Sarin et al. (9).

Ninhydrin reagents—stock solutions are prepared as follows:

Solution A. Mix 40 g of reagent grade phenol with 10 ml of absolute ethanol. Warm until dissolved. Stir with 4 g of Amberlite MB-3 resin for 45 min. Filter. Add this to a second filtered solution made by adding 2 mL of 100 mL of water containing 65 mg of KCN to 100 mL of pyridine, and also stirring with 4 g of Amberlite MB-3 resin.

Solution B. Dissolve 2.5 g of ninhydrin in 50 mL of absolute ethanol. Stopper and store in the dark under nitrogen.

Take enough suspended resin to half fill the stem of a small Pasteur pipet. Transfer to a plastic tube and wash three times with methanol. Dry the resin in a vacuum desiccator for 15 min, then weigh small samples of about 2 mg into plastic tubes, add 125 µL of solution A ninhydrin reagent and 32 µL of solution B. Heat in a boiling water bath for 5 min. Cool. The blue color present at this stage is generally sufficient for a visual qualitative assessment of coupling, but for quantitation proceed as follows. Add 1 mL of 60% ethanol and elute from a Pasteur pipet plugged with a small amount of glass wool. Rinse with 0.2 mL of 0.5M tetraethylamine hydrochloride in DCM, and then rinse with 1 mL of 60% ethanol. Dilute with 10 mL of 60% ethanol. Calculate the amount of unreacted group using the extinction coefficient of a free amino group of $1.5 \times 10^4/M/\text{cm}$ as a percentage of the amount of free amino group measured on piperidine-treated resin after coupling the C-terminal residue as described below.

4. Methods

4.1. Attachment of the First Residue

This method is appropriate for both acid- and base-labile resins. Calculate the amount of resin required for the amount of peptide you hope to get. About 0.6 mmol of peptide per g of resin is the maximum to expect from a synthesis after cleavage from the resin, but this may be reduced to 0.1 mmol after extensive chromatographic purification of a difficult sequence. A convenient amount of resin to work with is 2 g.

1. First clean the synthesis vessel with chromic acid, wash extensively with water, dry, and coat with a siliconizing agent. Dry again in an oven. Place the vessel on the shaker and add 2 g of resin.

2. Fully suspend the resin with DCM (about 30 mL for 2 g), shake for 2 min, and with a drying tube containing silica gel on the quickfit socket (Fig. 1) and the vessel in the vertical position, draw off the solvent by connecting the tubing from a water pump vacuum to the outlet tube via a trap consisting of a 2 l Buchner flask with a bung and outlet. This is repeated for all further washes.

3. Wash twice more with DCM and once with DCM/ DMF (1/1 mixture) and draw off the excess solvent.

4. Molar amounts of protected amino acids, coupling agents, and catalyst for addition of the first residue added are calculated from the number of moles of the alkoxy-substituent on the resin. Dissolve a sixfold molar excess of the Fmoc derivative of the C-terminal residue in a volume of DMF sufficient to swell the resin, i.e., for 2 g of the Bachem resin with 0.7 mmol/g substitution, 8.4 mmol of the Fmoc amino acid in 6 mL DMF, and add to the resin.

5. Add a sixfold molar excess of diisopropylcarbodiimide in 5 mL of DCM, shake for 2 min, and add a sixfold molar excess of N-methylmorpholine and 0.6-fold molar amount of dimethylaminopyridine in 1 mL of DMF, which act as catalysts for formation of the C-terminal ester with the resin, and shake for 1 h.

6. Wash once with DCM:DMF (1:1), (12 mL) and four times with DCM (16 mL each wash). If Asn, Gln, or Lys(Boc) is the C-terminal residue, they should be incorporated as the Fmoc pentafluorophenyl ester, using sixfold molar excess without the carbodiimide, and adding first a sixfold molar excess of N-hydroxybenzotriazole in 10 mL of DMF to the resin and shaking for 2 min, then adding a sixfold molar excess of the pentafluorophenyl ester in another 2 mL of DMF to the resin and shaking for another 2 min, and finally adding N-methylmorpholine (sixfold excess) and dimethylaminopyridine (0.6-fold amount) as catalysts and reacting for 3 h.

4.2. Measuring the Substitution of the C-Terminal Residue

1. To measure the extent of incorporation of the C-terminal amino acid, take a small amount of the resin, enough to half fill the stem of a short Pasteur pipet, transfer to a tube, remove excess DCM, and add about 0.5 mL of 20% piperidine in DMF.

2. Leave for 5 min, then remove excess reagent, and wash several times with DCM, then methanol.

3. Perform the monitoring ninhydrin reaction as described above.

4. To check by an independent method, hydrolyze a weighed sample of the deprotected resin with a 1:1 mixture of concentrated HCl and propionic acid in an evacuated tube at 130°C for 3 h, dry in vacuo, and

subject the residue to amino acid analysis. If the extent of substitution is less than 0.3 mmol/g, then repeat the coupling reaction with the first residue.

5. Finally, block remaining reactive groups having washed the resin three times with DCM by a 1-h incubation with 10-fold molar excesses of benzoyl chloride and pyridine in DCM. Wash out three times with DCM. The molar excesses of reagents described in the coupling methods below are relative to the amount of C-terminal residue incorporated.

4.3. Coupling Successive Residues

1. Deprotection and coupling cycles are initiated for addition of further residues, building toward the N-terminal end of the peptide. A general procedure suitable for most types of amino acid residues except Gln, Asn, and Lys(tBoc) will be described. It would be preferable to transfer the apparatus to a fume cupboard for handling DCM and piperidine, which are toxic.

2. The volume of each wash used below is enough to fully suspend the resin.

    Step 1. Wash with DMF, 3 x 2 min.

    Step 2. Wash with DCM, 3 x 2 min.

    Step 3. Wash with DMF, 3 x 2 min.

    Step 4. Deprotect with 20% piperidine, 2 x 5 min.

    Step 5. Wash with DMF, 3 x 2 min.

    Step 6. Wash with DCM, 3 x 2 min.

    Step 7. Wash with DCM:DMF (1:1 by vol), 2 x 2 min.

    Step 8. Add fivefold molar excess of Fmoc-amino acid over the amount of C-terminal residue on the resin in 6 mL of DMF, shake for 2

shake for 2 min, then add fivefold excess of diisopropylcarbodiimide in 6 mL of DCM. Shake for 1 h. Test for coupling with ninhydrin reagent. The reaction may be performed qualitatively, judging any blue color formed by eye.

Step 9. If the test gives no color, wash with DCM/DMF (1/1 by vol), 2 x 2 min, and proceed with the next cycle. If the test gives a blue color, then refer to Note 1 in section 5.1.

4.4. Coupling Fmoc-Asn, Fmoc-Gln, or Fmoc-Lys(Boc)

1. Side reactions occur if these residues are coupled by carbodiimide, so couple them instead as pentafluorophenyl esters (fivefold excess) (obtained from CRB).

2. At step 8 on the coupling cycle, first shake the resin for 2 min with fivefold excess of 1-hydroxybenzotriazole in about 10 mL of DMF, then add the pentafluorophenyl ester in 1–2 mL of DMF, and shake for 3 h.

3. Check for degree of coupling with the ninhydrin reagent, and repeat with a second batch of ester if the resin produces blue color.

4.5. Coupling the N-Terminal Amino Acid Residue

1. Since Boc-amino acids are cheaper that Fmoc-amino acids, savings can be made by adding the N-terminal amino acid residue at the final coupling as the α-N-Boc-amino acid with acid labile (e.g., Boc, O-t-butyl) side chain protecting groups. If Arg is the N-terminal residue, use Boc-Arg unprotected on the side chain.

2. Couple these residues using diisopropylcarbo-
    diimide as described for the Fmoc-amino acid resi-
    dues, and if Boc-Gln or Boc-Asn are at the N-term-
    inus, equilibrate the resin first with fivefold excess
    N-hydroxybenzotriazole, for several minutes be-
    fore adding Boc-Gln or Boc-Asn (fivefold excess)
    followed by diisopropylcarbodiimide. There are
    no problems with coupling Boc-Lys(Boc).

4.6. Cleavage and Deprotection of the Peptide

4.6.1. Acid-Labile Linker (Alkoxybenzyl-resin)

1. When the last residue has been coupled, and the
    resin washed, if a Fmoc-amino acid residue was
    added at the N-terminus, treat with 20% piperidine
    to step 4 to remove the Fmoc group on the N-term-
    inal residue as described above, then commence
    washing as follows (see section 4.2.2). If a Boc-
    amino acid residue was added, wash the resin after
    the final coupling as follows.
2. Wash three times with DMF, three times with
    DCM, three times with i-propanol, and three times
    with ether.
3. Leave the vessel with a drying tube in place on the
    water-pump vacuum for 10 min, then evacuate in a
    desiccator for 30 min on an oil pump.
4. Replace on the shaker and shake for 2 h with 50%
    (v/v) TFA, 5% (v/v) anisole in DCM (total 20 mL),
    and if the peptide contains Met, 0.5 mL of ethyl
    methyl sulfide. If the peptide contains Trp, include
    0.9% ethanedithiol along with anisole.
5. If Arg(Mtr) is present, then the cleavage/deprotec-
    tion conditions should be more drastic. Deprotect
    with 50% TFA in DCM for 6 h, or if the peptide has

Arg(Mtr), but not Met or Trp, deprotect with 50% (v/v) TFA plus 5% (v/v) thioanisole in DCM (total vol 20 mL) for 90 min. Collect the filtrate, add another 20 mL of 50% TFA (v/v) in DCM, shake for another 2 min, and combine the filtrates. TFA causes nasty burns, so wear gloves and goggles.

6. Rotary evaporate at 30°C and add ether to the remaining oil to give the peptide as a white solid. Wash the solid several times with ether, recovering the solid in a spark-free centrifuge. Vacuum desicate over NaOH pellets and weigh the recovered peptide.

4.6.2. Base-Labile Linker 4-(Hydroxymethy)-benzoyloxymethyl Resin

1. The peptide linked to the resin by the base-labile group must be treated with ammonia after TFA to yield a C-terminal amide. Treat the peptide-resin with 50% TFA as described in section 4.6.1. to remove protecting groups. Discard the filtrate and wash again with 50% TFA. Discard filtrate.

2. Wash the resin with DCM (three times), *t*-amyl alcohol (three times), 10% diisopropylethylamine in DCM (three times), DMF (three times), and ether (three times).

3. Evacuate first on a vacuum pump with the drying tube in place to let the ether evaporate, and then vacuum desiccate for 90 min over P_2O_5 on an oil pump.

4. In a fume cupboard, collect ammonia gas from a cylinder in a three-necked, round-bottomed flask containing 30 mL of HPLC-grade methanol cooled in ice, passing the gas first over KOH pellets in a gas bottle. The methanol should be protected from the

atmosphere by a KOH drying tube. Pass the gas until it no longer dissolves in the methanol, i. e., bubbles are seen.

5. Transfer the resin to a pressure bottle equipped with a magnet, pour on the ammonia-saturated methanol, stir for 24 h at room temperature. Cool again in ice before opening the bottle in a fume cupboard.

6. Filter off the resin, wash with further 10 mL of methanol, and rotary evaporate the combined filtrates. In case the peptide has precipitated, wash the resin further with 5% acetic acid, and rotary evaporate.

The use of the base-labile linker makes it possible to cleave the peptide with protecting groups still attached. This peptide may be then coupled to a free amino group in a second peptide, protein, or other substance, and the product finally deprotected by treatment in 50% TFA. This strategy, known as "fragment condensation," is often used to prepare peptides too long to put together by repetitive attachment of individual residues. The C-terminal fragment peptide with free amino group must be suitably protected on other reactive groups. This second peptide may be synthesized by the general procedure described in this chapter on the base-labile linker, using an Fmoc-amino acid for the N-terminal residue instead of a Boc residue. The N-terminal Fmoc group is removed by treatment with 20% piperidine and washed as described in section 4.6.1. The peptide is then cleaved as a C-terminal amide by treatment with ammonia as described in the preceeding paragraph, but without prior treatment with TFA.

7. To obtain from the base-labile resin a protected peptide fragment activated at its C-terminus for

coupling to an amino group in the C-terminal fragment, couple the α-N-Boc-protected amino acid
derivative as the protected N-terminal residue in
place of an Fmoc-derivative as described above.
Then instead of treating with piperidine or TFA,
wash the resin several times with DCM, then DMF.

8. Suspend the resin in DMF (5 mL/g), and add an-
hydrous hydrazine (Pierce) (wear gloves) (30 mol/
per mol of peptide). Stir in a closed container for 2
d.

9. Filter and wash with DMF.

10. Rotary evaporate the combined filtrate and washings under vacuum from a good oil pump.

11. The protected hydrazide formed may be converted
to an active azide before coupling to the amino
group of another peptide by the following procedure. To a solution of the hydrazide (100 mg in
about 4 mL DMF) cooled to –10°C, in an ice–salt
mixture, a solution of 1M HCl in DMF (2 mL) is
added, followed by isoamyl nitrite (0.26 mL). After
5 min at –10°C, the mixture is cooled to –30°C, in a
methanol-dry ice mixture, neutralized with trieth-
ylamine (0.26 mL), and added to the amino component in minimal volume of DMF. Leave for 2 d at
4°C. The product may be isolated by addition of
water. Protecting groups are then removed by incubation in 95% TFA and 5% water for 2 h. Rotary
evaporate at 30°C and precipitate the product with
ether.

4.7. Characterization and Purification

1. Some criteria of purity of the product must be
obtained. Amino acid analysis after acid (6M HCl)
or enzymic hydrolysis (aminopeptidase M) will
check the composition, and high-voltage electro-

phoresis of the product on Whatman 3MM paper will ascertain the heterogeneity in charge (*see* vol. 1 of this series). If HPLC is available, analysis by reverse-phase separation on a C8 300A silica column, eluting in 0.1% TFA with a gradient of acetonitrile, will indicate the degree of heterogeneity. Readers should refer to chapters (in vol. 1 of this series) on protein sequencing for detailed methods of peptide analysis and characterization.

2. If required, the peptide may be purified by ion-exchange chromatography, using either Whatman CM52 if the peptide is basic, eluting with ammonium acetate (pH 5.0), or if the peptide is acidic, chromatograph on Whatman DE52 resin, with ammonium bicarbonate (pH 8.0). Use the following procedure:

3. For 100 mg of crude peptide, place 40 g of CM52 on a sinter and slurry with $0.6M$ ammonium acetate (strong buffer). Adjust to pH 5.0 with acetic acid. Draw off and wash with $0.01M$ ammonium acetate (pH 5.0) (2.0 L) (dilute buffer). Slurry with this dilute buffer and pack into a 1.5 cm diameter column. Run the eluate through a UV monitor at 280 nm if the peptide contains Tyr or Trp, or 240 nm if not.

4. Pump dilute buffer at 30 mL/h until the conductivity of the eluate is close to that of the dilute buffer, then dissolve the peptide in water (5 mL), adjust pH to 5 with dilute acetic acid or ammonium hydroxide, and load onto the column. Collect fractions.

5. Pump dilute buffer for 1 h, then start a linear gradient with 200 mL of dilute buffer to 200 mL of strong buffer. Peptides are recovered by lyophilization and again checked for purity.

6. Acidic peptides are purified on DE52 in the same

way, using 0.01M ammonium bicarbonate (pH 8.0) as the dilute buffer and 0.6M ammonium bicarbonate as the strong buffer at pH 8.0.

7. If the peptide is insoluble at pH 5 or 8, it may be dissolve in 30% acetic acid and subjected to ion-exchange chromatography on SP-Sephadex (C25) suspended in 30% acetic acid, eluting with a gradient of NaCl up to 0.6M in 30% acetic acid using the same amounts and volumes as described for CM52. The resin is allowed to swell in excess 30% acetic acid (three times the expected suspended volume) overnight before use, and fines are removed by swirling up, allowing to settle for 20 min, and removing the excess supernatant. The resin is packed into a column and washed with 3 volumes of 30% acetic acid before loading the peptide. The peptides are desalted by gel filtration on Sephadex G-15 (4 x 60 cm) in 30% acetic acid, and the acetic acid removed by rotary evaporation.

4.8. Deprotection of Cysteine

If the acetamidomethyl group was chosen for protection of cysteine, it should be removed after purification. Dissolve the peptide in a minimum amount of water and adjust the pH to 4.0 with acetic acid or ammonia. Add a 10-fold molar excess over the acetamidomethyl content of mercuric acetate (BDH) and incubate for 3 h at room temperature. Then pass hydrogen sulfide from a gas bottle (obtainable from BDH) through the solution in a fume cupboard for 1 h. Centrifuge the suspension and isolate the product from the supernatant by chromatography on Sephadex G-15 in 2M acetic acid.

5. Notes

5.1. Incomplete Coupling

1.   If the ninhydrin test gives unacceptable blue color (>2% of total amino group prior to coupling), try adding fivefold molar excess of N-hydroxybenzotriazole in 1 mL of DMF, which will form an active ester *in situ*; incubate for a further hour and test again. If the resin still produces color, try adding 1 mL of 0.5M diisopropylethylamine in DCM to neutralize any protonated and therefore unreactive amino group. Shake for a further 15 min, then wash twice with DCM:DMF (1:1 by vol) (2×2 min). The peptidyl resin should not be left any longer with diisopropylethylamine for fear of removing the Fmoc group. If color with ninhydrin is still produced, then repeat steps 5–9 (section 4.3). If total coupling is still not achieved, then block by N-acetylation any remaining amino groups by 15-min incubation with 10-fold excess of N-acetylimidazole in DCM and proceed with the next cycle. This blocking reaction will terminate any unreacted peptide, which will thus not participate in further coupling steps. The resulting terminated peptide will be easier to separate out at the end of the synthesis than a form differing from the required peptide by the deletion of only one residue.

References

1.   Barany, G. and Merrifield, R. B. (1980) Solid-phase Peptide Synthesis, in *The Peptides* vol. 2 (Gross, E. and Meienhofer, J., eds) Academic, New York.

2. Meienhofer, J., Waki, M., Heimer, E. P., Lambros, T. J., Makofske, R. C., and Chang, C. -D. (1979) Solid-phase synthesis without repetitive acidolysis. *Int. J. Peptide Protein Res.* **13**, 35–42.

3. Arshady, R., Atherton, E., Gait, M. J., Lee, K., and Sheppard, R. C. (1979) Easily prepared polar support for solid phase peptide and oligonucleotide synthesis. Preparation of Substance P and a nonadeoxyribonucleotide. *J. Chem. Soc. Chem. Commun.* 423–425.

4. Carpino, L. A. and Han, G. Y. (1972) The 9-fluorenylmethoxycarbonyl amino-protecting group. *J. Org. Chem.* **37**, 3404–3409.

5. Kent, S. B. H. (1985) Difficult Sequences in Stepwise Peptide Synthesis; Common Molecular Origins in Solution and Solid Phase, in *Peptides; Structure and Function* Proc. 9th American Peptide Symposium. (Deber, C. M., Hruby, V. J., and Kopple, K. D., eds.) Pierce Chemical Co., Toronto, Ontario.

6. Wang, S. -S. (1973) p-Alkoxybenzyl alcohol resin and p-alkoxybenzyloxcarbonylhydrazide resin for solid-phase synthesis of protected peptide fragments. *J. Am. Chem. Soc.* **95**, 1328–1333.

7. Paquet, A. (1981) Introduction of protecting groups into O-unprotected hydroxyamino acids using succinimidyl carbonates. *Can J. Chem.* **60**, 976–980.

8. Green, N., Alexander, H., Olson, A., Alexander, S., Shinnick, T. M., Sutcliffe, J. G., and Lerner, R. A. (1982) Immunogenic structure of the influenza virus hemagglutinin. *Cell* **28**, 477–484.

9. Sarin, V. K., Kent, S. B. H., Tam, J. P., and Merrifield, R. B. (1981) Quantitative monitoring of solid-phase peptide synthesis by the ninhydrin reagent. *Anal. Biochem.* **117**, 147–157.

Chapter 23

Synthesis of a Series of Analogous Peptides Using T-Bags

Brian Austen

1. Introduction

To obtain a number of analogous peptides differing only in side chains in one or two positions, it may be convenient to perform the synthesis on resin confined to a number of porous bags (which are described as T-bags), so that when residues that are to be changed are coupled, a bag containing the peptidyl resin can be isolated from other bags and coupled to a different protected amino acid. At the completion of coupling, the bag may be returned to a bottle containing the other bags, and the synthesis continued. The principal and usefulness of this method was originally described by Houghton (1), who synthesized several hundred peptides at once in one step in order to explore the contribution of individual side chains in the specificity of antibody binding sites. The use of Fmoc chemistry and acid-labile resin means that peptides may be cleaved from the resin samples by individual reactions with TFA. If a method using tBoc chemistry is used as described by Houghton (1), then special apparatus to handle many individual reactions with HF to cleave the peptides from the resin at the end of synthesis is required. The main characteristics of solid-phase peptide synthesis are as described in the previous chapter, which should be referred to for description of methods and source of many of the materials, and solvents, and so on. Since it is not possible to monitor the completeness of coupling at each step in T-bags, the peptides obtained are unlikely to be very pure, but should be suitable for screening of the epitopes of antibodies.

2. Materials

1. *p*-Alkoxybenzyl resin (100–200 mesh) is obtained from Bachem or prepared from Merrifield resin (chloromethylated copolystyrene–1% divinylbenzene) (100–200 mesh) obtained from Pierce, as described in the previous chapter.
2. Fmoc-amino acids (CRB) (*see* previous chapter).
3. Polypropylene mesh (PP 74) for preparing T-bags can be obtained in meter lengths from Henry Simon Ltd., PO Box 3, Stockport, Cheshire, UK.
4. Heat-sealer may be obtained from Jencons Ltd.
5. Glass bottle with polypropylene cap; 100–200 mL (e.g., Sorval Corex centrifuge bottle).
6. For nature and source of solvents and reagents for solid phase synthesis, *see* previous chapter.

3. Methods

Before carrying out this procedure, please refer to previous chapter for a description of the main characteristics of solid phase peptide synthesis.

3.1. Preparing the Bags

1. Prepare the number of bags that you require, one for each peptide that you wish to make. For 20 bags, cut a strip of polypropylene mesh 30 x 4 cm. Draw with a pencil a line down the middle in the long dimension and at 1.5-cm intervals across the strip. Experiment with your heat sealer to find the correct time setting to completely seal two pieces of polypropylene together without destroying it. About 5 s heating followed by 10 s cooling is suitable with the Jencons heat-sealer.

2. Place identifying numbers using a graphite pencil in the middle of each 1.5 cm section, but close to the long edge so that when sealed down the long edge, the number will be sealed too. Fold the polypropylene down the center line, and heat seal the long edges together about a third of a centimeter from the edge. It helps to place clips along the fold to hold it in place while sealing. Take care to allow to cool before removing the mesh from the heat-sealer.

3. Heat-seal several more times along the length of the mesh toward the outside edge so that the sealed area extends over the full third of a centimeter and the number is completely sealed in. Heat-seal across the short dimensions of the folded strip, starting along close to one edge, then just below each pencil line. Cut along each pencil line across the short dimension. This should give 20 bags, each open along one edge.

3.2. Filling the Bags

Weigh out 30 mg of *p*-alkoxybenzyl-polystyrene into each bag and heat-seal along the remaining side

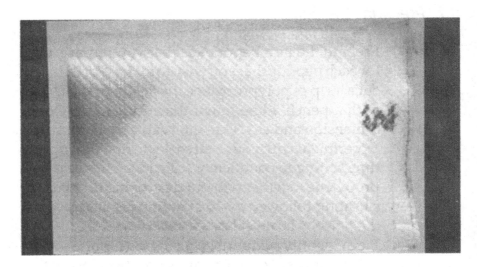

Fig. 1. A sealed polypropylene T-bag containing resin.

(Fig. 1). The yield from each bag should be about 10 µmol of peptide per bag.

3.3. Coupling the C-Terminal Residue

1. Place the bags in a beaker. Couple the C-terminal amino acid in at least 10 mL of reagents, which will take six bags. Scale up the volumes if more bags are used so that all the bags are covered with reagent.
2. The reagents consist of 0.25M Fmoc-amino acid in DMF and 0.25M diisopropylcarbodiimide in DCM added in equal amounts. If Fmoc-Asn, Fmoc-Gln, or Fmoc-Lys (Boc) are to be coupled (see Chapter 22) then first add 0.25M N-hydroxybenzotriazole in DMF, then an equal volume of an 0.25M solution of the Fmoc-amino acid pentafluorophenyl ester in DMF and omit the carbodiimide.
3. Cap the beaker with foil, shake on a vigorous rotary shaker for 2 min, then add N-methylmorpholine, so

that its concentration in the total volume is 0.125M, and 4-dimethylaminopyridine dissolved in a small volume of DMF so that its concentration in the reagents surrounding the T-bags is 0.0125M. Cover again and shake at room temperature for 2 h. If different residues are to be coupled at the C-terminus, then first sort bags according to their coded number, into different beakers, and couple with different protected amino acids.

3.4. Coupling Successive Residues

1. Transfer the bags together (up to 200 bags) to the centrifuge bottle and perform the following steps, shaking the bottle after placing the plastic seal and cap in place with each solvent, and decanting through a stainless steel sieve placed over the top to keep the bags from falling out. Use enough solvent to cover all the bags, i.e., between 30 and 100 mL per wash, and try not to let the bags dry out or there may be loss of resin from the bags. Perform the operations in a fume cupboard.

2. Carry out the following steps:

   Step 1. DMF, 3 x 2 min wash.
   Step 2. DCM, 3 x 2 min wash.
   Step 3. DMF, 3 x 2 min wash.
   Step 4. 20% Piperidine in DMF, 2 x 7 min wash.
   Step 5. DMF, 3 x 2 min wash.
   Step 6. DCM, 3 x 2 min wash.
   Step 7. DMF, 4 x 2 min wash.

3. After the washing in step 7, for each step in the synthesis, either the same amino acid residue will be coupled onto the resin in one or several of the bags

according to the sequences required. Sort the bags out according to number, then place in beakers appropriately marked for the amino acid residue to be added. Couple with Fmoc-amino acids using volumes of reagents that cover the bags. As before, the reagents consist of 0.25M Fmoc-amino acid in DMF and 0.25M diisopropylcarbodiimide in DCM added in equal volumes, but no tertiary amine catalyst is required. Add the Fmoc-amino acids first to the bags in a beaker followed by the diisopropylcarbodiimide, cover with aluminum foil, and leave stirring with a magnet for 2 h. For Fmoc-Asn, Fmoc-Gln, or Fmoc-Lys (Boc), add 0.25M 1-hydroxybenzotriazole in DMF first, then an equal volume of 0.25M solution of the pentafluorphenyl ester in DMF and omit the carbodiimide.

4. Wash with 100 mL of DCM:DMF (1:1 by vol) for 3 x 2 min. Repeat from step 1.

3.5. Deprotection and Cleavage from the Resin

1. As explained in the last chapter, Boc-amino acids can be coupled as the N-terminal residue using the same conditions as described above for Fmoc-amino acids. After the last residue has been added, wash three times each with DCM:DMF (1:1 by vol), DMF, DCM, and i-propanol.

2. Cut open the bags and transfer the resin in each to individual marked glass vials.

3. Wearing gloves and goggles and using a fume-cupboard, add 1 mL of 50% TFA in DCM with the appropriate added chemicals (to protect amino acids from side reactions as described in the last chapter) to each vial and leave for 2 h at room temperature.

4. Filter off each peptide from the resin through small sintered glass filters with side arms attached to a marked quickfit tube with slight vacuum from a water pump. Wash each resin with a further 2 x 1 mL of 50% TFA and also filter.
5. Evaporate the filtrates with streams of nitrogen in a fume cupboard and precipitate the peptides by addition of 3 mL of ether to each tube.
6. Vortex each tube, centrifuge in a spark-free centrifuge for 2 min, and dispose of the supernatant.
7. Wash another three times with ether and leave the tubes open on the bench for the remaining ether to evaporate. Then evacuate the tubes in a desiccator over KOH pellets.

4. Notes

1. Peptides may be characterized by HPLC or amino acid analysis as described in the last chapter, and tested for binding to antibodies by ELISA assay, measuring the competition exerted at various dilutions of peptide of binding of antibodies to immobilized antigens on the wells of PVC plates.

Reference

1. Houghton, R.A. (1985) General method for the rapid solid-phase synthesis of large numbers of peptides: Specificity of antigen-antibody interaction at the level of individual amino acids. *Proc. Natl. Acad. Sci. USA* **82**, 5131–5135.

Chapter 24

Production of Antisera
to Synthetic Peptides

W. J. Gullick

1. Introduction

The objective of immunization is to produce polyclonal or monoclonal antibodies that react with a chosen
molecule. Immunization may be in vivo by injecting
antigen into animals to produce antisera or to provide B-
lymphocytes for immortalization as hybridomas, or alternatively, in vitro using splenocytes from untreated
animals (*1*) for subsequent fusion with myeloma cells to
produce monoclonal antibodies (MAbs).

This chapter will be confined to the production of
antibodies to proteins using synthetic peptides as immunogens. Conventional immunization may involve
the injection of impure, semipure, or pure proteins to
produce polyspecific or monospecific antisera. Since
monospecific antisera are generally more useful, having
a pure antigen is desirable. The problem of impure anti-

gen can, however, be overcome by producing MAbs. This method relies on having an unambiguous screen in order to select the appropriate hybridomas. Frequently, this means a rather complicated analysis such as immunoprecipitation rather than by ELISA, which, when using impure immunogen, again requires purified protein for screening. Additional disadvantages are potential immunodominance by other antigens in the mixture over the desired immune response, and, importantly, there may be an immunodominant epitope on the antigen itself. If MAbs are required that recognize other regions of the molecule, this may involve time-consuming elimination of undesired MAbs of the same specificity produced against the dominant epitope. An example of this problem is work involving the acetylcholine receptor in which one region of the native molecule elicited many different MAbs and was subsequently called the main immunogenic region (2). In addition, mapping the binding sites of such MAbs on the surface or sequence of a protein is time-consuming and complicated (3).

Within the last few years, a different approach to generating protein-reactive antibodies has been developed that has several advantages over conventional immunization. This involves synthesizing short peptide sequences, coupling them to immunogenic carrier molecules, and immunizing animals with the conjugates. An immune response is usually generated both against the carrier and to the peptide acting as a hapten. The antipeptide antibodies produced, frequently, but by no means always, react with the protein from which the peptide sequence was selected. This technique is of course dependent upon having primary sequence information, either derived directly by sequencing a purified protein or quite commonly by using an amino acid sequence predicted from an open reading frame of cloned genomic or cDNA molecules. The latter situation does not require that any protein be purified, and in fact, in most cases the protein may never have been identified or characterized at all.

Immunization with synthetic peptides has the advantages that pure immunogens are used and an unlimited quantity of pure stable antigen is available for screening, often by ELISA, which is particularly useful in the production of MAbs when large numbers of hybridoma supernatants need to be screened quickly (*see* vol. 1 of this series). Perhaps more importantly, however, the region within the primary structure of the protein of interest to which the antibody binds is clearly defined with high resolution. The antibodies are, in fact, "site-directed" and can be generated against regions of proteins that are not normally immunogenic when the whole protein is employed as an antigen. The main problem of synthetic peptide immunogens is that although one almost always can make anti-peptide antibodies, perhaps on average only one in three or four of these sequences generates antibodies that react with this structure in the context of the whole molecule. It is therefore very important to choose a peptide immunogen carefully so as to improve the success rate, thereby minimizing the cost and the number of animals employed, and several techniques have now been developed for this end.

2. Choosing the Right Peptide

If, for instance, one only has the sequence of the 10 amino-terminal amino acids of a protein, or if two protein sequences are highly related except for a short sequence and one wishes to make antibodies that discriminate between them, the choice is already made. If, however, a longer sequence is available and you have no predisposition to select a particular region, there are several predictive methods that are considered helpful but not infallible.

2.1. Hydropathy Plots

Hydropathy is a measure of the polarity of a molecule. Each amino acid is assigned a hydropathy value, defined by its propensity to dissolve in the organic or aqueous phase of a two-phase solvent system. A plot of the local hydropathy of the sequence is generated by selecting groups of usually six or seven contiguous amino acids (the "window") and averaging their hydropathy value. By starting at the N-terminus of the protein and iteratively moving the window by one residue each time and performing the same calculation, one obtains a plot of the varying hydropathy along the sequence. The immunogenic epitopes of some proteins have been defined experimentally and are quite often composed of charged or hydrophilic residues that appear as peaks of local hydrophilicity on the plot (4,5). For a detailed discussion of this method, *see* ref. 6.

2.2. Protrusion Index

Water-soluble proteins are folded such that amino acids with hydrophilic side chains are generally disposed on the surface of the protein and are thus sterically available to interact with antibody combining sites. A different method of calculation called the "protrusion index" has been developed (7) that attempts to predict the relative protrusion of a sequence from the protein's globular surface. This method predicts that the more superficial the region, the more likely it is to be a good candidate as a synthetic peptide.

2.3. Secondary Structural Predictions

There is some evidence that turns in proteins are immunogenic, and the position of these within a sequence can be predicted mathematically (8). The great advantages of the above methods are that they only require primary sequence information and that the calculations involved are fairly simple, particularly that for hydropathy plots. Biochemical computer software is now commercially available that can perform one or more of these analyses.

2.4. Sequence Mobility

Recently two papers have suggested that the more flexible a stretch of amino acids (the atomic or segmental mobility), the more likely it is to be immunogenic (9,10). This analysis suffers from the fact that a quite detailed X-ray crystallographic structure of the protein is required, and this consequently limits its general usefulness.

All these methods are tested for their predictive ability by comparison with empirically determined antigenic sites of whole proteins. It should be emphasized that this is not the same as immunizing with short haptenic peptides that are attached to a carrier protein, usually at many different sites, and whose conformation will be influenced by their local environment on the surface of the carrier. By using these methods, however, one can hope to improve the chance of producing anti-protein reactive antibodies, and since all essentially lead to the choice of hydrophilic sequences, this produces the more practical advantage that the peptides are likely to be water-soluble and thus easier to use in aqueous coupling reactions (see below). One good bet if in doubt is to try the N- or C-termini of the protein, which are frequently, but not always, successful, possibly because they may more closely resemble the peptide immunogen than an internal, more conformationally constrained, sequence.

The choice of immunizing peptide may be influenced by other considerations. Functional significance of a particular region of a protein can sometimes be predicted from primary sequence motifs. An example of this is the sequence contributing to the nucleotide binding site of various protein kinases that can be rather well defined based upon conserved residues with characteristic spacing (11). Antibodies to such regions have been produced that are specific inhibitors of enzyme activity (12). Sometimes antibodies have been successfully raised to proteins homologous to that under study. Choosing the cognate region for synthesis may favor success, since, assuming the two proteins are folded in the same way, it is likely that the chosen region is superficial and available for antibody binding.

The length of peptides that have been synthesized has varied considerably, although 12–15 residues is most common. Longer sequences are more costly and do not seem to be more effective at generating antiprotein reactive antibodies, perhaps because they possess tertiary structure of their own that may not resemble that of the same sequence in the whole protein. Shorter peptides sometimes fail to stimulate a sufficiently high affinity and specific immune response for the native protein. One final point is that it is not wise to select a sequence that may be posttranslationally modified in the mature protein, such as a site of N-linked glycosylation or addition of lipid. Phosphorylation of serine, threonine, or tyrosine residues, however, does not seem to significantly influence the binding of polyclonal antibodies (13), although it might radically affect the binding of some MAbs.

3. Peptide Synthesis

There are two ways to obtain synthetic peptides; by synthesizing them oneself or by buying them. Several companies now perform custom synthesis, and some will also couple peptides to carrier molecules. The main disadvantage of this is that it is expensive. At the time of writing, a rough estimate of the cost of purchasing peptides is about $150 per residue for a partially purified product. The more peptides that are used, the more likely one is to obtain anti-protein reactive sera, but the cost increases. The average delivery time is 4–6 wk, and this may represent an undesirable delay. The great advantage of purchase is that no skill or effort is required from the experimenter, and the product is reliable.

Peptide synthesis, however, is not too daunting for the non-organic chemist. There are two basic chemistries in use. The more common is that developed by Merrifield (for which he won the Nobel prize), which employs t-BOC amino acid derivatives, and a newer approach, developed by Sheppard, using f-MOC amino acid derivatives, which is presently gaining in popularity. For references to the detailed procedure *see* chapters 22 and 23 and refs. 14 and 15. One very considerable advantage of the new f-MOC chemistry is the avoidance of liquid hydrogen fluoride, which is used in t-BOC syntheses to cleave off protecting groups and to remove the peptide from the solid support after synthesis is complete. This compound is potentially very hazardous, and although contained reaction vessels are available, great care must still be taken when using it.

Synthesis can be performed in instruments varying in cost from about $5–10 for a simple glass tube consisting of a scinter and two taps, to an essentially fully automated machine at about $100,000. There are now, however, more semiautomated machines appearing on the market in the $15,000–75,000 range. The advantages of the more automated machines are that, obviously, less time is required from the operator and, perhaps, a greater degree of reproducibility and purity of product. Clearly, however, the purchase of the more automated systems can only be justified over purchase of the peptides if many sequences are required. Peptide synthesis is described in detail in Chapters 22 and 23.

4. Materials

4.1. Glutaradehyde Coupling of the Peptide to the Carrier Protein

1. Keyhole limpet hemocyanin. Use as a slurry in 65% ammonium sulfate from Calbiochem (catalog no. 374811). It usually comes at about 30–40 mg/mL protein concentration.
2. Phosphate buffered saline. (PBS): 0.14M NaCl, 20 mM sodium phosphate buffer, pH 7.2.
3. 50 mM Sodium borate/HCl buffer, pH 9.0.
4. 100 mM Sodium phosphate buffer, pH 8.0.
5. 25% Solution of glutaraldehyde. It is not critical to use high-quality material such as EM grade, but this is quite cheap anyway. I have found Sigma (catalog no. G-6257) satisfactory, but almost any variety would do.
6. 1 M Glycine/HCl buffer, pH 6.0.

4.2. Immunization

1. Complete Freunds adjuvant.
2. Incomplete Freunds adjuvant.

5. Method

5.1. Coupling of Peptides to Carrier Proteins

Several carrier proteins have been used success-fully including bovine serum albumin, ovalbumin, thy-roglobulin, and keyhole limpet hemocyanin (KLH). A recent review of anti-peptide antibodies (16) lists almost two hundred examples, and one carrier clearly pre-dominates. Keyhole limpet hemocyanin has long been the immunologist's choice as a general antigen to pro-duce humoral antibodies, and it is frequently favored as a carrier for synthetic peptides. Hemocyanins are giant molecules, up to 9,000,000 mol wt, with quite repetitious structures. Many moles of peptide can be reacted per mole of KLH. All the other carriers have, however, been used with good results.

The methods available for coupling synthetic pep-tides to carriers are also quite varied, but can be clas-sified depending upon the functional group of the peptide that is employed for the reaction. Most, al-though not all, peptides used as immunogens are syn-thesized with free amino and carboxy termini, and these are most commonly employed for coupling. This is not only because there are several methods worked out for primary amines and carboxylic groups, but also because it is probably not a good thing to modify amino acids within a sequence since it will make the structure less like that of the parent molecule. Glutaraldehyde is the most commonly used reagent for coupling peptides via amine groups. Most papers refer to the method reported by Kagan and Glick (17), but I have found this reference rather difficult to find, so I present my method below, which is a minor modification of theirs.

This makes enough immunogen for a series of three injections into two rabbits.

1. Dialyze the KLH against PBS (normally 1 mL of KLH solution against 2 L of PBS for at least 2 h) without sodium azide, and then measure the protein concentration. A quick way of doing this is to dilute a small aliquot into 50 mM sodium borate buffer, pH 9.0, and measure the optical density of the solution at 280 nm. A 1.0 mg/mL solution of hemocyanin in this buffer gives 1.4 AU. Do not measure at neutral pH, since hemocyanin gives significant light scattering unless it is dissociated by acid or alkaline buffers.

2. Dissolve 7.5 mg of peptide in 100 mM sodium phosphate buffer, pH 8.0, mix with 7.5 mg of KLH, and make up to 0.5 mL with buffer.

3. Add 5 µL of 25% glutaraldehyde, mix well, and leave for 15 min at room temperature. The mixture may appear cloudy during this period.

4. Add a further 2.5 µL of glutaraldehyde and leave for 15 min.

5. Add 100 µL of 1 M glycine, pH 6.0, and leave for 10 min to quench the glutaradehyde. The reaction mixture will change to yellow to brown color. Store in aliquots at –20°C until use.

Other types of coupling reactions employ EDIC (1-ethyl-3-(-3-dimethylaminopropyl)carbodiimide), which reacts with amine and carboxylic acid groups (18), MBS (*m*-maleimidobenzoic acid *N*-hydroxysuccinimide ester), which reacts with free sulfhydryl groups (*19*), and *bis*-diazobenzidine, which reacts with tyrosine residues (*20*).

5.2. Immunization

1. Immunization protocols are very much the acquired preference of the experimenter. I suggest the following (for rabbits), since I have found it simple and reliable. I also use the same schedule for mice and rats, but employ about one tenth the amount of immunogen for mice and one fifth for rats.

2. First take a preimmune bleed. From a medium-sized rabbit (3–4 kg) about 20–25 mL of *serum* and from a mouse 50–150 µL would be ideal. Do not forget that this is the one thing that you will never be able to obtain again once you have immunized.

3. Day 1. Mix one third of the immunogen (200 µL from the glutaraldehyde protocol) with 0.8 mL of PBS and 1.0 mL of *complete* Freunds adjuvant to a stiff paste, and inject each animal at multiple sites (about 2–4) subcutaneously with a total of 1.0 mL of the mixture.

4. Day 14. Repeat as for day 1, but use *incomplete* Freunds adjuvant.

5. Day 28. Repeat as for day 14.

6. Day 35. Test bleed. 10–20 mL of *serum* for rabbits, 50–100 µL for mice.

7. Day 49. Test bleed.

6. Notes

1. I recommend ELISA for screening for anti-peptide antibodies, since the plates can be coated with

peptide alone, thereby allowing the antipeptide titer to be measured without any contribution from anti-carrier (usually a lot), or anti-crosslinker (sometimes present) response. The general procedure as applied to screening for hybridomas has been published (*see* vol. 1 of this series and ref. 21). For measuring serum titers, I normally measure in quadruplicate, starting at a dilution of serum of 1/50 and going down in five steps of fourfold dilution to 1/12,800. This will be a good range for most antisera, although high titer ones may need further dilutions. With most peptides, 50% of color development should be reached at a dilution of about 1/3200, although this is a considerable generalization. The anti-peptide antibodies can now be tested for their ability to recognize the protein of interest by immunoprecipitation or western blotting (*see* vol. 1 and this volume). Some will work in one or both assays, some in neither. The more peptides that have been used, the more likely it is that anti-protein reactive antibodies will be obtained.

2. Beware that complete Freunds adjuvant will cause just as good immune response in humans as it will in mice or rabbits. Wear gloves and eye protection whenever handling it.

3. Several companies sell KLH as a lyophilized powder. I have heard from others that this is virtually impossible to disolve without obvious large aggregates. To avoid this problem, I would recommend buying it as a solution or slurry.

References

1. Luben, R.A., Brazeau, P. Bolen, P., and Guillemin, R. (1982) Monoclonal antibodies to hypothalamic growth hormone releasing factor with picomoles of antigen. *Science* **218**, 887–889.

2. Tzartos, S.J. and Lindstrom, J.M. (1980) Monoclonal antibodies used to probe acetylcholine receptor structure: localization of the main immunogenic region and detection of similarities between subunits. *Proc. Natl. Acad. Sci. USA* **77**, 755–759.

3. Gullick, W.J. and Lindstrom, J.M. (1983) Mapping the binding of monoclonal antibodies to the acetylcholine receptor from *Torpedo californica. Biochemistry* **22**, 3312–3320.

4. Hopp, T.P. and Woods, K.R. (1981) Prediction of protein antigenic determinants from amino acid sequences. *Proc. Natl. Acad. Sci. USA* **78**, 3824–3828.

5. Kyte, J. and Doolittle, R.F. (1982) A simple method for displaying the hydropathic character of a protein. *J. Mol. Biol.* **157**, 105–132.

6. Hopp, T.P. (1986) Protein surface analysis. *J. Immunol. Meth.* **88**, 1–18.

7. Thornton, J.M., Edwards, M.S., Taylor, W.R., and Barlow, D.J. (1986) Location of "continuous" antigenic determinants in the protruding regions of proteins. *EMBO J.* **5**, 409–413.

8. Schulze-Gahmen, U., Prinz, H., Glatter, U., and Beyreuther, K. (1985) Towards assignment of secondary structures by anti-peptide antibodies. Specificity of the immune response to a turn. *EMBO J.* **4**, 1731–1737.

9. Westhof, E., Altschuh, D., Moras, D., Bloomer, A.C., Mondragon, A., Klug, A., and Van Regenmortel, M.H.V. (1984) Correlation between segmental mobility and the location of antigenic determinants in proteins. *Nature* **311**, 123–126.

10. Tainer, J.A., Getzoff, E.D., Alexander, H., Houghton, R.A., Olson, A.J., Lerner, R.A., and Hendrickson, W.A. (1984) The reactivity of anti-peptide antibodies is a function of the atomic mobility of sites in a protein. *Nature* **312**, 127–134.

11. Sternberg, M.J.E. and Taylor, W.R. (1984) Modelling the ATP-binding site of oncogene products, the epidermal growth factor receptor and related proteins. *FEBS Lett.* **175**, 387–392.

12. Gullick, W.J., Downward, J., Foulkes, J.G., and Waterfield, M.D. (1986) Antibodies to the ATP-binding site of human epidermal growth factor receptor as specific inhibitors of EGF stimulated protein tyrosine kinase activity. *Eur. J. Biochem.* **158**, 245–253.

13. Gullick, W.J., Downward, J., and Waterfield, M.D. (1985) Antibodies to the autophosphorylation sites of the epidermal growth factor receptor as probes of structure and function. *EMBO J.* **4**, 2869–2877.

14. Merrifield, B. (1986) Solid phase synthesis. *Science* **232**, 341–347.

15. Atherton, E. and Sheppard, R.C. (1985) Solid phase peptide synthesis using N-fluorenylmethoxycarbonylamino acid pentaflurophenyl esters. *J. Chem. Soc. Chem. Commun.* 165–166.

16. Palfreyman, J.W., Aitcheson, T.C., and Taylor, P. (1984) Guidelines for the production of polypeptide specific antisera using small synthetic oligopeptides as immunogens. *J. Immunol. Meth.* **75**, 383–393.

17. Kagan, A. and Glick, M. (1974) Oxytocin, in *Methods in Hormone Radioimmunoassay* (Jaffe, B.M. and Behrman, H. R., eds.) Academic, New York.

18. Kris, R.M., Lax, I., Sasson, I., Copf, B., Welin, S., Gullick, W.J., Waterfield, M.D., Ullrich, A., Fridkin, M., and Schlessinger, J. (1985) Synthetic peptide approach to the analysis of the kinase activities of avian EGF receptor and v-erbB protein. *Biochemie* **67**, 1095–1101.

19. Green, N., Alexander, H., Olson, A., Alexander, S., Shinnick, T.M., Sutcliffe, J.G., and Lerner, R.A. (1982) Immunogenic structure of the Influenza virus hemagglutinin. *Cell* **28**, 477–487.

20. Ratam, M. and Lindstrom, J. (1984) Structural features of the nicotinic acetylcholine receptor revealed by antibodies to synthetic peptides. *Biochem. Biophys. Res. Comm.* **122**, 1225–1233.

21. Gullick, W.J., Marsden J.J., Whittle, N., Ward, B., Bobrow, L., and Waterfield, M.D. (1986) Expression of epidermal growth factor receptors on cervical, ovarian and vulval carcinomas. *Cancer Res.* **46**, 285–292.

Chapter 25

Production of Antibodies Using Proteins in Gel Bands

S. A. Amero, T. C. James, and S. C. R. Elgin

1. Introduction

A number of methods for preparing proteins as antigens have been described (1). These include solubilization of protein samples in buffered solutions (2), solubilization of nitrocellulose filters to which proteins have been adsorbed (3), and emulsification of protein bands in polyacrylamide gels for direct injections (4–8). The latter technique can be used to immunize mice or rabbits for production of antisera or to immunize mice for production of monoclonal antibodies (9–11). This approach is particularly advantageous when protein purification by other means is not practical, as in the case of proteins insoluble without detergent. A further advantage of this method is an enhancement of the immune response, since polyacrylamide helps to retain the antigen in the animal and so acts as an adjuvant (7). The use of the protein directly in the gel band (without elution) is also helpful when only small amounts of protein are available. For instance, in this laboratory, we routinely immunize mice with 5–10 µg total protein using this method; we have not determined the lower limit of total protein that can be used to immunize rabbits. Since polyacrylamide is also highly immunogenic, however, it is necessary in some cases to affinity-purify the desired antibodies from the resulting antiserum or to produce hybridomas that can be screened selectively for the production of specific antibodies, to obtain the desired reagent.

2. Materials

1. Gel electrophoresis apparatus; acid-urea polyacrylamide gel or SDS-polyacrylamide gel.
2. Staining solution: 0.1% Coomassie Brilliant Blue-R (Sigma B-0630) in 50% (v/v) methanol/10% (v/v) acetic acid.
3. Destaining solution: 5% (v/v) methanol/7% (v/v) acetic acid.
4. 2% (v/v) glutaraldehyde (Sigma G-6257).
5. Transilluminator.
6. Sharp razor blades.
7. Conical plastic centrifuge tubes and ethanol.
8. Lyophilizer and dry ice.
9. Plastic, disposable syringes (3- and 1-mL).
10. 18-Gage needles.
11. Spatula and weighing paper.
12. Freund's complete and Freund's incomplete adjuvants (Gibco Laboratories, Grand Island, NY).
13. Phosphate-buffered saline solution (PBS) (50 mM sodium phosphate, pH 7.25/150 mM sodium chloride).

14. Microemulsifying needle, 18-gage (Becton Dickinson, Rutherford, NJ).
15. Female Balb-c mice, 7–8 wk old, or New Zealand white rabbits.

3. Method

1. Following electrophoresis (*see* Note 1 in section 4), the gel is stained by gentle agitation in several volumes of staining solution for 30 min. The gel is partially destained by gentle agitation in several changes of destaining solution for 30–45 min. Proteins in the gel are then crosslinked by immersing the gel with gentle shaking in 2% glutaraldehyde for 45–60 min (*12*). This step minimizes loss of proteins during subsequent destaining steps and enhances the immunological response by polymerizing the proteins. The gel is then completely destained, usually overnight.

2. The gel is viewed on a transilluminator, and the bands of interest are cut out with a razor blade. The gel pieces are pushed to the bottom of a conical plastic centrifuge tube with a spatula and pulverized. The samples in the tubes are frozen in dry ice and lyophilized.

3. To prepare the dried polyacrylamide pieces for injection, a small portion of the dried material is lifted out of the tube with a spatula and placed on a small square of weighing paper. In dry climates it is useful to first wipe the outside of the tube with ethanol to reduce static electricity. The material is then gently tapped into the top of a 3-mL syringe to which is attached the microemulsifying needle (Fig. 1A). Keeping the syringe horizontal, 200 μL of PBS solution is carefully introduced to the barrel of

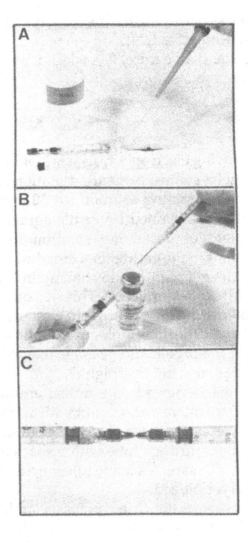

Fig. 1. Preparation of emulsion for immunizations. To prepare proteins in gel bands for injections, an emulsion of Freund's adjuvant and dried polyacrylamide pieces is prepared. (A) Dried polyacrylamide is resuspended in 200 µL of PBS solution in the barrel of a 3-mL syringe to which is attached a microemulsifying needle. (B). Freund's adjuvant is transferred into the barrel of a second 3-mL syringe. (C). An emulsion is formed by mixing the contents of the two syringes through the microemulsifying needle.

the syringe, and the plunger is inserted. Next, 200 µL of Freund's adjuvant is drawn into a 1-mL tuberculin syringe and transferred into the needle end of a second 3-mL syringe (Fig. 1B). This syringe is then atached to the free end of the microemulsifying needle. The two plungers are pushed alternatively to mix the components of the two syringes (Fig. 1C). These will form an emulsion within 15 min; it is generally extremely difficult to mix the material any further.

4. This mixture is injected intraperitoneally or subcutaneously into a female Balb-c mouse, or subcutaneously into the back of the neck of a rabbit (*see* refs. *13* and *14*). Since the emulsion is very viscous, it is best to use 18-gage needles and to anesthesize the animals. For mice, subsequent injections are administered after 2 wk and after 3 more wk. If monoclonal antibodies are desired, the animals are sacrificed 3–4 d later, and the spleen cells are fused with myeloma cells (*13* and *see* vol. 1 of this series). The immunization schedule for rabbits calls for subsequent injections after 1 mo or when serum titers start to diminish. Antiserum is obtained from either tail bleeds or eye bleeds from the immunized mice, or from ear bleeds from immunized rabbits. The antibodies are assayed by any of the standard techniques (*see* vol. 1 in this series).

4. Notes

1. We have produced antisera to protein bands in acetic acid–urea gels (*15*), Triton–acetic acid–urea gels (*16,17*), or SDS–polyacrylamide gels (*18*). In our experience, antibodies produced to proteins in one denaturing gel system will cross-react to those

same proteins fractionated in another denaturing gel system and will usually cross-react with the native protein. We have consistently obtained antibodies from animals immunized by these procedures.

2.   It is extremely important that all glutaraldehyde be removed from the gel during the destaining washes, since any residual glutaraldehyde will be toxic to the animal. Residual glutaraldehyde can easily be detected by smell. It is equally important to remove all acetic acid during lyophilization. Monoacrylamide is also toxic, whereas polyacrylamide is not. We do observe, however, that approximately 50 mm^2 of polyacrylamide per injection is the maximum that a mouse can tolerate.

3.   Freund's complete adjuvant is used for the initial immunization; Freund's incomplete adjuvant is used for all subsequent injections. The mycobacteria in complete adjuvant enhance the immune response by provoking a local inflammation. Additional doses of mycobacteria may be toxic.

References

1.   Chase, M.W. (1967) Production of Antiserum, in *Methods in Immunology and Immunochemistry* Vol. I (Williams, C.A. and Chase, M.W., eds.) Academic, New York.

2.   Maurer, P. H. and Callahan, H.J. (1980) Proteins and Polypeptides as Antigens, in *Methods in Enzymology* Vol. 70 (Van Vunakis, H. and Langone, J.J., eds.) Academic, New York.

3.   Knudson, K. A. (1985) Proteins transferred to nitrocellulose for use as immunogens. *Anal. Biochem.* **147**, 285–288.

4.   Tjian, R., Stinchcomb, D., and Losick, R. (1974) Antibody directed against *Bacillus subtilis* σ factor purified by sodium dodecyl sulfate slab gel electrophoresis. *J. Biol. Chem.* **250**, 8824–8828.

5. Elgin, S.C.R., Silver, L.M., and Wu, C.E.C. (1977) The in situ distribution of Drosophila non-histone chromosomal proteins, in *The Molecular Biology of the Mammalian Genetic Apparatus* Vol. 1 (Ts'o, P.O.P., ed.) North Holland, New York.

6. Silver, L.M. and Elgin, S.C.R. (1978) Immunological analysis of Protein Distributions in Drosophila Polytene Chromosomes, in *The Cell Nucleus* (Busch, H., ed.) Academic, London, New York.

7. Bugler, B., Caizergues-Ferrer, M., Bouche, G., Bourbon, H., and Alamric, F. (1982) Detection and localization of a class of proteins immunologically related to a 100-kDa nucleolar protein. *Eur. J. Biochem.* **128,** 475–480.

8. Wood, D.M. and Dunbar, B.S. (1981) Direct detection of two-cross-reactive antigens between porcine and rabbit zonae pellucidae by radioimmunoassay and immunoelectrophoresis. *J. Exp. Zool.* **217,** 423–433.

9. Howard, G.D., Abmayr, S.M., Shinefeld, L.A., Sato, V.L., and Elgin, S.C.R. (1981) Monoclonal antibodies against a specific nonhistone chromosomal protein of Drosophila associated with active genes. *J. Cell Biol.* **88,** 219–225.

10. Tracy, R.P., Katzmann, J.A., Kimlinger, T.K., Hurst, G.A., and Young, D.S. (1983) Development of monoclonal antibodies to proteins separated by two-dimensional gel electrophoresis. *J. Immunol. Meth.* **65,** 97–107.

11. James, T.C. and Elgin, S.C.R. (1986) Identification of a nonhistone chromosomal protein associated with heterochromatin in *Drosophila melanogaster* and its gene. *Mol. Cell Biol.* **6,** 3862–3872.

12. Reichli, M. (1980) Use of glutaraldehyde as a coupling agent for proteins and peptides, in *Methods in Enzymology* Vol. 70 (Van Vunakis, H. and Langone, J.J., eds.) Academic, New York.

13. Campbell, A.M. (1984) *Monoclonal Antibody Technology* Elsevier, New York.

14. Silver, L.M. and Elgin, S.C.R. (1977) Distribution patterns of three subfractions of Drosophila nonhistone chromosomal proteins: Possible correlations with gene activity. *Cell* **11:** 971–983.

15. Smith, B.J. (1984) Acetic Acid-Urea Polyacrylamide Gel Electrophoresis of Proteins, in *Methods in Molecular Biology* Vol. 1 (Walker, J.M., ed.) Humana, Clifton, New Jersey.

16. Alfagame, C.R., Zweidler, A., Mahowald, A., and Cohen, L.H. (1974) Histones of Drosophila embryos. *J. Biol. Chem.* **249**, 3729–3736.

17. Cohen, L.H., Newrock, K.M., and Zweidler, A. (1975) Stage-specific switches in histone synthesis during embryogenesis of the sea urchin. *Science* **190**, 994–997.

18. Smith, B.J. (1984) SDS Polyacrylamide Gel Electrophoresis of Proteins, in *Methods in Molecular Biology* Vol. 1 (Walker, J.M., ed.) Humana, Clifton, New Jersey.

Chapter 26

Purification of Immunoglobulins Using Protein A–Sepharose

Nicholas J. Kruger and
John B. W. Hammond

1. Introduction

    Most immunochemical techniques used in molecular biology rely on a specific class of antibodies, immunoglobulin G (IgG). In mammals, IgG antibodies are produced during the secondary humoral response and contribute about 80% of the serum immunoglobulin. Factors affecting the experimental production of antibodies have been discussed in detail (1). In this chapter we describe a method for the rapid and efficient purification of IgG to near homogeneity from rabbit serum. This is achieved by affinity chromatography using protein A.

    Protein A is a cell wall component of *Staphylococcus aureus* (Cowan 1 strain). This protein specifically binds the crystallizable fragment (Fc) region of IgG from many mammals (2). Because of this, protein A can be used to separate IgG from other serum proteins. The principal disadvantage of the technique is that it is not suitable for purifying IgG from all sources. The affinity of protein A for IgG varies among species and subclasses. In addition, other classes of immunoglobulin from some species may bind to this ligand, although in practice this is rarely a major problem. The interaction between protein A and immunoglobulins from several important species is summarized in Table 1.

There are obvious advantages to using purified IgG rather than crude serum as a source of antibodies, since other proteins present in antisera may interfere in many immunological methods. For example, contaminating proteins reduce the amount of antibody that can be immobilized on a solid support in enzyme-linked immunosorbant assays (*see* vol. 1 of this series, Chapter 38). This restricts the detection limit of the assay. In addition to such general protein effects, specific contaminants may interfere in several techniques. Digestion of proteins by proteases present in antisera may confound the identification of immunoprecipitated polypeptides. Alternatively, measurement of enzyme activity in immuno-inactivation studies is often hindered by contaminating enzymes, NADH oxidase in particular, that directly interfere with the enzyme assay. These examples indicate that the specific problems associated with using crude sera vary depending on the application. Most of these problems are overcome by using purified IgG.

2. Materials

1. Rabbit antiserum: The production of suitable antisera has been described in a previous chapter (*see* vol. 1 in this series, Chapter 32).
2. Protein A–Sepharose Cl-4B supplied commercially as freeze-dried beads.

Table 1
Affinity of Protein A for Immunoglobulins from Various Species[a]

Species	Serum IgG level, mg/mL	Binding		
		Strong	Weak	Unreactive
Rabbit	5	IgG		
Human	12	IgG1[b], 2, 4 IgA2 IgM (some)		IgG3
Guinea pig	6	IgG1, 2[b]		
Mouse	7	IgG2a, 2b, 3	IgG1[b] IgM	
Pig	18	IgG IgM (some) IgA (some)		
Goat		IgG2	IgG1	
Sheep		IgG2		IgG1
Dog	9	IgG IgM (some) IgA (some)		
Rat	16	IgG1, 2c	IgG2b	IgG2a
Cow	20	IgG2		IgG1

[a]The values for serum IgG levels are approximate, and significant variation may occur among individuals. Immunoglobulins failing to bind to protein A at pH 8.0 are described as unreactive. Immunoglobulins that bind to protein A, but are released above pH 6.0, are considered weakly bound. Strong binding indicates retention below pH 6.0 (data from refs. 2–7).
[b]Denotes the major IgG subclass.

3. Buffers

0.1M Sodium phosphate, pH 8.0 (1 L).
0.1M Sodium citrate, pH 3.5 (100 mL).
0.1M Sodium citrate, pH 3.0 (100 mL).
1.0M Tris-HCl, pH 8.5 (50 mL).

The pH of the above solutions are adjusted at 20°C. These buffers are stable, but are susceptible to microbial contamination. They should be stored at 4°C and discarded if contamination occurs.

4. Saturated ammonium sulfate: add 80 g of $(NH_4)_2$-SO_4 (enzyme grade—low in heavy metals) to 100 mL of water and stir overnight at 25°C. The solution, which should contain some undissolved crystals, is adjusted to pH 7.8 using NH_4OH (*see* Note 1 in section 4). This solution is stable for several months at room temperature. Do not store at 4°C.
5. Phosphate buffered saline (2 L): 10 mM NaH_2PO_4, 150 mM NaCl adjusted to pH 7.2 using NaOH. This solution is stable and may be stored at 4°C. It is susceptible to microbial contamination, however, and is usually made as required.

3. Methods

3.1. Preparation of Protein A-Sepharose Column

1. Add 1.5 g of protein A-Sepharose to 10 mL of phosphate buffer, pH 8.0, and allow the resin to swell at room temperature for 30 min. All subsequent steps are performed at 4°C.
2. Pour the resin into a small column (0.9 x 15 cm) and wash with phosphate buffer, pH 8.0, at 30 mL/h for about 30 min, until the resin has settled. The total volume should be about 5 mL.
3. Wash the column with 20 mL of citrate buffer, pH 3.0, to elute any bound material, and re-equilibrate the column using 50 mL of phosphate buffer, pH 8.0. The column is now ready for use (*see* Note 2 in section 4).

3.2. Purification of IgG

1. To 10 mL of serum at room temperature slowly add 5 mL of saturated ammonium sulfate, while stir-

ring constantly. Continue stirring for 1 h (*see* Note 3 in section 4), then centrifuge the mixture at 15,000g for 15 min at 25°C.

2. Discard the supernatant and dissolve the pellet in 5 mL of phosphate buffer, pH 8.0. Perform all subsequent steps at 4°C.

3. Dialyze the sample against 500 mL of phosphate buffer, pH 8.0, for 3 h to remove residual ammonium sulfate, then apply the dialyzed sample to the protein A–Sepharose column at 20 mL/h and discard the effluent. Most serum proteins pass straight through this column.

4. Wash the column with 50 mL of phosphate buffer, pH 8.0, to ensure complete removal of unbound proteins.

5. Elute the IgG from the column with 20 mL of citrate buffer, pH 3.5. Collect the eluate as 1-mL fractions in tubes containing 0.4 mL Tris-HCl, pH 8.5, to neutralize the buffer and prevent denaturation of IgG.

6. Combine the fractions containing the bulk of the IgG ($A_{280} > 0.2$) and dialyze twice, each time for at least 2 h against 1 L phosphate buffered saline. Divide the purified IgG into 1-mL aliquots and store at −20°C (*see* Note 4 in section 4).

7. Remove residual protein from the protein A–Sepharose by washing with 20 mL of citrate buffer, pH 3.0, and immediately re-equilibrate the column with 40 mL of phosphate buffer, pH 8.0 (*see* Note 5 in section 4).

4. Notes

1. When adjusting the pH of the saturated ammonium sulfate, monitor the change using a suitable narrow-range pH paper, then use an electrode to

check the final pH by diluting an aliquot of the solution about 20-fold in water. Do not use a combination glass electrode to measure directly the pH of the saturated solution, since the high salt and junction potentials may cause reading errors of greater than 1 pH unit.

2.   If the protein A–Sepharose column is not to be used immediately, it should be stored at 4°C in phosphate buffer containing 0.05% (w/v) NaN_3.

3.   The initial ammonium sulfate precipitation step in the purification of IgG is not essential. This procedure does extend the life of the protein A–Sepharose column, however, by removing lipids and proteases present in serum. If steps 1 and 2 and the dialysis of step 3 in section 3.2. are omitted, the serum should be diluted witth an equal volume of phosphate-buffered saline before being applied to the protein A–Sepharose column (*see* step 3).

4.   Avoid repeated cycles of freezing and thawing by dividing the purified IgG into smaller aliquots if necessary. For transport, IgG may be lyophilized. The powder is stable at room temperature for several days, but should be returned to –20°C when convenient.

5.   To prevent denaturation of protein A, the column should be washed and re-equilibrated with phosphate buffer as soon as possible after use. The column should be stored as described in note 2. If washed and stored correctly, this column may be used at least 50 times without noticeable deterioration in its ability to bind IgG.

6.   An initial indication of the purity of the IgG preparation may be obtained by measuring its absorbance at 280 and 250 nm. Although the exact value may vary between antibody clones, in general $A_{280}/$

A_{250} for mammalian IgG is about 2.5–3.0. In contrast, for other serum proteins this ratio is about 1.0–1.5. A more rigorous test of purity is provided by SDS-polyacrylamide gel electrophoresis of both reduced and nonreduced samples using the system described by Laemmli (*see* vol. 1 of this series, chapter 6). Typical results are shown in Fig. 1.

Provided the preparation is pure, the quantity of IgG may be estimated from its absorbance at 280 nm. A solution of IgG containing 1 mg/mL has an absorbance, in a 1-cm light path, of about 1.4.

7. With care this method can be used to purify IgG from sources other than rabbit. The amount of rabbit IgG purified, however, approaches the capacity of the affinity column. Other species may have higher serum IgG concentrations (Table 1). For such species, the volumes of antiserum and ammonium sulfate should be reduced proportionally. All other conditions should remain the same. Antisera in which most IgG binds poorly, or not at all, such as those from rat, sheep, and goat, should be purified by other methods (3).

8. The affinity of protein A for IgG subclasses, if present, may vary (Table 1). This variation has been exploited to identify and purify mouse IgG subclasses, which can be separated by selective desorption from the column (4).

References

1. Maurer, P.H., and Callahan, H.J. (1980) Proteins and Polypeptides as Antigens, in *Methods in Enzymology*, Vol. 70 (Vunakis, H.V. and Langone, J.J., eds.). Academic, New York.
2. Goding, J.W. (1978) Use of staphylococcal protein A as an immunological reagent. *J. Immunol. Meth.* **20**, 241–253.
3. Tijssen, P. (1985) *Practice and Theory of Enzyme Immunoassays.* Elsevier, Amsterdam.

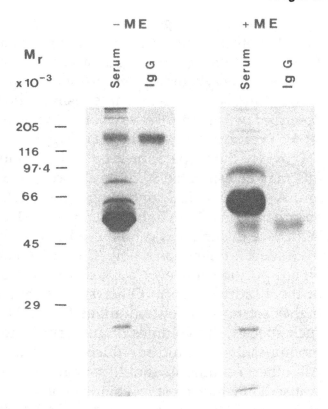

Fig. 1. Analysis of purified immunoglobulin G by SDS-poly-acrylamide gel electrophoresis. Samples of crude rabbit serum and purified IgG were fractionated by SDS-polyacrylamide gel electro-phoresis both before and after chemical reduction using 2-mercap-toethanol, and are labeled –ME and +ME, respectively. The molecu-lar weight standards used are myosin (205 kdalton), β-galactosidase (116 kdalton), phosphorylase b (97.4 kdalton), bovine serum albu-min (66 kdalton), ovalbumin (45 kdalton), and carbonic anhydrase (29 kdalton). The polypeptides were stained using Coomassie Blue.

Native IgG has a molecular mass of 150 kdalton and contains two heavy chains (50 kdalton) and two light chains (about 25 kdalton) held together by disulfide bonds. Following reduction by 2-mercaptoethanol, the heavy and light chains can be resolved. The light chain is often diffuse and stains poorly. No contaminants are visible in the purified IgG preparation.

4. Ey, P.L., Prowse, S.J., and Jenkin, C.R. (1978) Isolation of pure IgG1, IgG_{2a} and IgG_{2b} immunoglobulins from mouse serum using protein A-Sepharose. *Immunochemistry* **15**, 429–436.

5. Goudswaard, J., van der Donk, J.A., Noordzij, A., van Dam, R.H., andVaerman, J.-P. (1978) Protein A reactivity of various mammalian immunoglobulins. *Scand. J. Immunol.* **8**, 21–28.

6. Kessler, S.W. (1981) Use of Protein A-Bearing Staphylococci for the Immunoprecipitation and Isolation of Antigens from Cells, in *Methods in Enzymology* Vol. 73 (Langone, J.J. and Vunakis, H.V., eds.) Academic, New York.

7. Langone, J.J. (1980) ^{125}I-Labeled Protein A: Reactivity with IgG and Use as a Tracer in Radioimmunoassay, in *Methods in Enzymology* Vol. 70 (Vunakis, H.V. and Langone, J.J., eds.) Academic, New York.

Chapter 27

Antibody–Enzyme Conjugate Formation

G. B. Wisdom

1. Introduction

The ideal antibody–enzyme conjugate retains all the immunoreactivity and enzyme activity of its components, has defined and appropriate proportions, and is stable. There are many methods of labeling immunoglobulins with enzymes, but none meet all these criteria (1). Functional groups in the protein, such as amino, thiol, or saccharide groups, are exploited, and in nearly all cases bifunctional reagents are used to link the enzyme to the immunoglobulin G (IgG). The three most commonly used methods are presented here with minor modifications to the originally described procedures.

1.1. Alkaline Phosphatase Labeling Using Glutaraldehyde

The chemistry of the homobifunctonal reagent, glutaraldehyde, is complex. It reacts with amino groups of proteins, and when two proteins are mixed in its presence, stable conjugates are obtained without the formation of Schiff bases. This method (2) can be applied to most proteins, but self-coupling is a problem unless the proteins are at appropriate concentrations.

1.2. Horseradish Peroxidase Labeling Using Periodate Oxidation

Horseradish peroxidase labeling using periodate oxidation (3) is unusual in that an exogenous cross-linking reagent is not used. The peroxidase is a glycoprotein, and reaction of its saccharide residues with sodium periodate results in the formation of aldehyde groups. (The enzyme has few amino groups, and self-coupling can be avoided.) When IgG is added, the modified enzyme reacts with the immunoglobulin's amino groups to form Schiff bases. These are then reduced to give a stable conjugate.

1.3. β-Galactosidase Labeling Using m-Maleimidobenzoyl-N-hydroxysuccinimide Ester (MBS)

MBS is a heterobifunctional reagent, and it is of particular value when one of the proteins involved has no free thiol groups, e. g., IgG. In this method (4), the IgG is first modified by allowing the N-hydroxysuccinimide ester of MBS to react with amino groups in the protein. After the removal of unused reagent, the enzyme is added and the modified IgG reacts with thiol groups in the β-galactosidase via its maleimide groups to form thioether links.

The method employing alkaline phosphatase and glutaraldehyde is the easiest to carry out and gives useful conjugates, although the yields are relatively low; the other two methods give very active conjugates with similar levels of activity (5). The strength of the signal produced by a conjugate, however, will depend on the substrate and detection system used, as well as the specific activities of its components and the method of preparation. In all cases the conjugates formed are somewhat heterogeneous and their molecular weights may exceed 10^6. Removal of unconjugated IgG is important in most applications; this is usually achieved by gel filtration.

Antibody–enzyme conjugates can be used in several related techniques. The most common of these is the ELISA (1,2; and *see* vol. 1 of this series), and, in the "sandwich" forms of this technique, these conjugates are used for the measurement of antibodies and macromolecular antigens. The conjugates can also be applied to the detection of antigens or antibodies in tissue slices or cell layers and to the measurement of substances dotted or blotted on nitrocellulose or other membranes (*see* Chapters 31–34 in this volume).

2. Materials

Ideally, if a specific antibody conjugate is required, the IgG should be purified by immunoaffinity chromatography unless it is derived from a hybridoma and contamination by other proteins is minimal. For many purposes, however, the total IgG fraction of an antiserum may be used.

The enzymes should be free of other proteins and interfering substances. Several manufacturers provide suitable enzymes at very high specific activities, e. g., Boehringer Mannheim Corp., Sigma Chemical Co.

2.1. Preparation of the IgG Fraction of Serum

1. Ammonium sulfate [$(NH_4)_2SO_4$].
2. 20 mM Sodium phosphate buffer, pH 7.2, containing 0.15M NaCl (PBS).
3. 5 mM Sodium phosphate buffer, pH 6.5.
4. DE-52 cellulose (Whatman Ltd., Maidstone, UK) or similar DEAE ion exchanger

2.2. Alkaline Phosphatase Labeling Using Glutaraldehyde

1. Alkaline phosphatase from calf intestinal mucosa, 1000 U/mg or greater. (This is often available as an ammonium sulfate suspension.)
2. PBS.
3. Glutaraldehyde.
4. 50 mM Tris-HCl buffer, pH 7.5, containing 1 mM $MgCl_2$, 0.02% NaN_3, and 2% bovine serum albumin (BSA).

2.3. Horseradish Peroxidase Labeling Using Periodate Oxidation

1. Horseradish peroxidase, 250 purpurogallin units/mg or greater.
2. 0.1M sodium periodate ($NaIO_4$).
3. 1 mM Sodium acetate buffer, pH 4.4.
4. 10mM Sodium carbonate buffer, pH 9.5.
5. 0.2M Sodium carbonate buffer, pH 9.5.
6. Sodium borohydride ($NaBH_4$), 4 mg/mL.
7. Ultrogel AcA34 (LKB Produkter AB, Bromma, Sweden) or similar gel filtration medium.
8. PBS.
9. BSA.

2.4. β-*Galactosidase Labeling Using MBS*

1. β-Galactosidase from *Escherichia coli*, 600 U/mg or greater.
2. 0.1M sodium phosphate buffer, pH 7.0, containing 50 mM NaCl.
3. MBS.
4. Dioxan.
5. Sephadex G-25 (Pharmacia Fine Chemicals AB, Uppsala, Sweden) or equivalent gel filtration medium.
6. 10 mM Sodium phosphate buffer, pH 7.0, containing 50 mM NaCl and 10 mM MgCl$_2$.
7. 2-Mercaptoethanol.
8. DEAE-Sepharose (Pharmacia).
9. 10 mM Tris-HCl buffer, pH 7.0, containing 10 mM MgCl$_2$ and 10 mM 2-mercaptoethanol.
10. Item 9 containing 0.5M NaCl.
11. Item 9 containing 3% BSA and 0.6% NaN$_3$.

3. Methods

3.1. *Preparation of the IgG Fraction of Serum*

1. Stir serum gently at room temperature and add (NH$_4$)$_2$SO$_4$ slowly to give a concentration of 230 g/L. Adjust the pH to 7.8 if necessary and continue stirring for about 1 h.
2. Centrifuge the suspension at 1500g for 15 min at room temperature. Discard the supernatant.
3. Dissolve the pellet in sufficient PBS to give a volume equal to that of the original serum.
4. Repeat steps 1 and 2 in this section.
5. Dissolve the pellet in the minimal amount of PBS and dialyze against 5 mM sodium phosphate buffer, pH 6.5, at 4°C.

6. Add the dialyzed solution to a column of DE-52 (about 5 mL for each mL of serum) equilibrated at room temperature with the phosphate buffer and elute with the same buffer.

7. Pool the fractions with high A_{280}, concentrate by salt precipitation or ultrafiltration, and store the IgG at –20°C or at 4°C in the presence of an antimicrobial agent.

3.2. Alkaline Phosphatase Labeling Using Glutaraldehyde

1. Centrifuge a portion of the ammonium sulfate suspension, containing 1.5 mg of enzyme, at about 2000g for 10 min at 4°C; discard the supernatant.

2. Add 0.5 mg of IgG in 100 µL of PBS.

3. Dialyze the mixture against PBS overnight at 4°C.

4. Add 5% glutaraldehyde (about 10 µL) to give a final concentration of 0.2% (v/v), and stir the mixture for 2 h at room temperature.

5. Dilute the volume to 1 mL and dialyze, as in step 3 in this section.

6. Dilute the dialyzed solution to 10 mL with 50 mM Tris-HCl buffer, pH 7.5, containing 1 mM $MgCl_2$, 0.02% NaN_3, and 2% BSA; store at 4°C.

3.3. Horseradish Peroxidase Labeling Using Periodate Oxidation

1. Dissolve 2 mg of enzyme in 500 µL of water.

2. Add 100 µL of freshly prepared 0.1M $NaIO_4$ and mix for 20 min at room temperature.

3. Dialyze the modified enzyme against 1 mM sodium acetate buffer, pH 4.4, overnight at 4°C.

4. Dissolve 4 mg of IgG in 500 µL of 10 mM sodium carbonate buffer, pH 9.5.

5. Adjust the pH of the dialyzed solution to 9.0–9.5 by adding 10 µL of 0.2M sodium carbonate buffer, pH 9.5, and immediately add the IgG solution. Stir the mixture for 2 h at room temperature.

6. Add 50 µL of freshly prepared NaBH$_4$ solution (4 mg/mL), and stir the mixture occasionally over a period of 2 h at 4°C.

7. Fractionate the conjugate mixture on a column of Ultrogel AcA34 (1.5 × 85 cm) in PBS. Determine the A_{280} and A_{403}.

8. Pool the fractions in the first peak (both A280 and A403 peaks coincide), add BSA to a final concentration of 5 mg/mL, and store aliquots at –20°C.

3.4. β–Galactosidase Labeling Using MBS

1. Dissolve 1.5 mg of IgG in 1.5 mL of 0.1M sodium phosphate buffer, pH 7.0, containing 50 mM NaCl.

2. Add 0.32 mg of MBS in 15 µL of dioxan, mix, and incubate for 1 h at 30°C.

3. Fractionate the mixture on a column of Sephadex G-25 (0.9 × 30 cm), equilibrated with 10 mM sodium phosphate buffer, pH 7.0, containing 50 mM NaCl and 10 mM MgCl$_2$, and elute with the same buffer. Collect 0.5-mL fractions. Measure the A_{280} and pool the fractions in the first peak (about 3 mL).

4. Add 1.5 mg of enzyme to the modified IgG solution, mix, and incubate for 1 h at 30°C.

5. Stop the reaction by adding 1M 2-mercaptoethanol to give a final concentration of 10 mM (about 30 µL).

6. Fractionate the mixture on a column of DEAE-Sepharose (0.9 × 15 cm) equilibrated with 10 mM Tris-HCl buffer, pH 7.0, containing 10 mM MgCl$_2$ and 10 mM mercaptoethanol. Elute with the Tris buffer (50 mL), followed by Tris buffer containing

0.5*M* NaCl (50 mL). Collect 3-mL fractions in tubes containing 0.1 mL of Tris buffer containing 3% BSA and 0.6% NaN_3. Pool the major peak (this is eluted with the NaCl), and store at 4°C.

4. Notes

4.1. General Comments

1. The procedure described for the purification of IgG is suitable for rabbit serum. Other sera may require modifications such as an alteration of the pH or ionic strength used for ion exchange. In most cases a compromise between yield and purity must be reached.

2. The efficacy of an antibody–enzyme conjugate may be tested by immobilizing the appropriate antigen in the wells of a microtiter plate, incubating various dilutions of the conjugate for a few hours, washing, adding the substrate, and measuring the amount of product formed. This procedure may also be used to monitor the purification of conjugates by chromatography.

4.2. Alkaline Phosphatase Labeling Using Glutaraldehyde

1. If the alkaline phosphatase is available as a lyophilized powder or as an NaCl suspension, step 3 in section 3.2 should be omitted and step 1 in section 3.2 modified.

2. Dialysis of small volumes can be conveniently done in narrow dialysis tubing by placing a short glass tube, sealed at both ends, in the tubing so that the space available to the sample is reduced. Transfer losses are minimized by carrying out steps 2–5 in the same dialysis bag.

3. We have found that purification of these conjugates by gel filtration does not improve their activity.

4. BSA is routinely added to conjugate solutions to improve stability and minimize adsorbtion to container walls.

4.3. Horseradish Peroxidase Labeling Using Periodate Oxidation

1. Horseradish peroxidase is often specified in terms of its *RZ*. This is the ratio of the A_{403} to the A_{280} and provides a measure of the heme content of the preparation. A highly purified sample will have a value of about 3, but this is not a guarantee of high specific activity. Conjugates with an *RZ* of about 0.4 perform satisfactorily.

2. Preparations of the enzyme vary in their carbohydrate content, and this can affect the oxidation reaction. Free carbohydrate may be removed by gel filtration. Increasing the $NaIO_4$ concentration concentration to 0.2*M* can also help, but further increases lead to excess oxidation, which damages the activity of the enzyme.

3. During the course of the $NaIO_4$ treatment, the mixture changes color from orange to green; this is reversed during the dialysis.

4. NaN_3 should not be used with peroxidase conjugates because it inhibits the enzyme. If an antimicrobial agent is required, 0.02% merthiolate may be used.

5. This method may also be used with the glycoprotein enzymes, glucose oxidase and amyloglucosidase.

4.4. β-Galactosidase Labeling Using MBS

1. The thiol groups of β-galactosidase may become oxidized during purification and storage. Satisfac-

tory conjugates can be made from preparations with about 10 free thiol groups per molecule, and some manufacturers provide information about the thiol content. It is relatively easy to measure the number of thiol groups (6).

2. These conjugates may also be purified by gel filtration in Ultrogel AcA34 (or an equivalent medium) in 10 mM Tris-HCl buffer, pH 7.0, containing 10 mM MgCl$_2$, 10 mM 2-mercaptoethanol, and 50 mM NaCl.

3. MBS may be used to label IgG with enzymes lacking thiol groups. There are several procedures for introducing these groups into proteins (4,7).

References

1. Tijssen, P. (1985) *Practice and Theory of Enzyme Immunoassays* Elsevier, Amsterdam.

2. Engvall, E. and Perlmann, P. (1972) Enzyme-linked immunosorbent assay, ELISA. III. Quantitation of specific antibodies by enzyme-labelled anti-immunoglobulin in antigen-coated tubes. *J. Immunol.* **109**, 129–135.

3. Wilson, M. B. and Nakane, P. K. (1978) Recent Developments in the Periodate Method of Conjugating Horseradish Peroxidase (HRPO) to Antibodies, in *Immunofluorescence and Related Staining Techniques* (Knapp, W., Holubar, K., and Wick, G. ed.) Elsevier/North Holland Biomedical, Amsterdam.

4. O'Sullivan, M. J., Gnemmi, E., Morris, D., Chieregatti, G., Simmonds, A. D., Simmons, M., Bridges, J. W., and Marks, V. (1979) Comparison of two methods of preparing enzyme-antibody conjugates: Application of these conjugates for enzyme immunoassay. *Anal. Biochem.* **100**, 100–108.

5. Halliday, M. I. and Wisdom, G. B. (1986) A comparison of three methods for the preparation of enzyme antibody conjugates. *Biochem. Soc. Trans.* **14**, 473–474.

6. Ellman, G. L. (1959) Tissue sulfhydryl groups. *Arch. Biochem. Biophys.* **82**, 70–75.

7. Klotz, I. M. and Heiney, R. E. (1962) Introduction of sulfhydryl groups into proteins using acetylmercaptosuccinic anhydride. *Arch. Biochem. Biophys.* **96**, 605–612.

Chapter 28

Vacuum Blotting

An Inexpensive, Flexible, Qualitative Blotting Technique

Marnix Peferoen

1. Introduction

  The first system we used to transfer proteins from polyacrylamide gels onto nitrocellulose was a capillary blotting system, based on the work of Renart et al. (1). Although we were very enthusiastic about the wide range of possible applications created by the protein blotting technique, we were unsatisfied with the slow and inefficient transfer. Electroblotting (2,3) seemed to be much more efficient, but at that time there were serious financial limitations in the department.

  Instead of using capillary force to create a flow of buffer through the gel, we decided to speed up the process by using suction force to draw proteins out of the gel (4). Basically, the assembly for vacuum blotting is very similar to the set-up for capillary blotting, but the entire system is closed and connected to a vacuum pump. Vacuum blotting proved to be significantly faster than capillary blotting.

2. Materials

All buffers are prepared with analytical grade reagents that are dissolved in distilled water. Buffers, when stored at 4°C, are stable for at least 3 mo.

1. Transfer buffer: 3.03 g Tris (25 mM) and 11.26 g glycine (150 mM) are dissolved in 700 mL of water. To this solution, 200 mL of methanol is added, and the final volume is made up to 1000 mL with water. The pH of this buffer is 8.3.

2. Rinsing buffer: 1.14 g Na_2HPO_4 (16 mM), 0.27 g KH_2PO_4 (4 mM), and 3.36 g NaCl (115 mM) are dissolved in 400 mL of water followed by the addition of 1.5 mL of Tween 20. The buffer is made up to 500 mL with water, and the pH should be 7.3 without any adjustment.

3. Transfer membrane: A nitrocellulose membrane of 0.22 µm pore size (Millipore, HAWP 293 25) is used as immobilizing paper.

4. Vacuum blotting equipment. We use a commercial slab gel dryer from Bio-Rad (Model 1125), and vacuum is created with a suction pump (Heraeus, 250W). The pump is protected from buffer by inclusion of a side arm flask and a silica gel column in the vacuum line. The silica gel column is made from a plastic transparent tube (50 x 5 cm) filled with dry silica gel. One also needs a flexible plastic sheet and approximately 20 filter papers (e.g., Whatman No. 2, 16 x 16 cm). A porous polyethylene sheet on the bottom of the slab gel dryer creates an evenly distributed vacuum.

3. Method

3.1. Vacuum Blotting

1. Proteins are separated in SDS-polyacrylamide gels. After electrophoresis the gel is soaked for 20–30 min in the transfer buffer. The gel can be immediately processed or stored frozen in a plastic bag containing some 30 mL of transfer buffer.

2. A sheet of dry filter paper is laid on a porous polyethylene sheet on the bottom of the slab gel dryer. The nitrocellulose membrane is uniformly wet with transfer buffer and is placed on the dry filter paper. A window, some 5 mm smaller than the gel, is cut in the center of a plastic sheet (25 x 25 cm). The plastic sheet is then placed on the dry filter paper, and its window on the nitrocellulose membrane. The gel is put on the nitrocellulose so that its edges are resting on the plastic sheet (Fig. 1A). Air bubbles between the gel and the nitrocellulose membrane should be removed. Some 15 sheets of filter paper are saturated with transfer buffer and placed atop the gel. Again, trapping of air bubbles either between the filter papers and the gel or between the filter papers themselves, should be avoided. The four outer edges of the plastic sheet are folded on top of the saturated filter papers (Fig. 1 B,C).

3. The suction pump is connected via a silica gel column and a side arm flask to the slab gel dryer. The slab gel dryer is then closed with its rubber sheet, and the transfer buffer contained in the filter papers on top of the gel is driven through the gel by vacuum (Fig. 1D).

4. The gel is typically blotted for 1 h.

5. After the transfer, the nitrocellulose membrane is washed in rinsing buffer for 30 min at 37°C, fol-

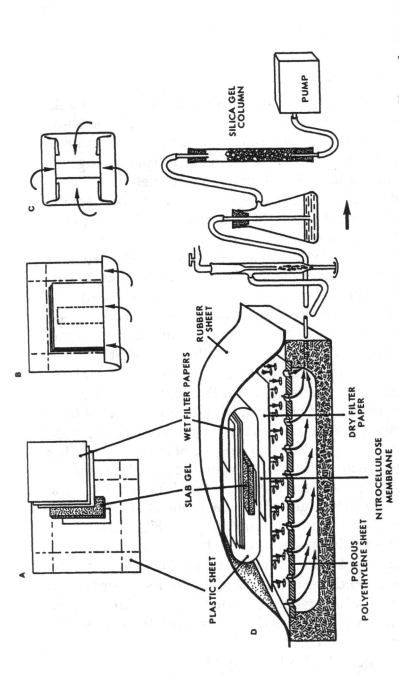

Fig. 1. Vacuum blotting set-up. (A) A nitrocellulose membrane is saturated in transfer buffer and put on a dry filter paper on the bottom of the slab gel dryer. A plastic sheet with a window 5-mm smaller than the gel is placed on the membrane. The gel is put on the nitrocellulose membrane so that its edges rest on the plastic sheet. The gel is covered with approximately 15 sheets of filter paper, saturated with transfer buffer. (B–C). The outer edges of the plastic sheet are folded on top of the filter papers. (D). The slab gel dryer is connected to the suction pump or a water jet pump and is closed with its rubber sheet.

lowed by three times 15 min at room temperature. The membrane is then stained with a general protein stain (Amido black, Indian ink, Aurodye, *see* Chapters 29 and 34 in this volume), or specific proteins are detected by immunostaining (*see* Chapters 31–33 in this volume).

4. Notes

1. In this study we used the transfer buffer originally described by Towbin et al. (2). Since we do not have to worry about high conductivity and high temperature, however, there is no limitation concerning the choice of buffer. One can use the buffer that provides the best conditions for transferring and binding proteins to the nitrocellulose membrane.

2. The rinsing buffer can also be used for immunostaining of membrane bound proteins. The concentration of Tween-20 is high enough to provide efficient blocking of protein binding sites on the nitrocellulose.

3. Nitrocellulose membranes with a pore size of 0.45 μm are less efficient in absorbing proteins of low molecular weight (lower than 20,000), which results in insufficient immobilization, causing leakage or diffusion of the blotted proteins (3,4,5). Vacuum blotting should also be compatible with nylon membranes or the Immobilon PVDF membrane (Millipore). Nylon membranes have higher binding capacities than nitrocellulose, and they are considerably stronger. In order to reduce background staining, nylon membranes require incubations with high concentrations of blocking agents and still give a high background with alkaline phosphatase and glucose oxidase conjugate detec-

tion systems. Nylon membranes are also incompatible with most protein stains (except Ferridye from Janssen Life Science Products and a biotin/avidin detection system). The Immobilon PVDF Transfer Membrane is reported to have high mechanical strength and high protein binding capacity and to be compatible with all commonly used protein and immunostaining techniques.

4. We use a commercial slab gel dryer, but any system that is suitable to dry gels should also be suitable for vacuum blotting. Even a mechanical suction pump is not indispensable, since a good water jet pump does work as well (results not shown). The plastic sheet between the stack of filter papers and the nitrocellulose membrane prevents the transfer buffer from passing around the gel instead of through the gel. It is therefore important to trim the window so that the edges of the gel rest on the plastic sheet. The 15 sheets of filter paper can absorb approximately 300–400 mL of transfer buffer.

5. Since AuroDye is a very sensitive stain, intense background staining can be a problem. A specific staining can substantially be reduced by inserting a sheet of nitrocellulose between the gel and stack of wet filter papers, so that the gel is sandwiched between two nitrocellulose sheets.

6. If the transfer is unsatisfactory after 1 h of blotting, the blotting time can be extended for several hours. In that case, it is necessary to add more transfer buffer to the filter papers. Every hour the blotting is interrupted by lifting the rubber sheet and unfolding the edges of the plastic sheet. Approximately 100 mL of transfer buffer is then poured on the filter papers. After folding the plastic sheet on top of the filter papers, the gel dryer is closed with

the rubber sheet, and the blotting continues. The addition of transfer buffer to the filter papers not only prevents the gel from drying, but also increases the transfer efficiency. The more transfer buffer used, the better the transfer.

7. The quality of the vacuum blotting is illustrated in Fig. 2. Female *Locusta migratoria* hemolymph is collected from a *Locusta* laboratory culture and separated in polyacrylamide-urea tubes by non-equilibrium pH gradient (3.5–10) gel electrophoresis (NEpHGE; 6) for 2000 Vh (volt-hours). The gel tubes are then electrophoresed at right angles on SDS-polyacrylamide gradient (5–15%) gels with the discontinuous buffer system (7). Proteins in the gel are silver stained according to Heukeshoven and Dernick (8). The duplicate gel is blotted for 1 h with 500 mL transfer buffer. Proteins on nitrocellulose are detected with AuroDye (9; Janssen Life Sci. Products). Figure 2a is part of the gel, stained with silver, containing two female specific yolk polypeptides (YP) with molecular weights in the range of 120,000. The blot is stained overnight with AuroDye (Fig. 2b). The vacuum blotting faithfully replicates the high resolution of the two-dimensional electrophoresis on the nitrocellulose membrane.

8. The transfer efficiency of vacuum blotting is determined by blotting [^{14}C]-methylated proteins of different molecular weight (M_r, 14,300–200,000, Amersham). After SDS-polyacrylamide gradient (5–15%) electrophoresis, the gel is cut in four strips, each containing two lanes of separated ^{14}C-methylated polypeptides. Three of these strips are then blotted onto nitrocellulose for 1, 2, or 4 h. Approximately 100 mL of transfer buffer is added every half hour. The proteins in the gel and on the nitrocellu-

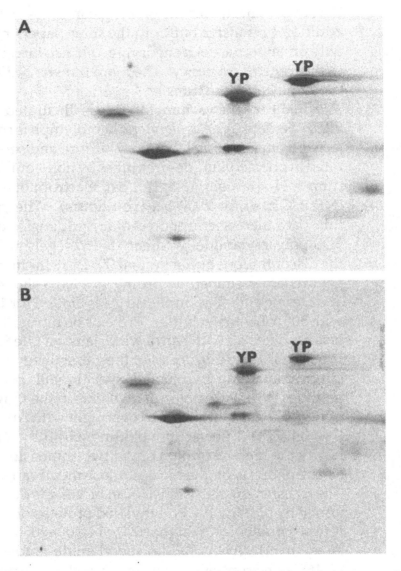

Fig. 2. NEpHGE followed by SDS-PAGE of female *Locusta migratoria* hemolymph. The gel is stained with silver (A) or blotted onto nitrocellulose for 1 h with 500 mL of transfer buffer. Proteins on the nitrocellulose are stained with AuroDye. (B). Only part of the gel and the corresponding piece of nitrocellulose containing two female specific yolk proteins (YP) are shown.

Table 1
Transfer Efficiency of Vacuum Blotting[a]

M_r of proteins	Percentage of ^{14}C-methylated proteins transferred to nitrocellulose		
	1 h	2 h	4 h
14,300	28	45	64
30,000	23	38	54
46,000	18	31	45
69,000	23	32	44
92,500	18	26	36
200,000	17	21	32

[a]After SDS-polyacrylamide gradient (5–15%) gel electrophoresis, ^{14}C-methylated polpeptides are transferred to nitrocellulose by vacuum blotting for 1, 2, and 4 h. Every half hour, approximately 100 mL of transfer buffer is added to the stack of filter papers, on top of the gel. After visualization by radioautography, radioactive polypeptides are cut from the gel and from the nitrocellulose membrane for liquid scintillation counting. Counts in polypeptides that are not blotted are referred to as 100%.

lose are visualized by radioautography, so that proteins can be cut in order to measure the radioactivity by liquid scintillation counting. Counts in electrophoresed polypeptides that are not blotted are referred to as 100%. Counts in transferred polypeptides are expressed as percentages of counts in polypeptides before blotting (Table 1). The transfer efficiency decreases with increasing molecular weight, and after 4 h there is twice as much lysozyme (M_r, 14,300) blotted as there is myosin (M_r, 200,000) (64 and 32%, respectively). These data show that the transfer efficiency of vacuum blotting is comparable to the efficiency of low-powered electroblotting (3,5). This also means that vacuum blotting, like low-powered electroblotting, is not very suitable for quantitative blotting. If, however, only a qualitative transfer is required, vacuum blotting is a very reasonable alternative to electroblotting. Apart from the fact that vacuum blotting is very inexpensive and easy to use, it may offer some particular advantages. There is for instance absolutely no limitation in the choice of buffers, since there is no increasing conductivity and temperature during transfer. In electroblotting, some proteins are at their isoelectric point and can therefore not be transferred. Vacuum blotting can be used to transfer proteins that otherwise would need buffers with very acidic or basic buffers. It is even possible to transfer proteins selectively from the gel, by adjusting the transfer buffer. Vacuum blotting is used in a wide variety of protein studies (10–14) and is also adapted to transfer DNA (15).

Acknowledgment

Figures are reprinted with permission from the *Handbook of Immunoblotting* (Bjerrum, O.J. and Heegard, N.H.H., eds.) CRC, Boca Raton, Florida.

References

1. Renart, J., Reiser, J., and Stark, G.R. (1979) Transfer of proteins from gels to diazobenzyloxymethyl-paper and detection with antisera: A method for studying antibody specificity and antigen structure. *Proc. Natl. Acad. Sci. USA* **76**, 3116–3120.
2. Towbin, H., Staehelin, T., and Gordon, J. (1979) Electrophoretic transfer of proteins from polyacrylamide gels to nitrocellulose sheets: Procedure and some application. *Proc. Natl. Acad. Sci. USA* **76**, 4350–4354.
3. Bittner, M., Kupferer, P., and Morris, C.F. (1980) Electrophoretic transfer of proteins and nucleic acids from slab gels to diazobenzyloxymethyl cellulose or nitrocellulose sheets. *Anal. Biochem.* **102**, 459–471.

4. Peferoen, M., Huybrechts, R., and De Loof, A. (1982) Vacuum-blotting: A new simple and efficient transfer of proteins from sodium dodecyl sulfate-polyacrylamide gels to nitrocellulose. *FEBS Lett.* **145**, 369–372.

5. Lin, W. and Kasamatsu, H. (1983) On the electrotransfer of polypeptides from gels to nitrocellulose membranes. *Anal. Biochem.* **128**, 302–311.

6. O'Farrell, P.Z., Goodman, H.M., and O'Farrell, P.H. (1977) High resolution two-dimensional electrophoresis of basic as well as acidic proteins. *Cell* **12**, 1133–1142.

7. Laemmli, U.K. (1970) Cleavage of structural proteins during assembly of the head of bacteriophage T$_4$. *Nature* **227**, 680–685.

8. Heukshoven, J. and Dernick, R. (1985) Simplified method for silver staining of proteins in polyacrylamide gels and the mechanism of silver staining. *Electrophoresis* **6**, 103–112.

9. Moeremans, W., Daneels, G., and De Mey, J. (1985) Sensitive colloidal metal (gold or silver) staining of protein blots on nitrocellulose membranes. *Anal. Biochem.* **145**, 315–321.

10. Ishida, I., Ichikawa, T., and Deguchi, T. (1983) Immunochemical and immunohistochemical studies on the specificity of a monoclonal antibody to choline acetyltransferase of rat brain. *Neurosci. Lett.* **42**, 267–271.

11. Briers, T. and Huybrechts, R. (1984) Control of vitellogenin synthesis by ecdysteroids in *Sarcophaga bullata*. *Insect Biochem.* **14**, 121–126.

12. Geysen, J., De Loof, A., and Vandesande, F. (1984) How to perform subsequent or "double" immunostaining of two different antigens on a single nitrocellulose blot within 1 day with an immunoperoxidase technique. *Electrophoresis* **5**, 129–131.

13. Peferoen, M. and De Loof, A. (1984) Intraglandular and extraglandular synthesis of proteins secreted by the accessory reproductive glands of the Colorado potato beetle, *Leptinotarsa decemlineata*. *Insect Biochem.* **14**, 407–416.

14. Levine, M.M., Nataro, J.P., Karch, H., Baldini, M.M., Kaper, J.B., Black, R.E., Clements, M.L., and O'Brien, A.D. (1985) The diarrheal response of humans to classic serotypes of enteropathogenic *Escherichia coli* is dependent on a plasmid encoding an enteroadhesiveness factor. *J. Infect. Dis.* **152**, 550–559.

15. Zaitsev, I.Z. and Yakovlev, A.G. (1983) Vacuum transfer of DNA to filters for detecting interindividual polymorphism by Southern blotting hybridization method. *B. Exp. B. Med.* **96**, 1442–1444.

Chapter 29

Blotting with Plate Electrodes

Marnix Peferoen

1. Introduction

Since its introduction by Towbin et al. (1), electro-blotting has become one of the most valuable analytical procedures in protein research. Conventionally, the gel and the immobilizing matrix are sandwiched between porous pads and inserted between two electrodes in a buffer tank. Electrodes of platinum wire are mounted in such a way that, with a minimum amount of material, they still generate a uniform electrical field in the center of the buffer chamber, with the minimum amount of material. On the other hand, transfer efficiency is largely dependent of the voltage gradient between the two electrodes. High voltages generate high currents and high temperatures, whereas reducing the distance between wire electrodes seriously disturbs the electrical field (2). Plate electrodes yield essentially homogeneous electrical fields, even when they are moved close together. Vaessen et al. (3) were the first to experiment with plate electrodes. Kyhse-Andersen (4) introduced horizontal

semidry blotting in which the transfer takes place in a
stack of buffer-saturated filter papers, inserted between
two graphite plates. Svoboda et al. (5) simplified the
buffer system from three to two different solutions and
used corrosion resistant electrodes. Recently Bjerrum
and Schafer-Nielsen (6) described a homogeneous buffer system with transfer efficiency equal to the efficiency
of the discontinuous buffer system. The major advantages of the horizontal semidry electroblottting are:
Simple assembly, reduced buffer consumption, uniform electrical field, high transfer efficiency, and simultaneous blotting of up to six full-size gels.

2. Materials

Buffers are made with analytical grade reagents
dissolved in distilled water. Buffers are stable for at least
3 mo when stored at 4°C.

1. Transfer buffer: 2.91 g of Tris base (48 mM), 1.46 g
    of glycine (39 mM), and 0.19 g of SDS (1.3 mM) are
    dissolved in 350 mL of water and adjusted to pH 9.2
    with HCl. After addition of 100 mL of methanol,
    the final volume is made up to 500 mL with water.
2. Rinsing buffer: 1.21 g of Tris base (10 mM) and 8.27
    g of sodium chloride (150 mM) are dissolved in 800
    mL of water and adjusted to pH 8.0 with hydrochlo-
    ric acid. The buffer is made up to 1000 mL, and 1 mL
    of Tween-20 is added.
3. Transfer membrane: Immobilon PVDF transfer
    membrane (Millipore, IPVH 000 10) is used as an
    immobilizing matrix.
4. Electroblotting equipment: Proteins are trans-
    ferred with a horizontal semidry multigel electro-
    blotter, produced by Ancos (Denmark) and distrib-
    uted in the US by Gelman Sciences Inc. The appa-
    ratus consists of a plastic base in which the anodic

carbon plate (17 x 17 cm) is fixed, whereas the cathodic carbon plate of equal dimensions is fixed in the lid of the apparatus. The transfer buffer is soaked in four sheets of absorbent paper.

5. Protein stain: Proteins in the gel are stained with Coomassie brilliant blue R250. Proteins on the Immobilon membrane are stained with Indian ink (0.1% v/v in rinsing buffer) or Amido black 10B (0.1% w/v in methanol and 2% acetic acid).

3. Method

3.1. Electroblotting

1. Proteins are separated in SDS-polyacrylamide slab gels. Four pieces of absorbent filter paper (Bio-Rad, 165-0921) and the transfer membrane, trimmed to the size of the gel, are saturated with transfer buffer.

2. After electrophoresis, two pieces of buffer-saturated absorbent paper are put on the anodic (bottom) plate electrode. The transfer membrane is pre-wet in a small volume of 100% methanol, followed by a brief rinse in transfer buffer and is then layered on the blotting papers. The gel is placed on the membrane and covered with the two other pieces of saturated absorbent paper. During assembly, trapping of air bubbles between any of the layers should be avoided. The semidry electroblotter is then closed with the lid containing the cathode.

3. The blotting apparatus is connected to a power supply and proteins are transferred for 1–2 h at a constant current density of 0.8–1.0 mA/cm^2 of gel (*see* Note 5 in section 4).

4. Following transfer, blots may be probed using immunodetection methods (*see* Chapters 31–34) or stained for protein as described below.

3.2. Protein Staining

1.   After the transfer, the membrane is incubated for 30 min at 37°C in rinsing buffer, followed by three times 15 min at room temperature. The membrane is then stained overnight with Indian ink and rinsed with rinsing buffer.
2.   Membranes can also be stained with Amido black for 15 min, followed by destaining for 15 min in 45% methanol, 7% acetic acid. The last traces of background stain are removed by a 1–2 min rinse in 90% methanol. The membrane is then dried for storage.

4. Notes

1.   This transfer buffer is one of the many that can be used with the semidry blotter. Originally, a discontinuous buffer system consisting of three different buffers, resulting in isotachophoretic transfer of the proteins, was developed (4). Svoboda et al. (5) use a Tris-glycine buffer (25 mM Tris-192 mM glycine, pH 8.3) with 20% methanol at the anodic side and with 0.1% SDS at the cathodic side. Bjerrum and Schafer-Nielsen (6) describe a continuous buffer system (48 mM Tris, 39 mM glycine, 1.3 mM SDS, 20% methanol, pH 9.2) with transfer efficiency as good as that obtained with the discontinuous buffer system of Khyse-Andersen (4).
2.   In this study we use an Immobilon PVDF membrane, which is a polyvinylidene difluoride matrix from Millipore. This membrane is not as brittle as nitrocellulose, is very resistant to high methanol concentrations, and has high protein-binding capacity and a superior retention. The only difference with this membrane is that it has to be pre-wet with

100% methanol. (The membrane should be kept wet at all times, and if it does dry out, it should be re-wetted in 100% methanol). The Immobilon PVDF membrane is compatible with all immuno-chemical stainings and with the AuroDye protein staining. It can also be stained with Coomassie blue and Amido black since the background staining can be removed by rinsing in 90% methanol.

3. Apart from the Ancos apparatus, there are several horizontal semidry electroblotters commercially available (e.g., Kem-En-Tec, Hellerup, Denmark; Sartorius, Gottingen, FRG; Biolyon, Lyon, France). The electroblotter from Idea Scientific Co. (Corvallis, Oregon) also uses plate electrodes, but is a vertical apparatus with a real buffer chamber. Most of the plate blotters use graphite electrodes, which are cheap, but need to be cleaned before each run and have a limited resistance to abrasion. The apparatus of Biolyon presents an elegant solution to that problem, without using the very expensive platinum plates. The anode is a surface conductive glass plate (Hortiplus from Glaverbel, Brussels, Belgium), whereas the cathode is a stainless steel plate (stainless steel does not resist anodic oxidation).

4. Most horizontal semidry blotters allow transfer of proteins from several (up to six) gels, stacked on top of each other. The transfer efficiency gradually decreases toward the cathode, however (6). Dialysis membranes must be interleafed to prevent contamination of the gels and membranes with substances being transferred from other gels in the same stack. When small gels (less than half the size of the plate electrode) are to be blotted, short circuiting should be prevented by inserting isolating material (e.g., slab gel spacers) between the electrodes.

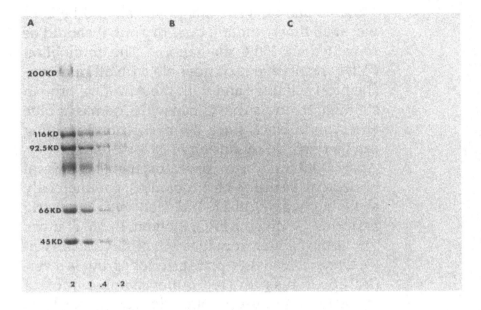

Fig. 1. SDS-polyacrylamide gradient (5–15%) gel electrophoresis of a high molecular weight mixture (45,000–200,000). The protein mixture is applied at 2, 1, 0.4, and 0.2 µg of individual protein per lane, stained with Coomassie brilliant blue (A) or first blotted for 60 (B) and 90 (C) min, and then stained.

5. The transfer can be conducted either at constant voltage or at constant current. If the transfer is performed at constant current, the size of the gel should be considered so that the current density ranges between 0.8 and 1.0 mA/cm². Under these conditions, the voltage will increase during the transfer, from about 10 V at the beginning to about 40 V after 2 h. Using constant voltage, the distance between the electrodes should be considered. As a rule of thumb, 20 V x h should result in an almost complete transfer. The power limit depends on the heat dissipation capacity of the apparatus. Bjerrum and Schafer-Nielsen (6) measured a core tempera-

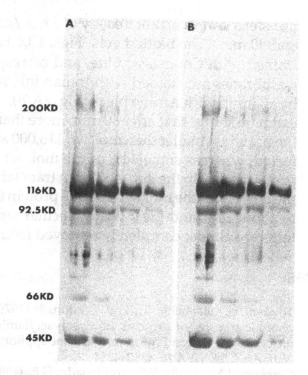

Fig. 2. Proteins transferred onto the Immobilon PVDF membrane from gels after 60 (A) and 90 (B) min or blotting and stained with Indian ink and Amido black. Blots (A) and (B) corresponds to gels 1 (B) and 1(C) respectively.

      ture of 40°C after 2 h of blotting at 18 W. At a 1 mA/cm² of gel (16 x 16 cm) and 40 V, the power is 10 W, well within the limit.

6. The transfer quality and efficiency is illustrated by blotting a protein mixture with molecular weights varying between 45,000 and 200,000 (Fig. 1). This protein mixture is electrophoresed at different concentrations (2, 1, 0.4, and 0.2 µg of individual protein) in a SDS polyacrylamide gradient (5–15%) gel. Part of the gel is stained with Coomassie brilliant blue (Fig. 1A), and two identical sets of gel lanes are

transferred at a current density of 1 mA/cm²for 60 and 90 min. The blotted gels (Figs. 1B,C) are then stained with Coomassie blue, and corresponding membranes are stained with Indian ink, followed by staining with Amido black (Fig. 2A,B). A rough estimate shows that after 60 min, more than 90% of the proteins smaller than mol. wt. 116,000 are transferred, whereas some 20% of the mol. wt. 200,000 protein is still in the gel. A 90-min transfer reduces the amount of the mol. wt. 200,000 protein in the gel to about 10%, and the lower molecular weigh proteins are almost completely removed from the gel.

References

1. Towbin, H., Staehelin, T., and Gordon, J. (1979) Electrophoretic transfer of proteins from polyacrylamide gels to nitrocellulose sheets: Procedure and some applications. *Proc. Natl. Acad. Sci. USA* **76**, 4350–4354.
2. Gershoni, J.M., Davis, F.E., and Palade, G.E. (1985) Protein blotting in uniform or gradient electric fields. *Anal. Biochem.* **144**, 32–40.
3. Vaessen, R.T.M.J., Kreike, J., and Groot, G.S.P. (1981) Protein transfer to nitrocellulose filters. A simple method for quantitation of single proteins in complex mixtures. *FEBS Lett.* **124**, 193–196.
4. Khyse-Andersen, J. (1984) Electroblotting of multiple gels: A simple apparatus without buffer tank for rapid transfer of proteins from polyacrylamide to nitrocellulose. *J. Biochem. Biophys. Meth.* **10**, 203–209.
5. Svoboda, M., Meuris, S., Robyn, C., and Christophe, J. (1985) Rapid electrotransfer of proteins from polyacrylamide gel to nitrocellulose membrane using surface-conductive glass as anode. *Anal. Biochem.* **151**, 16–23.
6. Bjerrum, O.J. and Schafer-Nielsen, C. (1986) Buffer Systems and Transfer Parameters for Semidry Electroblotting with a Horizontal Apparatus, in *Electrophoresis '86* Proceedings of the Fifth Meeting of the International Electrophoresis Society (Dunn, M.J., ed.) VCH Verlagsgesellschaft, Weinheim, FRG.

Chapter 30

Use of Dried Milk for Immunoblotting

Rosemary Jagus and Jeffrey W. Pollard

1. Introduction

Immunoprobing for specific peptides on solid supports such as nitrocellulose requires prior blocking of nonspecific protein binding sites. Many proteins, or protein mixtures, have been used for this, including bovine serum albumin, gelatin, and calf serum, but the cheapest and most efficient method makes use of dried milk protein.

The method was first described, and compared with other methods, by Johnson et al. (1), who refer to the blocking cocktail as BLOTTO (bovine lacto transfer technique optimizer). BLOTTO consists of a 5% solution of dried milk in Tris-buffered saline, along with antifoam and antimicrobial agents. This cocktail is used for blocking, as an incubation medium, and for subsequent washing to remove unreacted reagents. The initial description of the technique includes blocking protocols for immunoblots (Western blots), and the probing of RNA andDNA blots with radiolabeled DNA (Northern and Southern blots). However, the main usefulness of the procedure has been for immunoblotting, along with the probing of protein blots with radiolabeled RNA and DNA (North- and South-Western blots). The reagent is effective and easy to prepare, as well as being considerably cheaper than any available alternatives.

2. Materials

1.   10 x Tris-buffered saline (10 x TBS): $1.5M$ NaCl, 250mM Tris-HCl, pH 7.5. Solution should be autoclaved and stored at 4°C.
2.   1x TBS: 1:10 dilution of 10x TBS in sterile distilled water. Store at 4°C.
3.   BLOTTO: 5% (w/v) nonfat dried milk, 0.01% (v/v) Antifoam A (Sigma), 0.01% (v/v) sodium azide, in TBS. Store at –4°C.
4.   [^{125}I]-protein A, specific activity, 2–10 µCi/µg. Store at –4°C.
5.   Antibody directed against protein of interest. This may be a monoclonal antibody or antiserum. Immunoglobulins should be separated from whole serum by ion exchange chromatography (2). Antibodies should be stored at –70°C in TBS.

3. Methods

The following procedure is for processing two 13 x 15 cm blots.

1.   Rinse the nitrocellulose blots, to which proteins have been transferred, in 50 mL of TBS. Discard the TBS.

2. Add 50 mL of BLOTTO and incubate at room temperature (15–20°C), with shaking, for 1 h.
3. Discard the BLOTTO. Add 50 mL of fresh BLOTTO containing the first antibody.
4. Incubate at room temperature for 2–15 h, with shaking.
5. Rinse briefly in 50 mL of TBS. Discard the TBS.
6. Rinse three times in 50 mL of BLOTTO for 5 min each rinse, at room temperature, with shaking.
7. Add fresh BLOTTO containing the second antibody, if used (*see* section 4).
8. Incubate at room temperature for 2 h, with shaking.
9. Rinse briefly in 50 mL of TBS, then discard the TBS.
10. Rinse three times in 50 mL of BLOTTO for 5 min each rinse, at room temperature, with shaking.
11. Add fresh BLOTTO containing 20 µCi of [^{125}I] protein A, 0.1% (v/v) Triton X-100 and 0.1% (w/v) sodium dodecyl sulfate.
12. Incubate at room temperature for 2 h, with shaking.
13. Rinse briefly in 50 mL of TBS. Discard the TBS.
14. Rinse three times in 50 mL of BLOTTO containing 0.1% (v/v) Triton X-100 and 0.1% (w/v) sodium dodecyl sulfate, for 10 min each rinse, at room temperature, with shaking.
15. Rinse three times, briefly, in TBS.
16. Dry, mount blots on card, cover with plastic wrap, and expose to Kodak XAR-5 film, sandwiched between intensifying screens, at –70°C for 5–24 h.

4. Notes

1. Protein A, from *Staphylococcus aureus*, recognizes the Fc region of many, but not all, IgGs. For instance, rabbit and human IgGs are recognized by protein A, but mouse, sheep, and chicken IgGs are

not. If using a first antibody not recognized by protein A, it is necessary to use a second, "sandwich," antibody. For instance, if the available antibody is a mouse monoclonal antibody, then it is necessary to use, for instance, rabbit anti-mouse IgG as the second antibody. An additional advantage of using a sandwich antibody is that, since most antibodies are divalent, amplification of the signal occurs, increasing the sensitivity of the method.

2. The sensitivity of the method varies according to the avidity of the first antibody for the protein of interest, but can be sensitive down to 0.5 ng. Figure 1 shows a BLOTTO-processed immunoblot of 0.5–2 ng of the beta-subunit of eukaryotic initiation factor 2 (eIF-2), using affinity-purified sheep anti-beta-subunit as the first antibody. Details of the methodology can be found in Austin et al. (3).

3. Other methods of detection can be used. For instance, instead of being ^{125}I-labeled, protein A can be conjugated to horseradish peroxidase (4), alkaline phosphatase (5), or colloidal gold (4). Alternatively, second antibodies radiolabeled with ^{125}I or conjugated to the aforementioned components can be used (*see* Chapters 31–34 in this volume).

4. Even with the use of an antimicrobial agent, BLOTTO can turn sour. The use of putrid BLOTTO gives rise to high backgrounds, and the "smell" test is recommended. The nitrocellulose should be handled with gloved hands; finger "extracts" contain proteins and oils and can prevent wetting. Sodium azide inhibits horseradish peroxidase and therefore, if this method of detection is employed, the azide should be left out of the BLOTTO mix. Such a mix should be used for only a few days before it is discarded.

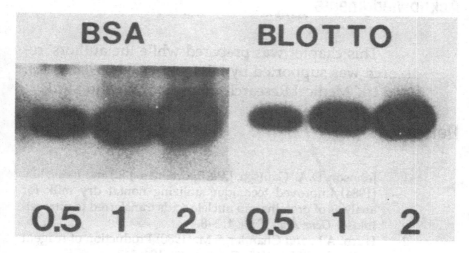

Fig. 1. Comparison of BLOTTO with BSA for blocking nonspecific protein binding sites on immunoblots. Purified eIF-2 (0.5–2 pmol) was subjected to electrophoresis by SDS-PAGE (15% polyacrylamide), and the separated subunits were transferred electrophoretically to nitrocellulose. Nonspecific protein binding sites were blocked with either BLOTTO or 3% BSA in TBS for 1 h. The blots were probed with affinity-purified antibody to the beta-subunit of eIF-2 and processed as described in this chapter. Autoradiographs are shown here.

5. Although BLOTTO has been described for use in the processing of Northern and Southern blots, such a technique gives rise to higher backgrounds than when using the conventional Denhardt's reagent. Nucleic acid binding to membrane-bound proteins (North- and South-Western blotting) appears to be significantly improved by the use of BLOTTO, however, and may require its use to guarantee specificity (6).

Acknowledgments

This chapter was prepared while the authors' research was supported by NIH grant GM 33905 to R.J. and UK Medical Research Council funding to J.W.P.

References

1. Johnson, D.A., Gautsch, J.W., Sportsman, J.R., and Elder, J.H. (1984) Improved technique utilizing nonfat dry milk for analysis of proteins and nucleic acids transferred to nitrocellulose. *Gene Anal. Tech.* **1**, 3–8.
2. Hurn, A.L. and Chantler, S.M. (1980) Production of reagent antibodies. *Methods in Enzymol.* **70**, 104–142.
3. Austin, S.A., Pollard, J.W., Jagus, R., and Clemens, M. (1986) Regulation of polypeptide initiation and activity of initiation factor 2 in Chinese hamster ovary cell mutants containing temperature-sensitive aminoacyl tRNA synthetases. *Eur. J. Biochem.* **157**, 39–47.
4. Brada, D. and Roth, J. (1984) Golden blot. Detection of polyclonal and monoclonal antibodies bound to antigens on nitrocellulose by protein A-gold complexes. *Anal. Biochem.* **142**, 79–83.
5. Burnette, W.N. (1981) Western blotting. *Anal. Biochem.* **112**, 195–203.
6. Miskimins, W.K., Roberts, M.P., McClelland, A., and Ruddle, F.H. (1985) *Proc. Natl. Acad. Sci.* **82**, 6741–6744.

Chapter 31

Immunodetection of Proteins on "Western" Blots Using ^{125}I-Labeled Protein A

Nicholas J. Kruger and John B. W. Hammond

1. Introduction

Immunoblotting provides a simple and effective method for identifying specific antigens in a complex mixture of proteins. Initially, the constituent polypeptides are separated using SDS-polyacrylamide gel electrophoresis, or a similar technique, and then are transferred either electrophoretically or by diffusion onto a nitrocellulose filter. The transfer of proteins from gels onto membranes is discussed in vol. 1 of this series and Chapters 28 and 29, this volume. Once immobilized on a sheet of nitrocellulose, specific polypeptides can be identified using antibodies that bind to antigens retained on the filter. In this chapter we describe conditions appropriate for binding antibodies to immobilized proteins and a method for locating these antibody–antigen complexes using ^{125}I-labeled protein A. These methods are based on those of Burnett (1) and Towbin et al. (2).

Although there are several different techniques for visualizing antibodies bound to nitrocellulose, radiolabeled protein A is both convenient and sensitive. Protein A specifically binds the Fc region of immunoglobulin G (IgG) from many mammals (*see* Chapter 26). Thus this compound provides a general reagent for detecting antibodies from several sources. Using this method, as little as 0.1 ng of protein may be detected, though the precise amount will vary with the specific antibody titer (3). The principal disadvantage of protein A is that it fails to bind effectively to major IgG subclasses from several experimentally important sources, such as rat, goat, and sheep. For antibodies raised in such animals, a similar method using iodinated protein G may be suitable (*see* section 4).

Both antigens and antisera can be screened efficiently using immunoblotting. Probing of a crude extract after fractionation by SDS-polyacrylamide gel electrophoresis indicates the specificity of an antiserum. The identity of the antigen can be confirmed using a complementary technique, such as immunoprecipitation of enzyme activity. This information is essential if the antibodies are to be used reliably. Once characterized, an antiserum may be used to identify antigenically related proteins in other extracts using the same technique. Examples of the potential of immunoblotting have been described by Towbin and Gordon (4).

2. Materials

1. Electrophoretic blotting system, such as Trans-Blot, supplied by Bio-Rad.
2. Nitrocellulose paper: 0.45 μm pore size.
3. ^{125}I-labeled protein A, specific activity 30 mCi/mg. Affinity-purified protein A, suitable for blotting, is available commercially (*see* Note 1 in section 4). ^{125}I

emits γ-radiation. Check the procedures for safe handling and disposal of this radioisotope.

4. Washing solutions: Phosphate buffered saline (PBS): Make 2 L containing 10 mM NaH$_2$PO$_4$, 150 mM NaCl adjusted to pH 7.2 using NaOH. This solution is stable and may be stored at 4°C. It is susceptible to microbial contamination, however, and is usually made as required.

The other washing solutions used are made by dissolving the appropriate weight of bovine serum albumin or Triton X-100 in PBS. Dissolve bovine serum albumin by rocking the mixture gently in a large, sealed bottle to avoid excessive foaming. The "blocking" and "antibody" solutions containing 8% albumin may be stored at –20°C and reused several times. Microbial contamination can be limited by filter-sterilizing these solutions after use or by adding 0.05% (w/v) NaN$_3$. Other solutions are made as required and discarded after use.

5. Staining solutions: These are stable at room temperature for several weeks and *may be reused*.

Amido black stain (100 mL): 0.1% (w/v) Amido black in 25% (v/v) propan-2-ol, 10% (v/v) acetic acid.

Amido black destain (400 mL): 25% (v/v) propan-2-ol, 10% (v/v) acetic acid.

Ponceau S stain (100 mL): 0.2% (w/v) Ponceau S, 3% (w/v) trichloroacetic acid.

Ponceau S destain (400 mL): distilled water.

3. Method

3.1. Immunodetection of Polypeptides

1. Following SDS-polyacrylamide gel electrophoresis, electroblot the polypeptides from the gel onto

nitrocellulose at 50 V for 3 h using a Bio-Rad Trans-Blot apparatus. This technique and other suitable transfer methods have been described in detail in vol. 1 of this series and Chapters 28 and 29 of this volume.

2.   After blotting, transfer the nitrocellulose filters individually to plastic trays for the subsequent incubations. Ensure that the nitrocellulose surface that was closest to the gel is uppermost. Do not allow the filter to dry out, since this often increases nonspecific binding and results in heavy, uneven backgrounds. The nitrocellulose filter should be handled sparingly to prevent contamination by grease or foreign proteins. Always wear disposable plastic gloves, and only touch the sides of the filter.

3.   Rinse the nitrocellulose briefly with 100 mL of PBS. Then incubate the blot at room temperature with the following solutions, shaking gently (*see* Note 2 in section 4).
(a) 50 mL of PBS/8% bovine serum albumin for 30 min. This blocks the remaining protein-binding sites on the nitrocellulose.
(b) 50 mL of PBS/8% bovine serum albumin containing 50–500 µL of antiserum for 16 h (*see* Note 3 in section 4).
(c) Wash the nitrocellulose at least five times, each time using 100 mL of PBS for 15 min, to remove unbound antibodies.
(d) 50 mL of PBS/4% bovine serum albumin containing 1 µCi ^{125}I-labeled protein A for 2 h.
(e) Wash the nitrocellulose at least five times, each time using 100 mL PBS/1% Triton X-100 for 5 min, to remove unbound protein A.

      To ensure effective washing of the filter, pour off each solution from the same corner of the tray and replace the lid in the same orientation.

4. If desired, stain the blot for total protein as described below.

5. Allow the filter to dry. Do not use excessive heat—nitrocellulose is potentially explosive when dry.

6. Mark the nitrocellulose with radioactive ink to allow alignment with exposed and developed X-ray film.

7. Fluorograph the blot using suitable X-ray film and intensifying screens (*see* vol. 1 of this series). Expose the film at –70°C for 6–72 h, depending on the intensity of the signal.

3.2. Staining of Total Protein

Either of the following stains is suitable for visualizing polypeptides after transfer onto nitrocellulose. Each can detect bands containing about 1 µg of protein. Coomassie blue is unsuitable, since generally it produces heavy background staining. An alternative staining method using colloidal gold is given in chapter 34.

3.2.1. Amido Black

Incubate the filter for 2–5 s in 100 mL of stain solution. Transfer immediately to 100 mL destain solution, and wash with several changes to remove excess dye. Unacceptably dark backgrounds are produced by longer incubation times in the stain solution.

3.2.3. Ponceau S

Incubate the filter with 100 mL of Ponceau S stain solution for 30 min. Wash excess dye off the filter by rinsing in several changes of distilled water. The proteins may be destained by washing the filter in PBS (*see* note 4 in section 4).

4. Notes

1. Iodination of protein A using Bolton and Hunter reagent labels the ε-NH$_2$ group of lysine, which apparently is not involved directly in the binding of protein A to the Fc region of IgG. This method is preferable to others, such as those using chloramine T or iodogen, which label tyrosine. The only tyrosine residues in protein A are associated with Fc binding sites, and their iodination may reduce the affinity of protein A for IgG (5).

2. Nonspecific binding is a common problem in immunoblotting. Several factors are important in reducing the resulting background.

    First, the filter is washed in the presence of an "inert" protein to block the unoccupied binding sites. Bovine serum albumin is the most commonly used protein, but others, such as fetal calf serum, hemoglobin, gelatin, and nonfat dried milk (*see* Chapter 30, this volume), have been used successfully. Economically, the latter two alternatives are particularly attractive.

    The quality of protein used for blocking is important, since minor contaminants may interfere with either antigen–antibody interactions or the binding of protein A to IgG. These contaminants may vary between preparations and can be sufficient to inhibit completely the detection of specific polypeptides. Routinely we use bovine serum albumin (fraction V) from Sigma Chemical Co. (product number A 4503), but no doubt albumin from other sources is equally effective. The suitability of individual batches of protein should be checked using antisera known to react well on immunoblots.

    Second, the background may be reduced further by including nonionic detergents in the appropriate solutions. These presumably decrease the hydrophobic interactions between antibodies and the nitrocellulose filter. Tween 20, Triton X-100, and Nonidet P-40 at concentrations of 0.1–1.0% have been used. In our experience, such detergents may supplement the blocking agents described above, but cannot substitute for these proteins.

Third, the nitrocellulose must be washed effectively to limit nonspecific binding. For this, the volumes of the washing solutions should be sufficient to flow gently over the surface of the filter during shaking. The method described in this chapter is suitable for 12 x 7 cm filters incubated in 14 x 9 cm trays. If substantially larger filters are used, the volumes of the washing solutions should be increased.

3. The exact amount of antibody to use will depend largely on its titer. Generally it is better to begin by using a small amount of antiserum. Excessive quantities of serum tend to increase the background rather than improve the sensitivity of the technique. Nonspecific binding can often be reduced by decreasing the amount of antibody used to probe the filter.

4. If desired, the nitrocellulose filter may be stained with Ponceau S immediately after electroblotting. This staining apparently does not affect the subsequent immunodetection of polypeptides, provided that the filter is thoroughly destained using PBS before incubation with the antiserum. Initial staining allows tracks from gels to be separated precisely and probed individually. This is useful when screening several antisera.

5. Protein G, a cell wall component of group G streptococci, binds to the Fc region of IgG from a wider

range of species than does protein A (6). Therefore antibodies that react poorly with protein A, particularly those from sheep, cow, and horse, may be detected by a similar method using ^{125}I-labeled protein G In principle, the latter is more versatile. At present, however, the limited availability of protein G means that, when appropriate, labeled protein A is still preferable.

6. Particular care should be taken when attempting to detect antigens on nitrocellulose using monoclonal antibodies. Certain cell lines may produce antibodies that recognize epitopes that are denatured by detergent. Such "conformational" antibodies are unlikely to bind to the antigen after SDS-polyacrylamide gel electrophoresis.

7. Even prior to immunization, serum may contain antibodies, particularly against bacterial proteins. These antibodies may recognize proteins in an extract bound to the nitrocellulose filter. Therefore, when characterizing an antiserum, control filters should be incubated with an equal amount of preimmune serum to check whether such preexisting antibodies interfere in the immunodetection of specific proteins.

8. Quantitation of specific antigens using this technique is difficult and must be accompanied by adequate evidence that the radioactivity bound to the filter is directly related to the amount of antigen in the initial extract. This is important, since polypeptides may vary in the extent to which they are eluted from the gel and retained by the nitrocellulose. Additionally, in some tissues proteins may interfere with the binding of antigen to the filter. Therefore the reliability of the technique should be checked for each extract.

Perhaps the best evidence is provided by determining the recovery of a known amount of pure antigen. For this, duplicate samples are prepared, and to one is added a known amount of antigen comparable to that already present in the extract. Ideally the pure antigen should be identical to that in the extract. The recovery is calculated by comparing the antigen measured in the original and supplemented samples. Such evidence is preferable to that obtained from only measuring known amounts of pure antigen. The latter indicates the detection limits of the assay, but does not test for possible interference by other components in the extract.

References

1. Burnette, W. N. (1981) "Western blotting": Electrophoretic transfer of proteins from sodium dodecyl sulfate-polyacrylamide gels to unmodified nitrocellulose and radiographic detection with antibody and radioiodinated protein A. *Anal. Biochem.* **112**, 195–203.

2. Towbin, H., Staehelin, T., and Gordon, J. (1979) Electrophoretic transfer of proteins from polyacrylamide gels to nitrocellulose sheets: Procedure and some applications. *Proc. Natl. Acad. Sci. USA* **76**, 4350–4354.

3. Vaessen, R. T. M. J., Kreike, J., and Groot, G. S. P. (1981) Protein transfer to nitrocellulose filters. *FEBS Lett.* **124**, 193–196.

4. Towbin, H. and Gordon, J. (1984) Immunoblotting and dot immunobinding—current status and outlook. *J. Immunol. Meth.* **72**, 313–340.

5. Langone, J. J. (1980) ¹²⁵I-Labelled Protein A: Reactivity with IgG and Use as a Tracer in Radioimmunoassay, in *Methods in Enzymology* Vol. 70 (Vunakis, H. V. and Langone, J. J., eds.) Academic, New York.

6. Akerstrom, B., Brodin, T., Reis, K., and Bjorck, L. (1985) Protein G: A powerful tool for binding and detection of monoclonal and polyclonal antibodies. *J. Immunol.* **135**, 2589–2592.

Chapter 32

Detection of Protein Blots Using the Avidin–Biotin System

Michael J. Dunn and Ketan Patel

1. Introduction

The high resolution capacity of polyacrylamide gel eletrophoresis for complex protein mixtures can only be fully exploited if suitable procedures exist for the characterization of the separated components. One approach to this problem is to investigate the interactions of the separated proteins with specific antibodies or other ligands such as lectins. This can be achieved by applying antisera directly to polyacrylamide gels after electrophoresis, but this technique of "immunofixation" is inefficient and time-consuming because of the slow rate of diffusion of antibodies into the gel matrix. This has resulted in the development of methods for transferring the pattern of separated proteins out of the gels onto thin substrates. This technique is known as "Western" blotting. The most popular substrate for proteins is currently nitrocellulose. Positively charged nylon membranes have a greater protein-binding capacity than nitrocellulose and can be used successfully with antibodies, but they have not proved popular because of the lack of an easy-to-use general staining procedure. Protein transfer can be achieved by capillary, contact-diffusion, or vacuum techniques, but more rapid and efficient transfer is achieved if the proteins are electrophoretically removed from the gel to the support by application of an electric field perpendicular to the plane of the gel. This technique, based on the method of Towbin et al. (1), is known as electroblotting. The methodology was described in vol. 1 of this series (2) and is reviewed in refs. 3–5.

After proteins have been transferred to nitrocellulose, all the unoccupied potential binding sites on the blot must be blocked before probing with a specific ligand. This is often achieved by soaking the blot in a solution of bovine serum albumin (1) or gelatin (6). These proteins do not always prove to be nonreactive during subsequent probing, however, particularly if lectins are used as probes, because of the presence of contaminating glycoproteins in the blocking reagent. An alternative strategy is to use a dilute solution of nonionic detergent for the blocking step, and polyethylene sorbitan monolaurate (Tween 20) has been found to be particularly suitable (7). It should be remembered, however, that certain proteins can be displaced from nitrocellulose by nonionic detergents.

Once the blot has been blocked, it can then be reacted with the specific antibody or ligand. Although it is feasible to label the primary antibody to give a direct visualization of the blot, indirect sandwich methods using a second antibody are generally employed because of the increased sensitivity that can be obtained.

The second antibody can be fluorescently labeled (e.g., fluorescein isothiocyanate, FITC), radiolabeled (usually with ^{125}I), or conjugated to an enzyme such as horseradish peroxidase or alkaline phosphatase. Fluorescent methods are not popular because of the requirement for UV illumination to visualize the result. Methods employing radiolabels are very sensitive, but since an autoradiographic step is required, there is a considerable time delay before the result is obtained. This has made methods dependent on enzyme activity very popular, as in the case of enzyme-linked immunoassays (ELISA), since they are very sensitive and yield rapid results. The secondary antibody can be replaced in some cases with *Staphylococcus aureus* protein A, which specifically binds to the Fc regions of IgG. A recent development is the use of gold-labeled secondary antibodies (8), which are very sensitive and have the added advantage that they are directly visible (red color) without further development. In addition, sensitivity can be increased by a silver enhancement procedure (*see also* Chapter 34).

A technique that has been found to increase the sensitivity of detection of blots exploits the specificity of the interaction between the small molecular weight vitamin biotin and the protein avidin. Many biomolecules (antibodies, lectins) can be readily conjugated with biotin and used as the second reagent in blot probing. The blots are then visualized using in a third step, avidin conjugated with an enzyme (e.g., peroxidase, alkaline phosphatase, β-galactosidase, glucose oxidase). Even greater sensitivity can be achieved at this stage by using preformed complexes of a biotinylated enzyme with avidin since many enzyme molecules are present in these complexes producing an enhanced signal. Egg white avidin (mol. wt. 68,000) is often used in these procedures, but has two distinct disadvantages: (1) it is highly charged at neutral pH so that it can bind to proteins nonspecifically, and (2) it is a glycoprotein that can interact with other biomolecules such as lectins via the carbohydrate moiety. It is, therefore, advantageous to use streptavidin (mol. wt. 60,000), isolated from *Streptomyces avidinii*, which has a p*I* close to neutrality and is not glycosylated.

2. Materials

1. Phosphate buffered saline (PBS) containing 0.05% (w/v) Tween 20 (PBS/Tween). For 1 L of solution, use:
   0.2 g KH_2PO4
   1.15 g Na_2HPO4
   0.2 g KCl
   8.0 g NaCl
   0.5 g Tween 20
2. Biotinylated affinity-isolated rabbit imunoglobulins to mouse immunoglobulins (DAKO) diluted 1:200 with PBS/Tween (biotinylated second antibody).
3. Peroxidase-conjugated avidin (DAKO), diluted 1:10,000 with PBS/Tween.
4. Diaminobenzidine (DAB) solution: 0.05% (w/v) Diaminobenzidine and 0.01% (v/v) Hydrogen peroxide (from 30% stock solution) in PBS/Tween. Prepare fresh when required.

3. Method

Note: All steps are carried out with gentle shaking.

1. The blot is blocked by incubation in PBS/Tween for 90 min at room temperature.
2. The blot is then incubated for 60 min at room temperature with the primary antibody diluted at

an appropriate concentration in PBS/Tween. In the example illustrated in Fig. 1 (left panel), mouse monoclonal antibody to desmin (Amersham) was diluted 1:250 (*see* Note 4 in section 4).

3. The blot is washed for 3 x 10 min with PBS/Tween at room temperature.

4. The blot is incubated for 60 min at room temperature with biotinylated secondary antibody diluted in PBS/Tween.

5. The washing step 3 is repeated.

6. The blot is incubated for 60 min at room temperature with the streptavidin-peroxidase diluted in PBS/Tween.

7. The washing step 3 is repeated.

8. The blot is visualized by the addition of DAB solution for 10 min at room temperature. Brown bands of peroxidase activity are detected (*see* Fig. 1, left panel). Other suitable substrates for peroxidase-linked antibodies are given in Chapter 33.

4. Notes

1. If lectins are used to detect glycoproteins, a similar method to that described here can be used if an antibody against the lectin is available. Alternatively, if the lectin is conjugated with biotin (many conjugates are available commercially), it can be used directly so that the secondary antibody procedure (step 4 and 5 in section 3) can be omitted. An example of this technique using biotinylated concanavalin A and alkaline phosphatase-conjugated avidin visualized with Fast Red is shown in Fig. 1 (right panel).

2. Other enzymes conjugated with avidin can be used with appropriate substrates at step 6 in section 3

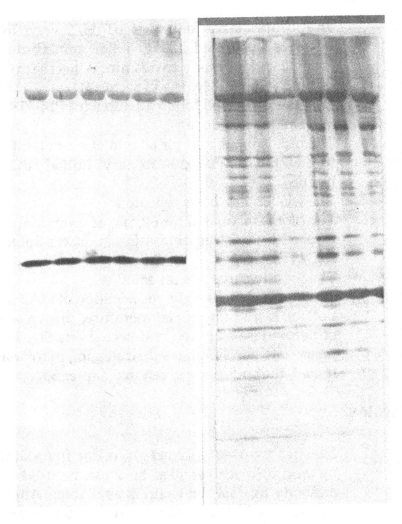

Fig. 1. Human skeletal muscle tissue proteins were separated by SDS-PAGE and transferred to nitrocellulose by electroblotting. The blots were blocked using PBS/Tween. (left) The blot was reacted sequentially with (1) mouse monoclonal antibody to desmin, (2) biotinylated antibody to mouse immunoglobulin, and (3) peroxidase-conjugated avidin, and (4) visualized using DAB. (right) The blot was reacted sequentially with (1) biotinylated Concanavalin A and (2) alkaline phosphatase-conjugated avidin, and (3) visualized with Fast Red.

(e.g., β-galactosidase, alkaline phosphatase (*see also* Chapter 33).

3. If an avidin–biotinylated enzyme complex (ABC) is required to enhance sensitivity, it is used at step 6 (section 3) suitably diluted in PBS/Tween. Reagents for these complexes are available from Vector Laboratories (avidin with biotinylated peroxidase, alkaline phosphatase, or glucose oxidase), DAKO (avidin with biotinylated peroxidase or alkaline phosphatase), and Amersham (streptavidin with biotinylated peroxidase or β-galactosidase).

4. No general guide can be given to the appropriate dilution for any particular primary antibody. This must be established for each antibody using serial dilutions, so that a strong specific signal is obtained, with minimal nonspecific and background staining.

References

1. Towbin, H., Staehlin, T., and Gordon, J. (1979) Electrophoretic transfer of proteins from polyacrylamide gels to nitrocellulose sheets. *Proc. Natl. Acad. Sci. USA* **76**, 4350–4354.
2. Gooderham, K. (1984) Transfer Techniques in Protein Blotting, in *Methods in Molecular Biology* Vol. 1 (Walker J.M., ed.) Humana, Clifton.
3. Gershoni, J. M. and Palade, G.E. (1983) Protein blotting: Principles and applications. *Anal. Biochem.* **131**, 1–15.
4. Symington, J. (1984) Electrophoretic Transfer of Proteins from Two-dimensional Gels to Sheets and Their Detetion, in *Two-dimensional Gel Electrophoresis of Proteins* (Celis, J.E. and Bravo, R., ed.) Academic, Orlando.
5. Beisiegel, U. (1986) Protein blotting. *Electrophoresis* **7**, 1–18.
6. Saravis, C.A. (1984) Improved blocking of nonspecific antibody binding sites on nitrocellulose membranes. *Electrophoresis* **5**, 54–55.
7. Batteiger, B., Newhall, W.J., and Jones, R.B. (1982) The use of Tween 20 as a blocking agent in the immunological detection

of proteins transferred to nitrocellulose membranes. *J. Immunol. Meth.* **55,** 297–307.

8. Moeremans, M., Daneels, G., Van Dijck, A., Langanger, G., and De Mey, J. (1984) Sensitive visualization of antigen-antibody reactions in dot and blot immune overlay assays with immunogold and immunogold/silver staining. *J. Immunol. Meth.* **74,** 353–360.

Chapter 33

Detection of Protein Blots Using Enzyme-Linked Second Antibodies or Protein A

J. M. Walker and W. Gaastra

1. Introduction

The immunodetection of a specific protein on a protein blot relies initially on the ability of a primary antibody to bind specifically to the protein of interest. This is simply achieved by washing the nitrocellulose paper in a dilute solution of the antibody. Before this step is carried out, however, unblocked protein binding sites on the nitrocellulose paper must be masked by washing with a general blocking solution (e.g., bovine serum albumin solution, dilute bovine serum, gelatin solution, hemoglobin solution, and so on). In this way, nonspecific binding of the antibody (which is itself a protein) to nitrocellulose is prevented, and any antibody that does bind to the paper will be doing so by a genuine antigen–antibody reaction alone. This antigen –antibody reaction must now be visualized. This can be achieved by washing the paper with an appropriately labeled ligand that binds specifically to the primary antibody. This ligand is generally of two possible types:

1. Protein A (which binds specifically to IgG molecules) either radiolabeled or linked to a marker enzyme (see below).
2. A second antibody (anti-IgG) that binds to the primary antibody. This second antibody is used linked to a marker enzyme. Alternatively this second antibody is linked to biotin. This biotinylated antibody is then treated with avidin linked to a marker enzyme (avidin binds with high specificity to biotin).

The general principles involved in detecting proteins on blots is shown in Fig. 1. The use of protein A is described in detail in Chapter 31, and the use of biotinylated antibodies is described in Chapter 32. With the exception of the use of radiolabeled protein A, however, all these approaches use the same principle for the ultimate detection of the required product— the use of a marker enzyme. Marker enzymes most commonly used are alkaline phosphatase or horseradish peroxidase. Both can be linked with high efficiency (while maintaining activity) to other proteins (such as antibodies, protein A, or avidin) without interfering with that protein's function. Such enzyme-linked second antibodies (or protein A or avidin) are commercially available, but can also be prepared quite easily in the laboratory (*see*, for example, Chapter 27). By washing the blot in a suitable substrate for the enzyme, the enzyme converts the colorless substrate into a colored product that

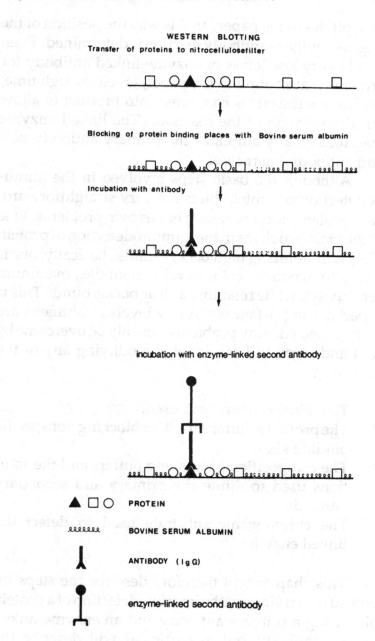

Fig. 1. Diagram showing the general principles involved in the immunodetection of protein blots.

precipitates in the paper. In this way the position of the original antigen–antibody reaction is determined. Even if only very low levels of enzyme-linked antibody (or protein A) are bound to the paper, given enough time, sufficient substrate is converted into product to allow for visualization of the reaction. The linked enzyme therefore greatly amplifies the primary antibody–second antibody interaction.

Although the basic steps involved in the immunodetection of protein blots are very straightforward, each protein blot experiment has its own problems. One of the main problems in the immunodetection of protein blots is minimizing high backgrounds. Basically, one is trying to maximize color development (i.e., maximum sensitivity), while retaining a clear background. This is especially important when low levels of antigens are being detected. Any problems can only be overcome by trial and error and can involve modifying any of the following:

1. The dilution of antisera used.
2. The protein solutions used for blocking nonspecific binding sites.
3. The composition of washing buffers and the solutions used to dilute the primary and secondary antibody.
4. The chromogenic substrate used to detect the linked enzyme.

This chapter will therefore describe the steps involved in carrying out the immunodetection of a protein blot using a primary antibody and an enzyme-linked second antibody, but in particular will describe the different blocking and washing solutions that can be used, and will examine the choice of substrates available for detecting linked enzymes (alkaline phosphatase and horseradish peroxidase).

2. Materials

1. Protein stain: 0.15% Amido black in 62:25:10, water:methanol:acetic acid. Destain: methanol:acetic acid:water; 90:2:100. Coomassie brilliant blue is not used, since it gives high background staining with nitrocellulose.

2. Washing buffer (PBS-Tween): NaCl (8.0 g), KCl (0.2 g), Na_2HPO_4 (1.15 g), KH_2PO_4 (0.2 g), Tween 80 (0.5 mL). Make to 1 L with distilled water (*see* note 6 in section 4).

3. Blocking solutions: The use of a range of different solutions has been reported in the literature. There is no suggestion that one is better than another. For any given experiment, the most suitable solution (i.e., that which gives lowest backgrounds) is found by trial and error. The following is an example of some of the solutions that have been used and could be tried.

a. 4% Bovine serum albumin in 0.05M Tris-HCl, 0.15M NaCl, 0.5% Tween 20, pH 10.3.

b. 2% Tween in PBS.

c. Non-fat dried milk (*see* Chapter 30).

4. Primary and secondary antibody solutions: These are usually used at dilutions of between 1:500 and 1:2000. The optimum dilution must be determined by trial and error, however. Antibodies are diluted in solutions containing protein and detergent to minimize nonspecific adsorption to the paper. Some workers recommend the use of relatively high pH buffers to minimize nonspecific binding, but this correspondingly decreases the binding of

antibody to antigen. This can be a problem when one is using low concentrations of antibodies (e.g., culture supernatants containing monoclonal antibodies), in which case more neutral buffers should be used. The following are some typical solutions used to dilute antisera:

a. 0.1% BSA, in 0.05M Tris-HCl, 0.15M NaCl, 0.5% Tween 20, pH 10.3 (or adjust to pH 8.5 if necessary).

b. 0.25% Gelatin, 0.05% Nonidet P40, 50 mM Tris-HCl, 5mM Na$_2$ EDTA, 0.15M NaCl, pH 7.5.

c. PBS containing 2% polyethylene glycol 6000 and 0.05% Nonidet NP-40.

5. Enzyme substrates:

a. BCIP substrate.
   1. 5-Bromo-4-chloro-3-indolyl phosphate (BCIP).
   2. Either 0.1M Tris-HCl, 1 mM MgCl$_2$, pH 8.6, or 1M ethanolamine, pH 9.8.

b. BCIP plus NBT substrate.
   1. Stock solutions: 5-bromo-4-chloro-3-indolyl phosphate (BCIP), 5 mg/mL in dimethylformamide. Store at –20°C.
   2. Nitro blue tetrazolium (NBT). 1 mg/mL in Veronal acetate buffer. Prepare fresh as required.
   3. 2M MgCl$_2$.
   4. 0.15M Veronal acetate, pH 9.6.

c. 4-methylumbelliferyl phosphate.
   1. 4-methylumbelliferyl phosphate.
   2. 1M Diethanolamine, 1 mM MgCl$_2$, pH 9.8.

d. Naphthol ASMX phosphate plus fast red TR salt.
   1. 3-Hydroxy-2-naphthoic acid 2,4-dimethylanilide phosphate (naphthol ASMX phosphate).
   2. Fast red TR salt. Fast red TR salt can be replaced by a range of other diazo dyes including fast blue BB salt, fast blue RR salt, fast garnet GBC, and so on.

6. 0.2M Tris-HCl, pH 8.2.

e.   3-Amino-9-ethylcarbazole substrate.
1. Stock solution: 3-amino-9-ethylcarbazole (1 mg/mL) in acetone. Store at 4°C.
2. 0.05M sodium acetate buffer, pH 5.0.
3. 30% (v/v) hydrogen peroxide.
f.   Diaminobenzidine.
1. Stock solution: diaminobenzidine (1 mg/mL) in acetone. Store at –20°C.
2. 0.05M Sodium acetate buffer, pH 5.0.
3. 30% (v/v) Hydrogen peroxide.
g.   4-Chloro-1-naphthol.
1. Stock solution: 4-chloro-1-naphthol (2%) in methanol. Store at –20°C.
2. 0.05M Tris-HCl, pH 7.4.
3. 30% (v/v) Hydrogen peroxide.

3. Method

All incubation and washing steps are carried out with gentle shaking.

3.1. Immunodetection of Protein Blots

1.   Gently remove the nitrocellulose filter from the gel after blotting.
2.   Be sure to indicate on which side of the filter the proteins are bound, and if not done so before, also indicate right, left, top, and bottom of the gel.
3.   Remove, if included, the lanes containing the molecular weight markers from the filter with a pair of scissors. The use of a scalpel on wet filters is not recommended, since this tends to damage the filter.
4.   Stain the strip with molecular weight markers in amido black solution for 5 min. Destain. This shows the efficiency of the electrophoretic transfer and gives the necessary molecular weight markers

if needed. Alternatively the filter strip can be
stained with Indian ink (1) or aurodye (*see* Chapter
34). Both these stains are considerably more sensitive than amido black.

5. If the transfer is satisfactory, place the rest of the
nitrocellulose filter in a small plastic dish and wash
it, with gentle shaking, in blocking solution (20 mL)
for 30 min or longer (overnight is often used). This
will block remaining binding sites on the nitrocellulose sheet. In all washing steps, take care that the
nitrocellulose sheet is mobile in the washing solution and not stuck to the bottom of the plastic dish.

6. Pour off the blocking solution and wash the nitrocellulose sheet in a solution of first antibody (20
mL) for 30 min (again, this incubation can be increased if necessary: some workers use an overnight incubation).

7. Remove the first antibody solution, and wash the
nitrocellulose sheet (4 x 5 min) in PBS-Tween (4 x 20
mL). This should remove all traces of any unbound
first antibody.

8. Wash the nitrocellulose sheet in the appropriate
second antibody solution (20 mL) for 30 min, then
pour off this solution and wash the sheet (4 x 5 min)
in PBS-Tween (4 x 20 mL). Finally, wash the sheet
in the buffer used for the enzyme substrate (20 mL)
for 5 min.

9. Pour off the buffer solution and wash the sheet in
freshly prepared substrate solution (see below). A
colored band (blue or red, depending on the enzyme and substrate used) should appear on the
nitrocellulose sheet within 5–30 min. When sufficient color has developed, wash the sheet in distilled water and blot dry between two sheets of
filter paper.

3.2. Preparation of Substrates

All papers should be washed in the appropriate buffer for a few minutes before adding substrate solution to bring the paper to the optimum pH for enzyme reaction.

3.2.1. Substrates for Alkaline Phosphatase (AP)

3.2.1.1. 5-Bromo-4-chloro-3-indolyl Phosphate (BCIP)

In this method (2), the enzyme converts the BCIP to the corresponding indoxyl compound, which precipitates, tautomerizes to a ketone, oxidizes, and dimerizes to form a blue indigo, which is deposited in the nitrocellulose. This substrate has been used by workers at concentrations between 0.5 and 2.0 mg/mL in various buffers (*see* note 1 in section 4). Two typical solutions are described below. Dissolve the BCIP in a small volume of DMSO, then add to the appropriate buffer.

1. 2.0 mg/mL in 1M ethanolamine, pH 9.8.
2. 0.5 mg/mL in 0.1M Tris-HCl, 1 mM MgCl$_2$, pH 8.6.

3.2.1.2. BCIP with Nitro Blue Tetrazolium (NBT)

In this method (3), the hydrogen ion released from dimerization (see above) reduces the NBT salt to the corresponding diformazan, which is intensely purple. This compound is therefore deposited along with the indigo giving an enhanced signal. To produce the substrate solution, mix the following:

1. 9 mL of Veronal-acetate buffer.
2. 0.1 mL of BCIP solution.
3. 1 mL of NBT solution.
4. 20 µL of MgCl$_2$ solution.

3.2.1.3. 4-Methylumbelliferyl Phosphate

This substrate is converted to 4-methylumbellifer-one, which is detected as a blueish-white fluorescent band using UV (360 nm) light. The substrate solution is dissolved at 0.1 mg/mL in $1M$ diethanolamine, 1 mM MgCl$_2$, pH 9.8.

3.2.1.4. Naphthol ASMX-Phosphate Plus Fast Red TR Salt

Mix equal volumes of naphthol ASMX phosphate (0.4 mg/mL in distilled water) and Fast Red TR salt (6 mg/mL in 0.2M Tris-HCl, pH 8.2). Both compounds are poorly soluble, and each should be stirred for 20 min before mixing. When mixed, filter the solution, then use directly. The reaction is terminated by washing the nitrocellulose sheet in distilled water (5).

3.2.2. Substrates for Horseradish Peroxidase (HP)

All HP substrates are susceptible to oxidation and should be stored in dark bottles. When preparing solutions, rigorous stirring should be avoided to prevent unwanted oxidation.

3.2.2.1. 3-Amino-9-ethylcarbazole

Mix the following:
1. Stock solution (0.8 mL).
2. Sodium acetate buffer, pH 5.0 (20 mL).
3. 30% Hydrogen peroxide (30 µL).

This substrate produces a red-brown insoluble deposit.

3.2.2.2. Diaminobenzidine

Mix the following:
1. Stock solution (2 mL).
2. Sodium acetate buffer, pH 5.0 (50 mL).

3. 30% H_2O_2 (30 µL).
This substrate produces a red-brown insoluble deposit.

3.2.2.3. 4-Chloro-1-naphthol

Mix the following:
1. Stock solution (1.5 mL).
2. Tris-HCl buffer, pH 7.4 (50 mL).
3. 30% H_2O_2 (30 µL).
This substrate produces a blue deposit.

4. Notes

1. Knecht and Diamond (2) reported that for alkaline phosphatase, BCIP substrate gave higher backgrounds, with no increase in sensitivity, when diethanolamine buffer was used in place of Tris.
2. Although sensitivity limits using HRP or AP have been reported, they vary considerably (in the ng–pg range), and comparison of different methods is not possible since each study was carried out using different antigens, primary antibodies, and protocols. *In general*, however, AP appears to be more sensitive than HRP when direct comparisons have been made (*see*, for example, ref. 4). This may be caused in part by the fact that with AP, color development can continue for many hours with extended incubation time, whereas HRP is substrate-inactivated, resulting in an effective reaction time of the order of 30 min. Also AP-developed blots are less susceptible to fading or photobleaching and thus maintain their color for longer periods of time. HP-developed blots should be kept in the dark and preferably photographed as soon as possible after development of the color.

3. Prolonged reaction times are possible with AP using the BCIP substrates and methylumbelliferyl phosphate since when prepared, the reagents are stable and do not precipitate. When using ASMX-phosphate, however, nonspecific precipitation is a problem when using prolonged reaction times.

4. A minor disadvantage of the methylumbelliferyl substrate for AP is the need to immediately UV photograph the blot to preserve the result.

5. For all staining methods, the reaction is terminated by washing the blot in distilled water, blotting the nitrocellulose between two sheets of filter paper, and air-drying.

6. If nonspecific adsorption, leading to high backgrounds, continues to be a problem, increase the protein and detergent concentrations in the blocking and antibody solutions, and include protein in the washing buffer. The use of protein to concentrations as high as 10% is not uncommon. In addition to gelatin and BSA, many other blocking proteins have been used, e.g., hemoglobin, fetal calf serum. The only requirement of a blocking protein is that it does not cross-react with either the primary or secondary antibody. Increasing the numbers of washing steps and the volume of washing solution used can sometimes reduce background staining.

7. When using horseradish peroxidase, do not use buffers containing azide [many workers include this in low concentrations (0.05%) in buffers as an antimicrobial agent]. Azide inhibits HRP.

8. When using horseradish peroxidase, 4-chloro-1-naphthol is often the substrate of choice since both diaminobenzidine and 3-amino-9-ethylcarbazole are reported to be mutagens.

9. At no stage allow your nitrocellulose paper to dry

out. This will lead to heavy, uneven background staining.

10. Nonionic detergents are used in all solutions to minimize hydrophobic interactions between antibodies and nitrocellulose that would lead to high backgrounds. Tween 20, Triton X-100, and Nonidet NP-40 can all be used at concentrations between 0.1 and 2%.

11. In our own work we have found 2% Tween in PBS to be an excellent blocking agent, but others have reported that this is not the case. This just stresses the point that optimal conditions must be found by trial and error for each experiment.

12. Although this is stating the obvious, it is of course essential that the enzyme-linked second antibody is against IgG of the *species* in which the primary antibody was raised.

13. High background can be a particular problem when studying proteins from *Escherichia coli*. This is because the animal in which the primary antibody is raised has, at a particular time in its life, been through an infection with *E. coli*. This gives rise to anti-*E. coli* antibodies that can give substantial background staining when screening for the production of a particular protein in *E. coli*. In those cases the antiserum should be absorbed with an *E. coli* extract before use.

References

1. Hancock, K. and Tsang, V.C.W. (1983) India ink staining of proteins on nitrocellulose paper. *Anal. Biochem.* **133**, 157–162.
2. Knecht, D.A. and Diamond, R.L. (1984) Visualisation of antigenic proteins on Western blots. *Anal. Biochem.* **136**, 180–184.
3. Blake, M.S., Johnson, K.H., Russell Jones, G.J., and Gotschlich, E.C. (1984) A rapid sensitive method for detection of alkaline-

phosphatase conjugated anti-antibody on Western blots. *Anal. Biochem.* **136**, 175–179.

4.   Promega Notes No. 3. *Promega Biotech.* Dec., 1985.

5.   O'Connor, C.G. and Ashman, L.K. (1982) Application of the nitrocellulose transfer technique and alkaline phosphatase conjugated anti-immunoglobulin for determination of the specificity of monoclonal antibodies to protein mixtures. *J. Immunol. Meth.* **54**, 267–271.

Chapter 34

Colloidal Gold for the Detection of Proteins on Blots and Immunoblots

Alan Jones and Marc Moeremans

1. Introduction

Immunoblotting techniques, in which electrophoretically separated proteins are transferred to the surface of nitrocellulose or nylon membranes ("Western" blots), are becoming increasingly popular in the analysis of protein systems. Transfer techniques in protein blotting have been covered in this series (Chapter 20 in vol. 1, and Chapters 28 and 29 in this volume). In this chapter we will describe the use of the colloidal gold marking system for the detection of proteins on blots.

Immunodetection of specific bands within a separated "field" of proteins on a blot can be done in a number of ways. Protein A-^{125}I or peroxidase are common labels used for the visualization of these immunoblotted proteins (see Chapters 31 and 33). These labels require further processing by autoradiography or chromogenic development, respectively, in order to localize the antigen to the naked eye. Using immunogold staining (IGS), a red signal is produced without further development, thus simplifying the process (1). In contrast to autoradiography, the technique is faster and operator safety is improved since the handling of shorter shelf-life radioactive substances is avoided (2–4). Nevertheless, autoradiography has the advantage that major bands and minor bands can be detected on the same blot by adapting the exposure time, without the necessity of repeating the whole experiment.

The sensitivity of the red color produced by indirect immunogold staining yields a sensitivity comparable to that of indirect peroxidase. We would, however, now strongly recommend the additional and more sensitive silver intensification step. Amplification of the signal by silver enhancement, in which the red bands of the gold-marked antibodies on the antigen become intensely black, increases the sensitivity by at least one order of magnitude, making the immunogold-silver staining (IGSS) system even more sensitive than the PAP and ABC immunoperoxidase systems (1). IGSS also produces a highly contrasted and stable signal with practically no background. Enzymatic methods also usually require possibly carcinogenic developers, and the signal tends to fade with time. The possibility of evaluating the result first with immunogold alone and then after with silver enhancement is yet another advantage of immunogold over the enzymatic methods. Figures 1 and 2 illustrate diagrammatically the IGS and particularly the IGSS techniques, and Fig. 3 compares these various detection methods.

Because of their convenience and reliability, the IGS and IGSS techniques have proved very useful for

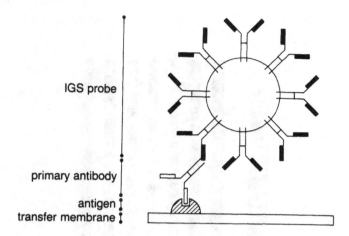

IGS probe

primary antibody

antigen
transfer membrane

Fig. 1. Principle of the indirect visualization of antigens with primary antibodies recognizing the immobilized antigen and gold-labeled secondary antibodies (or IGS probes) directed against the primary antibody. Colloidal gold particles of about 20 nm linked to the affinity-purified secondary antibodies give the best results. The actual pattern of binding of the antibodies to the gold particle is unknown, and the illustration above is simply pictorial.

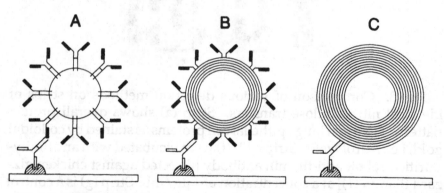

A   B   C

Fig. 2. Schematic representation of the silver-enhancement process. In the initial phase (A) the gold-probe piggybacks on the primary antibody, which in turn sticks to the antigen. During the silver enhancement step layers of silver selectively precipitate on the colloidal gold suface (B). The result is a significantly larger particle (C), a silver surface that generates a much more intense macroscopic signal.

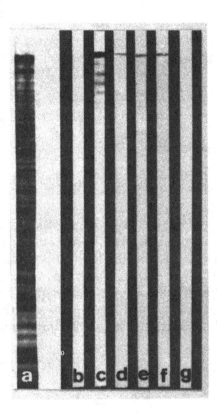

Fig. 3. Comparison of various detection methods on strips of identical nitrocellulose transfers. Strip (a) shows overall protein pattern of chicken lung epithelial cell proteins as stained by colloidal gold, i.e., AuroDye™. Strips (b)–(f) were incubated with an affinity-purified rabbit anti-filamin antibody [directed against chicken gizzard filamin (1)] at a concentration of 1 µg/mL. Strip (g) is a control of the immunogold/silver staining in which the primary antibody was omitted. Visualization method: (b) and (c) immunogold and immunogold/silver staining using AuroProbe BL GAR; (d) indirect peroxidase using GAR-HRP; (e) avidin—biotinylated horseradish peroxidase complex method (according to the instructions of the manufacturer); and (f) peroxidase anti-peroxidase. For the enzymatic methods the strips were developed with 4-chloro-1-naphthol.

screening monoclonal antibodies on strips of Western-blotted protein mixtures. It can also be expected that these techniques should become useful for serological testing, e.g., in tests of HTLV-III virus proteins. The IGS and IGSS techniques are also applicable to the specific detection of translation products from cloned c-DNA fragments in overlay assays of phage plaques or lysed bacterial colonies.

The sensitivity of immunoblotting techniques can be very high (as low as pg levels). This has created the need for an ultrasensitive staining method on a duplicate or even on the same protein blot, to make possible direct correlation between the total electrophoretogram (1-D or 2-D) and detected bands or spots in the immunoblot. Such a procedure would also be very useful for assessing the quality of the transfer from the separating gel to the blotting membrane. Conventional stains on nitrocellulose membranes are known to be very insensitive, and they do not work at all on nylon membranes.

An India ink staining method has been described (5) and in our hands has given good results. It is, however, not always possible to find a good batch of India ink. Alternative methods are based on the immunochemical detection of protein bands after derivation with 2,4-dinitrofluorobenzene, pyridoxal 5'-phosphate, or TNP (6–8).

We (9) have found that colloidal gold particles, appropriately buffered at a pH of 3.0 [and stabilized with Tween 20 (9)], can be used as a highly sensitive total protein stain on nitrocellulose membranes (Fig. 4). A similar application of colloidal gold has also been developed independently by others (10). We have called this gold stain AuroDye™, and it is available from Janssen Life Sciences Products. Gold staining of proteins transferred to nitrocellulose has subsequently become very

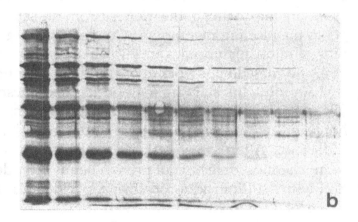

Fig. 4. Comparison of Amido Black and colloidal gold staining on nitrocellulose (b) with silver staining of SDS-polyacrylamide gel (e) and the same for FerriDye™ transferred to either nitrocelulose (c) or Zetaprobe (d). Nine dilutions (twofold) of Bio-Rad high molecular weight standards separated on a 7.5% gel. Amount per protein: (Lane 1) 1000 ng; (Lane 2) 500 ng; (Lane 3) 250 ng; (Lane 4) 125 ng; (Lane 5) 62.5 ng; (Lane 6) 31.3 ng; (Lane 7) 15.6 ng; (Lane 8) 7.8 ng; (Lane 9) 3.9 ng. The last well (10) was loaded with sample buffer alone.

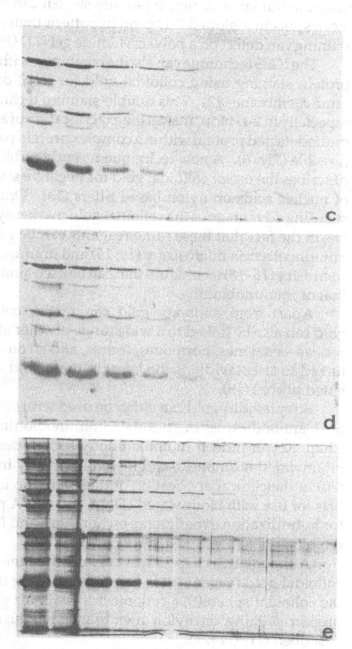

Fig. 4 (*continued*)

useful in our laboratories, especially since it can detect more spots in a blotted 2-D protein pattern than silver staining can detect on a polyacrylamide gel (11) (Fig. 5).

The IGSS technique can also be combined with total protein staining using colloidal gold (or iron) on the same membrane (12). This double-staining technique, especially in 2-D blots, makes the exact location of an immunodetected protein within a complex protein pattern possible (Fig. 6). A new technique recently published, describes the use of colloidal gold for negative staining of nucleic acids on nylon based filters (13). One outstanding advantage of the colloidal gold marker system lies in the fact that these same reagents can be used in immunoelectron microscopy (14, 15) and immunocytochemistry (16–18) techniques that can be used along side that of immunoblotting.

Apart from antibody–gold conjugates, colloidal gold can also be linked to a wide range of other affinity probes—enzymes, hormones, lectins, and so on. Gold linked to streptavidin is the latest in these gold conjugated probes (19).

Streptavidin-gold can either be used with biotinylated antibodies, as in "normal" immunoblotting (*see* Chap. 32), or indeed in nonantibody techniques (e.g., biotinylated hormones), and there is currently interest (20) in the clinical application of these streptavidin labels for use with biotinylated cDNA and cRNA probes for hybridization experiments on Southern and Northern blots.

Although this chapter will only describe the use of colloidal gold reagents, it is worth mentioning that, in the colloidal sphere(!), a cationic iron stain for general protein staining on nylon membranes has just been described (21) (Fig. 7).

FerriDye™ is supplied by Janssen Life Sciences Products and contains a colloidal iron sol supplemented

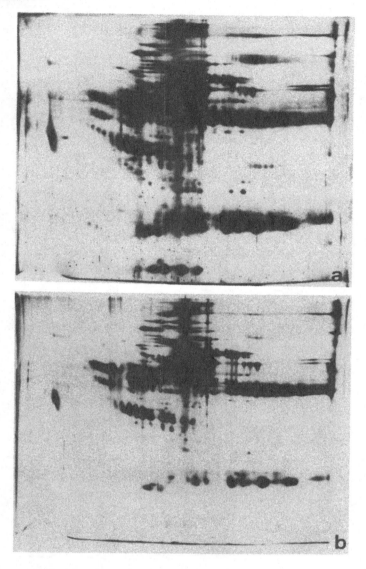

Fig. 5. Comparison of AuroDye™ on nitrocellulose (a) with silver staining of polyacrylamide gel (b). Human pleural fluid (30 mg) was run on a 2-D gel. Proteins separated in the first dimension with IEF using 2% LKB ampholines, pH 3.5–10 and 5–8, in a ratio 3:1 and in the second dimension with SDS-electrophoresis using a 10% T 2.6% C polyacrylamide gel. A duplicate gel was electrotransferred to nitrocellulose. Courtesy of Prof. Rabaey and Dr. Segers (State University of Gent, Belgium).

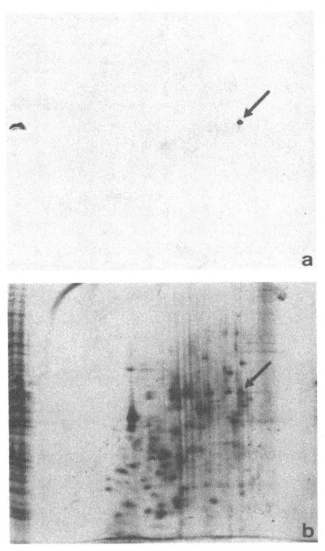

Fig. 6. A whole cell extract of rat hepatoma cells was prepared for 2-D electrophoresis according to Garrels (24). After NEPGHE in the first dimension, proteins were separated according to their molecular weight with SDS-electrophoresis on a 12% gel and electrotransferred to nitrocellulose. Immunodetection (indirect immunogold-silver staining) of tubulin (arrow) with rabbit anti-tubulin and AuroProbe™ BL GAR (a). Sequential overall protein staining of the same blot with AuroDye (b).

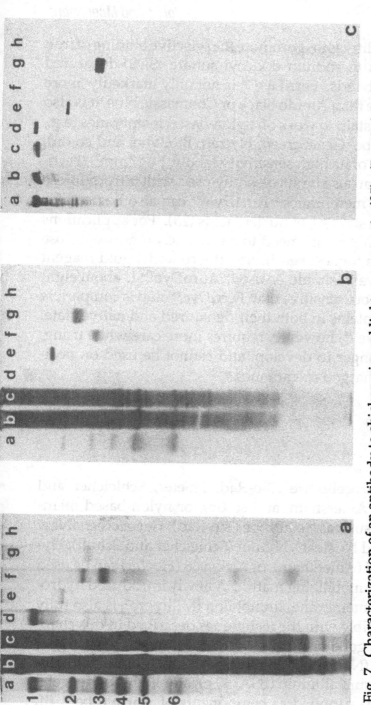

Fig. 7. Characterization of an antibody to chicken gizzard light meromyosin (25). (Panel a) Silver staining of SDS-polyacrylamide gel. (Panel b) Visualization of total protein pattern on ZetaProbe by colloidal iron, i.e., FerriDye™. (Panel c) Immunodetection of antigen after incubation with antibody: (a) purified reference proteins at 0.5 mg per band: (1) Chicken gizzard myosin heavy chain; (2) Chicken gizzard alpha-actinin; (3) bovine serum albumin; (4) rat brain tubulin; (5) chicken gizzard actin; (6) pig stomach tropomyosin. (b) Chicken heart fibroblast. (c) Chicken lung epithelial cells. (d) Chicken gizzard myosin heavy chain (0.25 mg). (e) Chicken gizzard myosin rods (0.25 mg). (f) Chicken gizzard myosin, Subfragment 1 (g) Chicken gizzard light meromysin. (0.25 mg). (h) Chicken gizzard L20 (0.25 mg).

with additives that promote the selective binding of iron particles to sodium dodecyl sulfate (SDS)-denatured protein bands. FerriDye™ is not only markedly more sensitive than Amido black or Coomassie blue, it is also the first stain to work on nylon-based membranes (e.g., ZetaProbe, GeneScreen, Nytran, Biodyne, and so on). Because of its high sensitivity (about 1 ng/mm^2), Ferri-Dye™ can also be successfully used with nitrocellulose blotting membranes. FerriDye™ can also be used for positive staining of nucleic acids (13). For applications in which proteins need to be stained on nitrocellulose with the highest sensitivity, the colloidal gold reagent AuroDye™, should be used. AuroDye™ is about eight times more sensitive than FerriDye™ and is compatible with proteins in both their denatured and native state. AuroDye™, however, requires more care while using, takes longer to develop, and cannot be used on positively charged membranes.

2. Materials

2.1. General

Nitrocellulose (Bio-Rad, Hoefer, Schleicher and Schull, Amersham, and so on), or nylon-based membranes such as ZetaProbe (Bio-Rad), GeneScreen (New England Nuclear), Nytran (Schleicher and Schull), Hybond-N (Amersham), or Nylon 66 (Hoefer) can be used as an immobilizing matrix. A newly introduced hydrophobic membrane, Immobilon (Millipore), is also fully compatible with the techniques described in this chapter. For good results we would recommend normal goat serum (NGS, heat-inactivated, 5 mL lyophilized), bovine serum albumin (BSA, 50 g), and high-quality gelatin (5 g) supplied as components of AuroProbe™ BL

Kits (Janssen Life Sciences Products). Recommended total protein-staining reagents are Ponceau red (reversible) or Amido black (low-sensitivity nitrocellulose), FerriDye™ (intermediate-sensitivity nylon/nitrocellulose), or AuroDye™ (high-sensitivity nitrocellulose). The latter two are available from Janssen Life Sciences Products. Reagents for the silver enhancement step are now in a kit form (IntenSE™, Janssen Life Sciences Products).

Analytical grade chemicals should be used throughout, and water should be deionized and glass-distilled. If silver enhancement is used, the glassware used in all the steps of the procedure from immunoincubation to silver enhancement should preferably be machine-washed, including an acid wash step. It should be rinsed with the same quality of water used in the silver enhancement. If washed manually, the use of brushes with metallic parts should be avoided since they give off metallic ions (such as zinc) that interfere with the enhancement process.

2.2. Immunogold/Silver Staining Procedure

The standard procedure refers to BSA-Tris. This is a buffer composed of 20 mM Tris-base, 150 mM NaCl, adjusted to pH 8.2 with 1N HCl, supplemented with 20 mM NaN$_3$, and containing a (w/v)% BSA indicated by the preceding figure, e.g., 0.1% BSA-Tris.

1. Tris buffered saline (TBS): 20 mM Tris-HCl, 150 mM NaCl, 20 mM NaN$_3$, pH 8.2.
2. Blocking solution (nitrocellulose): 5% (w/v) BSA in TBS.
3. Blocking solution 2 (for charge-modified nylon): 10% (w/v) BSA in TBS.

4. Washing buffer: 0.1% (w/v) BSA in TBS.
5. Gelatin buffer: (Washing buffer supplemented with gelatin.) TBS containing 0.4% (w/v) gelatin and 0.1% (w/v) BSA. Heat to 50°C to dissolve the gelatin.
6. AuroProbe™ BL: AuroProbe™ BL reagent 1% (v/v) in gelatin buffer.
7. IntenSE™ kit (enhancer, initiator, and fixing solution)
a. Enhancer
    Enhancer powder, 2.35 g
    Bidistilled water, to 50 mL
b. Initiator
    Initiator powder, 3.9 g
    Bidistilled water, to 50 mL
c. Fixing solution
    Fixing solution, 10 mL
    Bidistilled water, to 100 mL

2.3. Alternative Silver Enhancement Method
(Light Sensitive)

The principal solutions required for this step are a citrate buffer, the silver enhancer, and the fixing solution.

1. A 2M stock solution of citrate buffer is prepared by dissolving 23.5 g of trisodium citrate • 2H_2O and 25.5 g of citric acid • H_2O in 100 mL of water. Dilute half of it ten times.
2. Next prepare the silver enhancement developing solution (77 mM hydroquinone and 5.5 mM silver lactate in 200 mM citrate buffer, pH 3.85). This solution is light-sensitive. It should be prepared in a container wrapped in aluminum foil immediately prior to use. Mix 10 mL of the citrate buffer stock solution with 60 mL double distilled water (solu-

tion A). Dissolve 0.85 g hydroquinone in 15 mL water (solution C). The solutions B and C are also light sensitive, and they should be protected by aluminum foil. Add Solution B to solution A. Add solution C and mix well. The developer should always be freshly prepared immediately before use.

2.4. Staining Blotted Proteins with Gold Particles

1. Phosphate buffered saline, pH 7.2 (PBS):
    a. Potassium dihydrogen phosphate, 0.2 g.
    b. Disodium hydrogen phosphate, 1.15 g.
    c. Sodium chloride, 8.00 g.
    d. Potassium chloride, 0.2 g.
    e. Distilled water, to 1000 mL.
2. Washing and blocking buffer: 0.3% (v/v) Tween 20 in PBS.
3. Tween 20 stock solution: 2% (v/v) Tween solution in bidistilled water.

2.5. Double Staining Using IGSS and AuroDye™ Total Protein Staining

1. Phosphate buffered saline, pH 7.2.
2. Blocking solution: 0.3% (v/v) Tween 20 in PBS.
3. Washing buffer: 0.05% (v/v) Tween 20 in PBS.
4. Enhancer, initiator, fixing solution (for the light-insensitive method, using an IntenSE™ kit).
5. Citrate buffer, silver lactate, hydroquinone (for the light sensitive method, *see above*).

3. Methods

3.1. Immunogold/Silver Staining Procedure

3.1.1. Immunogold Staining

1. Cut a narrow strip off the separating gel to silver stain the protein pattern after electrophoresis. We

recommend the use of 0.75-mm thick gels. Fix this strip with fixing solution for 30 min, then silver stain (see vol. 1 of this series, and Chapter 13 of this volume).

The silver staining method matches the sensitivities of the IGSS technique. Alternatively the gel strip can be stained with Coomassie Brilliant Blue R-250 using the standard procedure. This is not particularly recommended, however, because of its nonmatching relative low sensitivity.

2. Transfer the proteins from the remainder of the gel to nitrocellulose or nylon-based membranes, preferably using electroblotting. One can either use conventional methods (e.g., Hoefer's Transphor) or the recently introduced semidry blotter (22; *see* Chapter 29 in this volume). Recommendations for obtaining optimal transfer and low background are given in section 4. *To avoid skin contact with the transfer membrane, handle blots by their edges with clean plastic tweezers.*

3. Silver stain the gel afterward to evaluate the efficiency of transfer.

4. When nitrocellulose is used, a strip of the transfer membrane can also be stained with Amido black or Ponceau Red. A dramatically increased sensitivity can be obtained, however, by using the newly developed colloidal gold stain, AuroDye™. Its sensitivity matches that of silver staining of the gel and immunodetection with IGSS on the blot membrane. [When nylon-based membranes are used, FerriDye™ (Janssen Life Sciences Products) is the only direct protein stain available for this purpose.]

5. Saturate the nonoccupied protein-binding places by incubating the transfer membrane in blocking solution 1 at 37°C for 45 min. Nylon-based mem-

branes, with their higher binding capacities, should be incubated overnight at 37°C in blocking solution 2. Do not throw away the blocking solution afterwards; it can be reused several times. After quenching excess binding sites, the blot should then be washed in 100 mL of washing buffer in a Petri dish for 2 x 5 min.

6. Incubate the saturated transfer membrane, or strips of it, under constant agitation in washing buffer suppplemented with 1% normal serum and containing 1–2 µg of the primary antibody per mL (if purified). If unpurified serum is used, a dilution of preferably higher than 1/500 should be used. Use about 1 mL for every 10 cm² of transfer membrane surface. To reduce the amount of antibody, one can prepare tailored plastic bags that, after addition of the incubation fluid, are heat sealed. In this case a volume of 15–20 mL is sufficient for a 10 x 15 cm membrane. A typical incubation time, at room temperature, is 2 h. To ensure that the whole membrane is flooded by the antibody solution, the plastic bag must be placed horizontally on a tilting apparatus. A negative control consists of replacing the primary antibody solution by washing buffer plus 1% normal serum (prepared from the same species as the secondary antibody).

7. Wash the transfer membrane in washing buffer three times for 5 min. All washing steps are performed in Petri dishes with 100 mL of washing fluid.

8. Dilute the Immunogold staining (AuroProbe™ BL) reagent 1:100 in gelatin buffer. Gelatin comes with AuroProbe™ BL kits or can be purchased from Difco (Bacto gelatin 0143-01); it has been shown to inhibit nonspecific binding of gold probes to cer-

tain proteins, especially tropomyosin, without affecting the immunoreactivity.

9. Incubate the transfer membrane overnight in the diluted AuroProbe™ BL reagent under constant agitation. When a shorter secondary incubation (2 h) is required, use a more concentrated solution (1:25 dilution). To reduce the amount of Auro-Probe™ BL reagent used, the incubation can again be performed in tailored plastic bags. About 15–20 mL gold reagent is sufficient for a 10 x 15 cm blot.

10. Do not throw away the incubation fluid afterward; it can be reused several times. Store at 4°C.

11. Wash the transfer membrane for 2 x 10 min in washing buffer.

12. Evaluate the result. If no signal amplification is needed, the membrane can be air-dried. More often, however, it is best to stain the transfer membrane by the silver enhancement procedure as described below. This will give the highest possible sensitivity.

3.2. Silver Enhancement (Light Insensitive) Using Janssen Life Sciences Products IntenSE™ Kit

Before proceeding with this step, it is highly recommended to carefully read the additional information in section 4.

1. Perform the IGS method as outlined above.

2. Wash the gold-stained transfer membrane for 2 x 5 min in excess of double-distilled water to remove chloride ions.

3. Just before use, prepare the solutions required for silver enhancement. These can be made from scratch or from IntenSE™ kits, produced by Jans-

sen Life Sciences Products. Make an appropriate volume (about 100 mL for a 10 x 15 cm blot) of developer solution by mixing 50 mL of enhancer solution with 50 mL of initiator solution.

4. Pour the mixed enhancer and initiator solution into a glass Petri dish. Put the gold-stained transfer membrane in this silver enhancer *immediately* afterward.

5. Enhance the gold signal on the transfer membrane. The enhancement should be monitored visually. Typical development times range from 5 to 15 min. For a further description, please read the additional information in section 4.

6. Put the transfer membrane in a Petri dish with the diluted fixing solution for 5 min (watch the signal intensity).

7. Wash the membrane for 3 x 5 min in excess water, after which it can be air-dried.

3.3. Alternative Silver Enhancement Method

The above method has been given for silver enhancement using the light-insensitive kit, IntenSE™, from Janssen Life Sciences Products. Alternatively, the silver enhancement step can be performed by a light-sensitive procedure. The principal solution required for this step are a citrate buffer, the silver enhancer, and the fixing solution.

1. The immunogold-labeled blotting membranes are further washed in an excess of distilled and deionized water for 5 min twice. It is essential that all chloride ions be removed, since chloride impurities will induce a precipitate. Then the strips or sheets are washed again, but only for 2 min, in 100 mL of the diluted citrate buffer.

2. The blotting membranes are then incubated in a glass Petri dish or the like, filled with developer, for 5–15 min. Here, too, aluminum foil should be used to protect the reaction from daylight. Strips can be developed in a dish of 15-cm diameter and sheets in a dish of 25-cm diameter. The silver enhancement can be visually checked by regularly taking a look. Avoid, however, exposing the developer to strong daylight. Possible turbidity of the developing fluid could indicate contamination with chloride ions.

3. After the reaction has been completed, the strips or sheets are transferred into the fixing solution for 5 min and then washed three times in an excess of water. Use a photographic fixer (Agfa-Gevaert), diluted 1:4, for 2 min, or alternatively, a 5% aqueous solution of sodium thiosulfate.

The treatment with fixing solution should not take too much time because the enhancement obtained then diminishes. If this seems to occur before the 5 min have passed, this period of time can be shortened. To conclude the procedure, the immunogold-silver stained transfer membranes are air-dried on a piece of filter paper.

3.4. Staining Blotted Proteins with Gold Particles

Because of the very high sensitivity obtained with AuroDye™, any protein impurities will also appear, and it is necessary to work under very clean conditions during the electrophoretic run as well as during electrotransfer. When using conventional electrotransfer in a buffer tank, putting an additional nitrocellulose membrane at the cathodic side of the gel is very helpful in filtering and adsorbing impurities out of the system that would otherwise adsorb on the protein blot. This is not necessary when using "semidry" electrotransfer (*see* Chapter 29).

1. After transfer, that part of the blot to be stained with gold particles (AuroDye™) is cut off the transfer membrane. It usually contains a duplicate protein pattern that is to be probed with an (immune) overlay assay. The membrane is washed extensively on a gyratory shaker with excess washing and blocking buffer at 37°C for 30 min and three times for 15 min at room temperature. Use 0.4 mL of buffer per cm² of blotting membrane. About 100 mL (four times) is sufficient for a blot of 15 x 10 cm. This extensive washing is necessary to remove adherent polyacrylamide gel particles and residual SDS. The blot is then rinsed in excess distilled water (200 mL) for 3 min.

2. The washed blot is drained off and incubated in AuroDye™, undiluted, at a minimum 0.2 mL AuroDye™/cm² blotting membrane. With high protein loads, 0.5 mL/cm² may be needed. Sheets are incubated in sealed plastic bags and strips in stoppered tubes under constant agitation on a tilting apparatus (e.g., Hoefer's Red Rocker). If a tilting apparatus is not available, best results are obtained by using excess Aurodye™ in a clear glass or polycarbonate tray on a reciprocal or rotary shaker. Incubation should continue until optimum color formation has been obtained. In general, incubation time is in the order of 4 h to overnight. Overstaining does not occur. Background only forms on proteinaceous impurities.

    The colloidal gold stain may turn blue, probably by reaction with proteins released from the blot.

Table 1
Staining Procedure with Colloid Gold

Wash extensively in excess PBS with 0.3% Tween 20
30 min at 37°C
⇓
Wash extensively in excess PBS with 0.3% Tween 20
15 min at room temperature
⇓
Wash extensively in excess PBS with 0.3% Tween 20
15 min at room temperarue
⇓
Wash extensively in excess PBS with 0.3% Tween 20
15 min at room temperarue
⇓
Incubate in AuroDye™
4 h to overnight
⇓
Wash blot in excess H$_2$O for 15 min and air dry

This happens when the proteins are present in high concentrations and can be relieved by diminishing the protein load, if possible. The capacity of nitrocellulose is limited to about 0.8 µg/mm^2. Otherwise, it is recommended to replace the gold sol immediately after a brief wash with excess bidistilled water.

3. After staining, the blot is washed several times with excess water and air dried. The stain turns darker upon drying and does not fade. These steps are summarized in Table 1.

3.5. Immunogold Staining of Protein Dots

Materials and procedures are identical to the above, except for the spotting procedure, which is given below.

1. Soak the transfer membrane in water.
2. Dry it in air for 15 min.
3. Dilute the antigen in an appropriate buffer containing 50 μg of BSA per mL.
4. Apply 1 μL of the diluted antigen on the transfer membrane with a pipet or a microsyringe (for example the Drummond Microdispenser).
5. Dry it in air for 15 min.

Recently several companies (e.g., BRL, Bio-Rad, and so on) have introduced small manifolds that enable the user to apply 96 dots on a sheet of transfer membrane. Using these manifolds generally results in smooth and regular dots. Immunochemical manipulations such as incubation and wash steps are also greatly facilitated.

3.6. Combined Immunogold-Silver Staining and AuroDye™ Total Protein Staining of Electrophoretically Separated Proteins Transferred to Nitrocellulose

This double staining method results in a combination of immunodetection with the staining of the complete electrophoretogram on the same membrane. Such a technique makes an exact localization of an immunodetected band within a complex protein pattern possible. It is based on the use of Tween 20 (polyoxyethylene sorbitan monolaurate) as a blocking agent and is particularly interesting for 2-D blots. The procedure requires silver enhancement materials.

1. Transfer the electrophoretically separated proteins to a nitrocellulose membrane. For optimum results, follow the recommendations regarding clean conditions as explained above. One of these recom-

mendations is the use of an additional blotting membrane at the cathodic site in the sandwich to absorb impurities from the blotting system.

2. After transfer, saturate the blotting membrane in the blocking solution for 30 min at 37°C.

3. Wash for 3 x 5 min in the blocking solution at room temperature under constant agitation. About 100 mL is sufficient for a blot of 10 x 15 cm. The extensive washing is necessary to remove adherent polyacrylamide gel particles and residual SDS. All subsequent steps are also to be carried out at room temperature. After quenching of excess binding sites with blocking buffer, the blot is washed again with washing buffer in a Petri dish for 2 x 5 min.

4. Incubate the saturated and washed transfer membrane under constant agitation in washing buffer containing 1–2 µg/mL of the primary antibody (or a suitable dilution of the primary antiserum) for 2 h. Use about 1 mL for every 10 cm² of transfer membrane. Normal serum is not used in this instance in order to reduce the protein loading and to keep any background staining to a minimum.

5. Wash for 3 x 10 min in washing buffer.

6. Dilute the AuroProbe™ BL reagent (up to 1:100) in washing buffer (solution 3). Gelatin cannot be used in this AuroProbe™ dilution because gelatin is not compatible with the AuroDye™ step that follows later. Since gelatin blocks the nonspecific binding of gold probes with certain proteins like tropomyosin, its obligatory omission may result in some nonspecifically immunostained bands. This can be controlled by omitting the primary antibody.

7. Incubate the transfer membrane for 2 h (or overnight) in the diluted AuroProbe™ BL gold reagent under constant agitation.

8. Wash for 3 x 5 min in washing buffer.
9. Wash for 2 x 5 min in double-distilled water.
10. Perform a silver enhancement step on the immuno-gold-stained transfer membrane as described previously.
11. Wash for 3 x 5 min in a large excess of distilled water.
12. Wash for 5 min in blocking solution.
13. Take a photograph of the result.
14. Incubate the immunogold-silver transfer membrane overnight in AuroDye™.
15. Rinse the double-stained membrane in excess water and dry it in air before storing.
16. Take another photograph of the result.

4. Notes

4.1. Use of Colloidal Gold for General Protein Staining

1. The stabilized colloidal gold sol AuroDye™, containing gold particles of 20 nm in size, has been adjusted to a pH of about 3, at which the negatively charged particles will now bind very selectively to proteins by hydrophobic and ionic interactions. The latter may result in different staining intensities of proteins differing in isoelectric point. This is to our knowledge a common feature of all protein staining methods (e.g., silver staining, Coomassie Blue, and so on). However, AuroDye™ stains more proteins on a 2-D blot that can be shown in a gel with silver staining. Indeed, some proteins are not detected with silver staining, but become clearly visible with AuroDye™. It cannot be used on nylon-based membranes since, because of these interactions, the whole membrane will be stained. The

method works well on blots of both denatured and nondenatured proteins.

2. Use of filter papers in the sandwich set-up. Whatman 3MM (No. 3030917) or Schleicher and Schull (No. 604A-24343) are recommended. Other brands should be checked for the presence of substances that are released during the transfer and produce a very disturbing speckled background.

3. The high sensitivity of AuroDye™ used in the protein staining procedures requires special care and handling to avoid background staining. High-quality chemicals should be used throughout. Many researchers routinely work with far less sensitive stains such as Amido black or with relatively insensitive overlay detection techniques. As a result, they are used to rather heavy protein loads.

    Heavy protein loads on gels, when transferred to nitrocellulose, will not only be heavily stained, but will leak protein off the membrane from saturated sites. Excessive protein leakage will cause an aggregation of the gold particles and destroy the AuroDye™ reagent. The problem will be more severe with 1-D gels than with 2-D gels, in which the proteins are distributed as spots over the entire blotting membrane. In general, for 1-D gels it is recommended to use protein loads that, in the case of a complex mixture, give resolvable banding patterns after silver staining. Single bands in 1-D gels should not exceed 1000 ng/band. The ideal load for molecular weight standards is 200 ng/band. Lower loads lead to better separations and help save precious samples. For immunovisualization, AuroDye™ is best used in conjunction with a detection system such as the AuroProbe™ BL IGSS kit, which matches its high sensitivity.

Because it is practically impossible to avoid proteinaceous impurities in the transfer apparatus, it is advisable to sandwich the gel between two nitrocellulose membranes. One will give you the exact replica of the gel; the other will adsorb these impurities. As mentioned previously, however, this is unnecessary when using "semidry" blotting.

4. It is important to wear gloves when handling gels and blots. For extra care, handle blots by their edges with clean plastic (not metallic!) forceps, since gloves can leave smears. All current laboratory electrophoresis protocols may be used. 2-D gels should be thoroughly washed in several changes of excess transfer buffer, in order to remove the ampholytes, unless the first-dimension gels have been fixed selectively for proteins before use in the second dimension.

5. In a 1-D gel system, blots contain one to several lanes. Identical blots can react in parallel with various probing agents (e.g., antibodies). Usually one blot with an additional lane containing molecular weight standards is stained with AuroDye™.

For 2-D gels, the combination of an immunogold-silver overlay assay with subsequent AuroDye™ staining is particularly effective, because it facilitates correlation of detected polypeptides with the total electrophoretogram. Note that Tween 20, not BSA, should be used when double staining.

6. It is advisable not to reuse the AuroDye™ solution. The concentration of the particles will probably be too low to give optimal results.

7. Low ionic strength transfer buffers are recommended for blotting in conventional apparatus (i.e., 25 mM Tris, 192 mM glycine, 20% methanol, pH 8.3). When SDS is used in the transfer buffer to

SANDWICH SET-UP

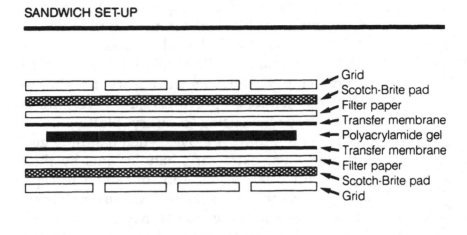

Fig. 8. Schematic representation of the protein blotting sandwich setup.

increase transfer efficiency, the following procedure is recommended. An initial transfer is done overnight at 4 V/cm with initial transfer buffer (25 mM Tris, 192 mM glycine, 20% methanol, 0.05% SDS, pH 8.3). The initial transfer buffer is then removed (can be reused once), and the apparatus, cassettes, and cooling coils are rinsed with distilled water. The apparatus is filled with secondary transfer buffer (25 mM Tris, 192 mM glycine, 20% methanol, pH 8.3), and the transfer continued at 4 V/cm for 4 h to remove SDS from nitrocellulose.

8. For obtaining optimal AuroDye™ protein staining with minimal background, it is recommended to use the following sandwich set-up for conventional electrotransfer: The cathode-oriented grid—a Scotch-Brite pad—two sheets of filter paper—a nitrocellulose sheet—the gel—a nitrocellulose sheet —two sheets of filter paper—a Scotch-Brite pad— the anode-oriented grid (Fig. 8).

We have found that the use of an extra nitrocellulose sheet at the cathodic side of the gel and of a high-quality filter paper leads to a sharp, high-contrast staining with negligible background. Wash the Scotch-Brite pads extensively with distilled water after each run to remove protein contaminants. In special cases, the AuroDye™ signal can be silver enhanced with Janssen's IntenSE™ silver-enhancement kit. This is not, however, to be considered a standard procedure, since the use of the stain by itself already gives very sensitive detection.

4.2. Double-Staining Method

1. By using Tween 20 as blocking agent, instead of BSA, it is possible that some background staining may occur because of higher nonspecific interactions. It is, therefore, essential to include a negative control. In addition, because of the subsequent protein staining, it is not possible to include gelatin in the immunogold reagent, which may result in nonspecific binding of the gold probe. Normally this is no problem because nonspecific interactions (if they occur) are shown by the negative control and can be easily distinguished from the positive signal.

2. It is sometimes difficult to recognize an immunodetected band or spot after overall protein staining. It is advisable then to take a photograph after immunodetection or to mark the detected spot before performing the overall protein staining.

4.3. Summary of the Use of Colloidal Gold General Staining

1. Colloidal gold may be used for staining the proteins on duplicate blots of overlay assays to make corre-

lation of detected bands with the total protein transfer accurate and simple. For quick reference, blots may also include an additional lane with molecular weight markers.

2. Colloidal gold may be used for staining proteins on blots that have previously been subjected to immunogold-silver staining. This is particularly useful for Western blots of 2-D gels.

3. Colloidal gold may be used for high-sensitivity staining of Western blots of 2-D gels to maximize the number of stained spots. In cases in which an extremely low background occurs, the AuroDye™ signal can be further amplified with IntenSE™. This is not, however, to be considered as a standard procedure.

4. Colloidal gold may be used to negatively stain transfers of nucleic acids on nylon-based membranes prior to hybridization. This facilitates correlation of hybridized bands with the total electrophoretogram (13).

Should problems arise during the use of colloidal gold in general protein staining, the trouble-shooting guide in Table 2 may be useful.

4.4. Use of Colloidal Gold for Immunoblotting

1. Immunoblotting combined with immunogold staining (IGS) or immunogold-silver staining (IGSS) can be applied in a variety of experiments. The most common field of application is the study of antigens and antibodies. The technique can be used for antigen detection as well as for antibody characterization in both a qualitative and a quantitive way. Blots from electrophoretically fraction-

Table 2
Trouble-Shooting

Observations	Probable cause	Remedy
AuroDye™ turns purplish during staining	Agglutination of gold particles by proteins released from the blotting membrane	Wash blot brieflywith excess water. Replace AuroDye™. If possible, use lower protein loads
AuroDye™ turns colorless during staining	Adsorption of all the gold particles by excess protein on blotting membrane	Replace AuroDye™ and double the volume/cm². If possible, use lower protein loads
Spotty background	Impurities released from filter paper adsorbed onto the blotting membrane during blotting	Use high-quality filter paper in electrotransfer apparatus during blotting
High background	Interference with contaminants of chemicals	Contact your distributor for a list of high quality chemicals
	Interference with proteinaceous contaminants	Use sandwich set-up as outlined above. If possible, use lower protein loads. Replace Scotch-Brite pads and follow recommendations as outlined above.
Smears on background	Incorrect handling	Handle blot with plastic forceps and avoid contact with gloves

ated antigens, for example, are most helpful in the evaluation of monoclonal antibodies. The dot blot technique on the other hand is very convenient for large-scale screening assays or for measuring the activity of an antibody. In the latter case a serial dilution series of the antigen is applied as spots (e.g., 250–0.1 ng/spot) on transfer membrane strips. The least detectable amount of antigen is a measure of the activity of the primary antibody. Similarly, protein dot blots are used to assess cross-reactions of antigens or relative sensitivities of detection systems. Cross-reaction can be expressed as the proportion between the least detectable amount of homologous antigen over the least detectable amount of heterologous antigen multiplied by 100. Another useful way is to spot, e.g., in a mini-manifold, constant amounts of antigen and to test a dilution series of the immunogold reagent.

2. The incubation time in the immunogold-reagent is influenced by the concentration of the reagent. A dilution of 1:100 (OD_{520} = 0.05) or even 1:200 (OD_{520} = 0.025) can be used overnight. When a dilution of 1:25 (OD_{520} = 0.20) is used, however, optimal results are obtained within 2 h.

The immunogold staining reagent can be used several times, certainly when it is diluted 1:25. Of course by reusing the reagent, longer incubation times may become necessary. The number of times the reagent can be used will be dependent on the amount consumed during the assay. This can be reduced by diminishing protein loads and relying on the stronger possible silver enhancement for obtaining satisfactory signals.

3. Controls are important. It is advisable to incubate one strip without primary antibody or if possible

with a nonrelated primary antibody. A positive control, i.e., the use of an antibody of known reactivity, is helpful to judge the experimental conditions. When nonpurified antigens give multiple bands, absorption with excess pure antigen will indicate which detected band corresponds to antibodies against it.

4. It is essential to add gelatin to the immunogold reagent. In our hands this prevented the nonspecific binding between the immunogold-reagent and tropomyosin, as demonstrated by the negative control. The source of the gelatin is equally important. If other brands of gelatin are used than the one supplied with the kit, it may be necessary to include a negative control.

4.5. Silver Enhancement

1. The enhancement reagents are extremely sensitive to the purity of the water used. Low-quality water results in the formation of precipitates that lower the reactivity of the enhancement reagent and in most cases produce a background. We therefore advise using double-distilled water throughout the procedure.

   As well as the high purity of the water used, the glassware used with the silver enhancement reagents should be very clean. Contact with metallic objects and the silver enhancer or its components should be avoided.

2. Using the Janssen Life Sciences IntenSE™ kit, the enhancer and initiator solutions have to be prepared on the same day the actual silver enhancement will take place. Using solutions that are kept longer than a day will result in a deteriorated silver

Table 3
Variations in Enhancement Time Depending
on Ambient Temperature

Temperature, 0°C	Typical enhancement time, min	Typical silver enhancer stability, min
16	13.0	18.0
18	11.0	16.0
20	10.0	13.5
22	9.0	11.5
24	8.0	9.5

enhancer. The enhancer and initiator solutions do not have to be shielded from normal (day) light. We advise that the user, however, not expose them to extreme heat or light. Use care: The silver enhancer or its components can stain the skin. The actual silver enhancer is prepared by mixing equal volumes of enhancer and initiator solution. Since this silver enhancer is only usable for about 15 min (see below), its components should be mixed immediately prior to the actual silver enhancement. The silver enhancer does not have to be shielded from normal (day) light. Note: At the time of press, an advanced IntenSE™-II kit was being prepared that obviates the need for such stringency regarding water purity and will be ready to use in liquid form.

Both the time required for enhancement and the time the silver enhancer can be used are considerably sensitive to the temperature of the environment. Table 3 illustrates this.

Both the typical enhancement time and typical silver enhancer stability are somewhat arbitrary in that for some applications it may be necessary to enhance longer than suggested, and for others, shorter. Since the enhancement can be monitored visually, the researcher can judge this personally. A comfortable margin between enhancement time and silver enhancer stability time exists around room temperature. For certain very demanding applications, in which a very strong signal amplification is required, it is possible to repeat the silver enhancement. Enhance the gold signal for 5 min or less, rinse with distilled water, and put in fresh silver enhancer for another 5 min before fixing.

After enhancement, put the sample in the fixing solution. Fixing the sample too long may cause bleaching of the signal obtained. It is also necessary, therefore, to remove any remaining fixing solution afterward by thoroughly washing the sample in distilled water.

3. Because both metallic gold and silver catalyze the reduction of silver ions, the reaction, which is initially catalytic, soon becomes autocatalytic. Antigen spot tests show that high concentrations of antigen (100 and 50 ng/mm^2) result in the same signal intensity as judged with the naked eye. We would, therefore, advise monitoring the enhancement from time to time. Long incubations can result in the visualization of bands that are of minor importance compared to the main signal.

As already explained in the procedure, the incubation in fixing solutions must be monitored. If the signal intensity diminishes before the 5 min have passed, the fixation time can be shortened. This is especially true for weak signals. As also indicated in the procedure, it is necessary to remove the fixative by thoroughly washing with distilled water.

The light-insensitive developer makes it possible to monitor the development continuously in normal daylight. The IntenSE™ kit can be stored at room temperature.

4. Bovine serum albumin is expensive. Because of the addition of 20 mM NaN$_3$ to the blocking solutions, these solutions can be reused several times.

5. Recently we have formulated a new buffer that increases the speed of the immunogold color development as well as the sensitivity. Further, after silver enhancement a sharper contrast is obtained. This is a BSA-phosphate-Tween 20 buffer composed of 50 mM phosphate buffer, pH 7.5, 0.5% Tween 20 (v/v), 0.5% BSA (w/v), and 0.05% sodium azide (w/v). This buffer can be used through the whole procedure, i.e., for blocking, washing, and incubating with immunoreagents. In addition no normal serum is needed to dilute the antibody, and no gelatin is used to dilute the AuroProbe™ BL reagent.

4.6. Preparation of Colloidal Metals and Immunoprobes

For the sake of convenience in this review, the use of commercially available gold probes and colloidal metals has been referred to throughout. Please refer to ref. 23 for the preparation of gold sols and probes, ref. 9 for the preparation of colloidal gold for protein staining, and ref. 21 for the preparation of colloidal iron for general protein staining.

Acknowledgments

The authors wish to thank Barbara Oliver for her great help and patience in the preparation of this manuscript, and particular thanks are due to Dr. Jan De Mey for his time in reviewing the original manuscript. Guy

Daneels, Marc De Raeymaeker, Bart De Wever, and Jan De Mey are acknowledged for their contribution to the methods described in this chapter.

References

1. Moeremans, M., Daneels, G., Van Dijck, A., Langanger, G., and De Mey, J. (1984) Sensitive visualisation of antigen-antibody reactions in dot and blot immune overlay assays with immunogold and immunogold-silver staining. *J. Immunol. Meth.* **74**, 353–360.
2. Surek, B. and Latzko, E. (1984) Visualisation of antigenic proteins blotted onto nitrocellulose using the immunogold staining (IGS) method. *Biochem. Biophys. Comm.* **121**, 284–289.
3. Brada, D. and Roth, J. (1984) "Golden-blot"—detection of polyclonal and monoclonal antibodies bound to antigens on nitrocellulose by protein A-gold complexes. *Anal. Biochem.* **142**, 79–83.
4. Hsu, Y. (1984) Immunogold for detection of antigen on nitrocellulose paper. *Anal. Biochem.* **142**, 221–225.
5. Hancock, K. and Tsang, V.C.W. (1983) India ink staining of proteins on nitrocellulose paper. *Anal. Biochem.* **133**, 157–162.
6. Wojtkowiak, Z., Briggs, R.C., and Hnilica, L.S. (1983) A sensitive method for staining proteins transferred to nitrocellulose sheets. *Anal. Biochem.* **129**, 486–489.
7. Kittler, J.M., Meisler, N.T., Viceps-Madore, D., Cidlowski, J.A., and Thanassi, J.W. (1984) A general immunochemical method for detecting proteins on blots. *Anal. Biochem.* **137**, 210–216.
8. Wolff, J.M., Pfeifle, J., Hollmann, M., and Anderer, A. (1985) Immunodetection of nitrocellulose-adhesive proteins at the nanogram level after trinitrophenyl modification. *Anal. Biochem.* **147**, 396–400.
9. Moeremans, M., Daneels, G., and De Mey, J. (1985) Sensitive colloidal metal (gold or silver) staining of protein blots on nitrocellulose membranes. *Anal. Biochem.* **145**, 315–321.
10. Rohringer, R. and Holden, D.W. (1985) Protein blotting: Detection of proteins with colloidal gold, and of glycoproteins and lectins with biotin-conjugated and enzyme probes. *Anal. Biochem.* **144**, 118–227.

11. Segers, J. and Rabaey, M. (1985) Sensitive Protein Stain on Nitrocellulose Blots, in *Protides of the Biological Fluids* (Peeters, H., ed.) Pergamon, Oxford.

12. Daneels, G., Moeremans, M., De Raeymaeker, M., and De Mey, J. (1986) Sequential immunostaining (gold/silver) and complete protein staining on Western blots. *J. Immunol. Meth.* **89**, 89–91.

13. Saman, E. (1986) A simple and sensitive method for detection of nucleic acids fixed on nylon-based filters. *Gene Anal. Tech.* **3**, 1–5.

14. Romano, E.L. and Romano, M. (1984) Historical Aspects, in *Immunolabelling for Electron Microscopy* (Polak, J.M. and Varndell, I.M., eds.) Elsevier, Amsterdam.

15. Beesley, J.E. (1985) Colloidal gold: A new revolution in marking cytochemistry. *Proc. Roy. Micr. Soc.* **20**, 187–196.

16. Holgate, C.S., Jackson, P., Cowen, P.N., and Bird, C.C. (1983) Immunogold-silver staining: A new method of immuno-staining with enhanced sensitivity. *J. Histochem. Cytochem.* **31**, 938–944.

17. Springall, D.R., Hacker, G.W., Grimelius, L., and Polak, J.M. (1984) The potential of the immunogold-silver staining method for paraffin sections. *Histochemistry* **81**, 603–608.

18. Hacker, G.W., Springall, D.R., van Noorden, S., Bishop, A.E., Grimelius, L., and Polak, J.M. (1985) The immunogold-silver staining method: A powerful tool in histopathology. *Virchows Arch. (Pathol. Anat.)* **406**, 449–461.

19. Liesi, P., Julien, T., Vilja, P., Grosveld, F., and Rechardt, L. (1986) Specific detection of neuronal cell bodies: In situ hybridization with a biotin-labeled neurofilament c-DNA probe. *J. Histochem. Cytochem.* **34**, 923–926.

20. Editorial (1986) Blotting ones' copy. *Lancet* ii, 21–22.

21. Moeremans, M., De Raeymaeker, M., Daneels, G., and De Mey, J. (1985) FerriDye™: Colloidal iron binding followed by Perl's reaction for the staining of proteins transferred from SDS-gels to nitrocellulose and positively charged nylon membranes. *Anal. Biochem.* **153**, 18–22.

22. Kyshe-Anderson, J. (1984) Electroblotting of multiple gels: A simple apparatus without buffer tank for rapid transfer of proteins from polyacrylamide to nitrocellulose. *J. Biochem. Biophys. Meth.* **10**, 203–209.

23. De Mey, J. (1986) The Preparation and Use of Gold Probes, in *Immunocytochemistry, Modern Methods and Applications* 2nd

ed. (Polak, J.M. and Van Noorden, S., eds.) Wright, Bristol, London, Boston.

24. Garrels, J.I. (1979) Two-dimensional gel electrophoresis and computer analysis of proteins synthesized by clonal cell lines. *J. Biol. Chem.* **254**, 7691.

25. Langanger, G., Moeremans, M., Daneels, G., Sobieszek, A., De Brabander, M., and De Mey, J. (1986) The molecular organization of myosin in stress fibers of cultured cells. *J. Cell. Biol.* **102**, 200–209.

Note: A range of publications concerning the colloidal metal marker system in immunochemistry, immunocytochemistry, and immunoelectron microscopy is available free on request from the authors of this chapter. These include:

Volumes No. 1 and 2 of the colloidal gold marker bibliography
A booklet, *Colloidal Gold Sols for Macromolecule Labelling*
A booklet, *A Guide to the Immunogold-Silver Staining (IGSS) of Transferred Proteins*
A booklet, *Colloidal Gold Reagents for the Staining of Protein Blots on Nitrocellulose Membranes*

Chapter 35

Enzyme Immobilization by Adsorption

M. D. Trevan

1. Introduction

Much has already been written on the subject of biocatalyst (cell/enzyme) immobilization, and it is not the function of a text such as this to go into all the various aspects of the field. For general reviews on the area, the reader is referred to refs. 1–3. The introduction to this chapter is also meant, however, to serve as an introduction to the succeeding three chapters, and is thus more general than is usual in this series.

Biocatalyst immobilization has been defined as the imprisonment of an enzyme (or cell) in a phase (the biocatalyst phase) that is distinct from, but communicates with, the phase (the bulk phase) in which substrate product or effector molecules are dispersed and monitored. Thus the usual form of the biocatalyst phase is some type of "polymer" matrix made to retain the biocatalyst by interacting with it either physically or chemically; the "polymer" may be soluble or insoluble in the bulk phase.

Immobilization may alter the behavior of a biocatalyst. The effect of immobilization on enzymes has been studied in depth and extensively reviewed (1,4,5). The effect of immobilization on cells has hardly been studied. Immobilization has been shown to affect all aspects of an enzyme's behavior: stability may be affected; K_m, V_{max}, and pH optimum may go up or down; sensitivity to inhibitors or activators may be altered; normal kinetic patterns may be totally disrupted. The changes may be attributed to a direct real effect of immobilization on the enzyme's intrinsic characteristics *per se* and/or to apparent changes in the enzyme's characteristics. This latter is usually caused by disturbances in solute concentrations in the immediate vicinity of the enzyme molecule, brought about by the imposition on the enzyme of a microenvironment determined by the physicochemical nature of the immobilizing "polymer" matrix.

Methods for biocatalyst immobilization probably number several hundred and present a bewildering choice to one intent upon a first attempt at immobilization. Broadly, however, immobilization methods can be divided into three major categories; adsorption, covalent bonding, and entrapment (Fig. 1). By far the greatest number of methods fall into the covalent bonding catergory, and it is for that reason that several protocols are given in Chapter 37.

Immobilization of biocatalysts is carried out for a number of reasons: to facilitate reuse of the biocatalyst; to aid retention of the biocatalyst in a continous reactor; to enhance the biocatalyst's useful characteristics, for example stability or productivity; to facilitate higher

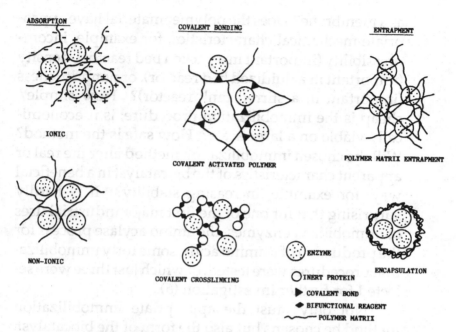

Fig. 1. Forms of biocatalyst immobilization.

concentrations of biocatalysts (in particular, viable cells) within bioreactors than would be possible with the unimmobilized biocatalyst; to protect large cells from high shear forces often found in bioreactors; or to study enzyme reactions or cellular physioloy in defined heterogeneous systems. No one immobilization method is therefore of universal use, and most of the art of biocatalyst immobilization lies in selecting the least inappropriate method. A variety of factors may be involved in the choice of an appropriate method of immobilization. For example, does the method of immobilization inactivate the biocatalyst or detrimentally alter its characteristics? Is the "bond" between polymer matrix and biocatalyst strong enough to retain the molecules/cells under the conditions of their use? Does the polymer material exist in the desired physical form; for example, as a membrane? Does the polymer material have appropriate mechanical characteristics, for example, incompressibility (important in a packed bed reactor), density (important in a fluidized bed reactor), or nonbrittleness (important in a stirred tank reactor)? How simple/cheap is the immobilization procedure; is it economically viable on a large scale? How safe is the method? Will the chosen immobilization method alter the real or apparent characteristics of the biocatalyst in a beneficial way; for example, increasing stability? It is hardly surprising that for one of the first major industrial uses of immobilized enzymes, the amino acylase process for the production of L-amino acids, some forty immobilization procedures were tested, of which just three were selected for further investigation (6).

Not only must the appropriate immobilization method be chosen, but also the form of the biocatalyst. At one extreme, biocatalysts are highly purified enzymes; at the other end of the spectrum, they are viable, growing cells. Somewhere in between there is a gray area in which enzyme becomes dead cell, becomes nongrowing viable cell. The selection of the appropriate form of biocatalyst will invariably depend upon individual circumstance. Thus in certain applications, e.g., biosensors, purified enzyme may have to be used in order to maintain either very high activity in the immobilized form or precise specificity. The expense of enzyme purification may be avoided by the use of "dead" cells or crude cell extracts, providing that the immobilized biocatalyst does not display any unwanted side reactions. Thus in most industrial-scale immobilized enzyme processes involving a single hydrolytic or isomeric reaction with a pure substrate stream, e.g., the conversion of glucose to fructose by xylose isomerase (E.C. 5.3.1.5), dead cell preparations rather than purified enzyme are usually used, since significant side reactions are unlikely to occur. When the desired conversion is a complex series of reactions or involves the use/regeneration of coenzymes, then whole cells (nongrowing) may well provide a more useful system than purified enzymes. Thus, the form of the biocatalyst will also in part dictate the type of immobilization method chosen. In general, covalent binding is usually reserved for purified enzymes, with the notable exception of glutaraldehyde cross-linking used to immobilize floculated "dead" cells containing xylose isomerase activity. Adsorption techniques have been applied to both enzymes and cells, but work more effectively with the former. Entrapment techniques are usually reserved for cell immobilization, for the reasons given in Chapter 38.

The rest of this introductory section is concerned specifically with immobilization by adsorption, probably the oldest and simplest form of enzyme immobilization. The adsorption forces involved may be nonspecific, hydrophobic interactions or ionic interactions (the most common). Typically adsorbent materials used have included clay, charcoal, hydroxyapatite, and, most frequently, ion-exchange materials. Of the latter, ion-exchange celluloses and like materials have proved more suitable than traditional ion-exchange resins, since the high degree of ionic substitution of the resins too often results in protein denaturation. A number of characteristics of the absorbent material must be considered. For the highest possible enzyme loadings, the material must have a high available surface area-to-volume ratio. This can be achieved in one of two ways: finely dividing the absorbent material or providing a macroporous structure so that the inner core of the material, not just its external surface, is available to the enzyme. Neither approach is without its drawbacks. Very small particles are likely to be mechanically fragile, thus rendering them vulnerable to attrition by the stirrer mechanism in a stirred tank reactor, and are likely to present practical problems in maintaining true plug-flow conditions in a packed-bed reactor (because of overpacking and channeling of the bed material). Conversely, although larger particles are likely to be mechanically tougher and exhibit better flow characteristics in packed bed reactors, the distribution of the majority of the enzyme throughout the core of the adsorbant particle enhances the likelihood that significant mass transfer (diffusion) limitations will occur. This in turn will reduce the overall catalytic efficiency of the immobilized enzyme (see, however, ref. 7). Another significant, potential disadvantage of using physical adsorption to immobilize enzymes is that the binding forces between enzyme and polymer are relatively weak, and hence the loss of enzyme, by desorption, is highly likely. Thus, pH, ionic strength, and temperature must all be closely controlled. Even so the addition of substrate to the immobilized enzyme occasionally produces sufficient conformational change in the enzyme to cause its desorption! Assuming that all these factors can be controlled, however, the potential desorption of the enzyme can be put to good use. Thus, in their evaluation of immobilization methods for the aminoaclyase process, Chibata and Tosa (6) chose DEAE-Sephadex-bonded enzyme, because it allowed them to regenerate the packed-bed enzyme reactor's activity every so often (without unloading the reactor), by flushing the system with high salt concentrations and low pH to remove the denatured enzyme, subsequently adding fresh enzyme directly to the reactor to allow immobilization to take place *in situ*. The original DEAE-Sephadex was left in their reactors for at least 10 yr. This case of immobilized enzyme regeneration, coupled to the almost perpetual reuse of the otherwise expensive polymer support material, added significantly to the overall economy of the process. The major advantage of this type of method is, of course, the ease with which it can be performed and the mild conditions employed during the immobilization procedure. The only other limitation to this technique is the necessity that the enzyme must be active within the narrow range of environmental conditions (particularly pH) required to keep the enzyme bound to the polymer matrix. The alteration of an enzyme's apparent pH optimum as a consequence of immobilization onto a polyelectrolyte matrix does not matter in this context, because the apparent pI value of the enzyme will, in all probability, also be altered!

The method given below is for the binding of amyloglucosidase (E.C. 3.2.1.3) to DEAE-Sephadex. Within the constraints mentioned above, almost any enzyme can be used.

2. Materials

1. Amyloglucosidase solution (5 g in 100 mL of 0.02M acetate buffer, pH 4.2).
2. DEAE-Sephadex Type A 25 (10 g).
3. 0.02M Acetate buffer, pH 4.2.
4. 0.5M HCl.
5. 0.5M NaOH.

3. Method

1. Suspend 10 g of DEAE-Sephadex in 50 mL of 0.5M HCl and stir for 20 min.
2. Filter, resuspend in 50 mL of 0.5M NaOH, and stir for 20 min.
3. Filter, wash with distilled water until washings are neutral, suspend in 100 mL of acetate buffer, pH 4.2,

and stir for 10 min.

4. Filter, resuspend in 100 mL of acetate buffer, and leave to equilibrate overnight. Filter dry.

5. Add the prepared DEAE-Sephadex to 100 mL of the amyloglucosidase solution, stir at room temperature for 2 h, and then filter.

6. Reslurry the enzyme–Sephadex complex in 200 mL of acetate buffer pH 4.2, stir for 15 min, and filter.

7. Repeat step 6 until the filtrate is free of protein as judged by its absorbance at 280 nm.

8. Resuspend the amyloglucosidase–Sephadex complex in 50 mL of acetate buffer and store at 4°C.

4. Notes

1. This method is virtually foolproof — it is on occasion run by the author as a second-year undergraduate class practical, and it invariably works.

2. When using other enzymes, care should be taken to check that the pH used for adsorption and also for assay is at least 1.5 units higher than the pK_a values for DEAE-Sephadex.

3. Maximum loadings obtainable will depend in part on the molecular weight of the enzyme and the net charge density on the enzyme molecule. In practice with this (and any other methods of immobilization), the law of diminishing returns applies to enzyme loading; after a given point the polymer becomes effectively saturated with enzyme, and up to that point the relationship between the quantity of enzyme used and the amount bound is nonlinear.

4. In general, it is also better to use a small volume of concentrated enzyme solution (rather than a large volume of dilute enzyme), since this enhances the

degree of substitution of enzyme molecules onto the polymer.

5. The addition of substrate (or inhibitor) during the immobilization process may increase the final activity of the immobilized enzyme.

Acknowledgment

The materials and methods used in this chapter are copyrighted by John Wiley & Sons (1980), and are reproduced, with permission, from ref. *1*.

References

1. Trevan, M. D. (1980) *Immobilized Enzymes: An Introduction and Their Applications in Biotechnology.* Wiley, Chichester.
2. Rosevear, A. and Lambe, C. A. (1983) Immobilized plant and animal cells. *Top. Enzyme Fermen. Biotechnol.* **7**, 13–37.
3. Chibata, I. and Tosa, T. (1981) Uses of immobilized cells. *Ann. Rev. Biophys. Bioeng.* **10**, 197–216.
4. Klibanov, A. M. (1983) Immobilized enzymes and cells as practical catalysts. *Science* **219**, 722–727.
5. Engaser, J-M. and Horvath, C. (1976) Diffusion and Kinetics with Immobilized Enzymes, in *Applied Biochemistry and Bioengineering* Vol. 1 (Wingard, L. B., Katchalski-Katzir, E., and Goldsmith, L., eds.) Academic, London.
6. Chibata, I. and Tosa, T. (1976) Industrial Applications of Immobilized Enzymes and Immobilized Microbial Cells, in *Applied Biochemistry and Bioengineering* vol. 1 (Wingard, L. B., Katchalski-Katzir, E., and Goldstin, L., eds.) Academic, London.
7. Trevan, M. D. (1987) Diffusion limitation—friend or foe? *Trends Biotechnol.* **5**, 7–9.

Chapter 36

Enzyme Immobilization by Entrapment

M. D. Trevan

1. Introduction

Entrapment methods of immobilization are mostly used in cell immobilization procedures, but have found some application with enzymes, particularly when the enzyme is essentially a dead cell or crude homogenate(*1*). In theory the entrapped enzyme is not attached to the polymer; its free diffusion is merely restrained. In practice, however, some or all of the enzyme may be an integral part of the polymer matrix; for example, chain termination in polyacrylamide gel formation may occur onto the enzyme (*see* also ref. *2*).

The great virtue of entrapment methods for immobilizing enzymes is that they are fast, cheap, very easy, and usually involve fairly mild conditions. The most used method seems to be that based upon entrapment in polyacrylamide gels, although the concentrations of acrylamide and quantity of crosslinking reagent used vary from method to method. The major drawback with this form of immobilization for pure soluble enzymes is the tendency of the enzyme to leak from the gel. This may be overcome by increasing the degree of cross-linking of the gel matrix, thus reducing pore size. This will in turn, however, increase the likelihood of mass transfer/diffusion limitation occurring, a particular problem with entrapment of biocatalysts. The effect of mass transfer/diffusion limitation may be ameliorated to a large extent by reducing the size of the entrappping polymer particle. The method given here easily produces a very finely divided, highly cross-linked polyacrylamide gel. It should be noted that mass transfer/diffusion limitation will be increased with increasing enzyme loading, such that the apparent specific activity of the entrapped enzyme falls as the amount of enzyme entrapped is increased (3).

When the enzyme is already insoluble or is part of a dead cell, leakage of the entrapped enzyme from the gel matrix is seldom a problem (*see* section 4). Clearly, enzymes immobilized by entrapment are likely to be inactive with high molecular weight substrates. The method given below is for the entrapment of urease in an acylamide gel, although almost any enzyme may be used.

2. Materials

1. Urease (EC 3.5.1.5).
2. Tris buffer, 0.1M, pH 7.0, containing $10^{-4}M$ EDTA.
3. Tris buffer, 0.1M, pH 7.0, containing $10^{-4}M$ EDTA and 0.5M NaCl.
4. Methylene-*bis*-acrylamide (MBA).

5. Acrylamide (ACR).
6. Dimethylamino-propionitrile.
7. Potassium persulfate solution 10 mg/mL.

3. Method

1. Make up a solution of 110 mg MBA and 10 mg ACR dissolved in 10 mL of Tris buffer, pH 7.0, and place in a stoppered flask.
2. Dissolve 15 mg of urease in the MBA:ACR solution and gas with nitrogen for 20 min.
3. Add 0.2 mL of dimethylamino-propionitrile and mix gently.
4. Add 0.5 mL of potassium persulfate solution, mix gently, stopper the flask, and allow to stand at room temperature for 20 min. An opaque gel will form.
5. Disrupt the gel by shaking the flask and passing the contents through a fine-needle syringe.
6. Filter on a Buchner funnel and wash the gel with 50 mL of 0.1M Tris buffer, pH 7.0, containing 0.5M NaCl.
7. Repeat step 6 three times.
8. Resuspend the enzyme gel in 15 mL of 0.1M Tris buffer, pH 7.0.

4. Notes

1. The method given here is routinely used as an undergraduate student laboratory exercise, largely because of the speed of preparation.
2. In practice, little leakage of enzyme occurs from the gel. The gel particles are not particularly robust, however, and thus vigorous stirring may break them down and release enzyme.
3. At the suggested level of addition, about 20% of the enzyme is actually retained in the gel. Using less

enzyme will result in a higher percentage being entrapped.

4. This method can be easily adapted for use with dead cells or nonpurified enzymes. In those cases in which diffusion of the "enzyme" out of the gel is likely to be less of problem, however, it would be better to increase the quantity of acrylamide in the gel in order to increase the gel strength. The exact proportions and concentrations of acrylamide to methylene-bis-acrylamide are best determined by experimentation.

5. Soaking the formed immobilized enzyme in a 2.5% solution of glutaraldehyde for 30 min at pH 7.5 will reduce any enzyme leakage to virtually zero, but may inactivate the enzyme. Inactivation may be overcome by adding a high concentration of competitive inhibitor (or substrate).

Acknowledgment

The materials and methods used in this chapter are copyrighted by John Wiley and Sons (1980), and are reproduced, with permission, from ref. 3.

References

1. Poulsen, P. B. (1985) Current Applications of Immobilized Enzymes for Manufacturing Purposes, in *Biotechnology and Genetic Engineering Reviews* Vol. 1 (Russel, G., ed.) Intercept, Newcastle.

2. Chibata, I. and Tosa, T. (1983) Continous production of L-malic acid by immobilized cells. *Trends Biotechnol.* **1**, 9–11.

3. Trevan, M. D. (1980) *Immobilized Enzymes: An Introduction and Their Application in Biotechnology* Wiley, Chichester.

Chapter 37

Enzyme Immobilization by Covalent Bonding

M. D. Trevan

1. Introduction

Enzyme immobilization by covalent bonding to an insoluble polymer has the supposed advantage of irreversible binding of the enzyme to the support matrix. Many of the bonds commonly employed are not 100% stable in use, however. Unless some care is taken to ensure that the enzyme can withstand the reaction conditions, substantial loss of enzyme activity will occur. Even under favorable conditions, 60% loss of activity is not uncommon.

Covalently bound immobilized enzymes are not only suitable for use in large-scale industrial processes [although their use in this field is in practice limited (1)] and in the construction of biosensors, but also as the immobile phase in affinity chromatography for the purification of enzyme inhibitors, activators, or substrates.

A wide variety of binding chemistries has been studied, and it is possible to bind an enzyme to almost any polymer or solid material (e.g., cellulose, agarose,

polyacryl-type compounds, glass, nylon, and so on).
The choice of chemistry and support is still largely an
empirical matter, and few studies have been performed
to compare the efficacy of the various methods. In a recent study, McManus and Randerson (2) compared the
effect of chemistry and support on the uptake, stability,
and immunoreactivity of a covalently immobilized
antibody. They found that both the support and chemistry affected the uptake, the former largely because of
variations in surface area, the latter in an unpredictable
manner. For example, aldehyde groups were effective
in binding the antibody when attached directly to the
polymer support or via support-bound glutaraldehyde,
but not in the form of support-phenylaldehyde. The
nature of the support was shown not to affect the stability of the antibody–polymer bond, whereas the chemistry involved in producing that bond had a marked effect
on stability. In general they concluded that the order of
decreasing stability was binding via glutaraldehyde to
amino groups, N-hydroxysuccinide esters, tresyl
bonds, and, least stable, isourea bonds formed in cyanogen bromide reactions. None of the methods affected
the K_s for binding of antibody to antigen.

Many commercially prepared activated supports
for enzyme immobilization are available with a wide
range of support materials and chemistries. There is obviously a premium to be paid for this however; cyanogen bromide-activated Sepharose 4B costs some 10
times more than the Sepharose 4B, and with an enzyme
ready immobilized to Sepharose 4B, the cost soars by
another 100 times!

There is obviously room for the do-it-yourself en-
thusiast. The four different chemistries that are detailed
here are among the most commonly employed and are:
coupling to cyanogen bromide-activated agarose; cou-

pling to tresylchloride-activated agarose; coupling to 2-amino-6-chloro-s-triazyl-cellulose, and coupling by glutaradehyde.

The covalent coupling of enzymes, proteins, and other ligands to polymer-support matrices activated by cyanogen bromide (3) is probably the most widely used immobilization method. The coupling relies on the formation of an isourea bond between a hydroxyl group on the polymer and an amino group on the protein. The immobilization method is essentially a two-stage process. The first stage, the activation of the polymer with cyanogen bromide, is potentially hazardous, and it is possible to buy (at some cost) ready-activated polymers from many suppliers of chromatography materials. The second stage, the coupling of protein to activated polymer, is carried out under very mild conditions (neutral pH, low temperature), and even sensitive proteins can be coupled with very little loss of activity. Any hydroxyl-containing polymer can be used, but among the most useful, from the view point of enzyme loading, high surface area-to-volume ratio, mechanical stability, and flow characterisics, are the beaded cross-linked agaroses such as Sepharose 4B. The major problem with this technique is the relative instability of the isourea bond causing loss of enzyme.

Activation of hydroxyl-containing polymers by tresyl chloride (2,2,2-trifluoroethyl sulfonate chloride) has been proposed as an alternative to bromide activation (4). Although tresylchloride is not exactly innocuous (a corrosive, lachrymator), it is certainly less potentially hazardous than cyanogen bromide. Coupling of the activated polymer to enzyme occurs under very mild conditions, and, in addition, the tresylchloride-activated polymer is stable for several months at pH 3 in aqueous suspension, whereas cyanogen bromide-activated polymers rapidly lose their enzyme-binding ability in aqueous suspension. Activation by tresylchloride is thought to occur by tresylation of a hydroxyl group on the polymer (with elimination of HCl). At neutral pH, this then is available for nucleophilic attack by an NH_2 group on the enzyme. At pH values above 8.5, it is possible that substitution of individual fluorines occurs. The resulting bond between the polymer and enzyme seems in practice to be 100% stable. The degree of activation of the polymer must be carefully controlled, because there appears to be an optimum level above which percentage yields of enzyme activity markedly decrease. This optimum level probably varies with both support material and enzyme.

Coupling of enzymes to cyanuric chloride (trichloro-s-triazine)-activated polymers again relies on the presence of hydroxyl groups on the polymer and amino groups on the enzyme. The original method (5) first reacted the trichloro-s-triazine to cellulose and then coupled the enzyme to the dichloro-s-triazinyl cellulose compound, quenching the third chloro-group with ammonium salts. The highly reactive nature, however, of both the tri- and dichloro compounds means that reaction times for the activation of the cellulose are measured in seconds, and for coupling the enzyme, in minutes. The apocryphal story of the exact reaction time for the celluose-activation step being achieved by dropping a flask containing the reaction mixture down the well of a four-story staircase remains one of the more curious of scientific anecdotes—that the story originated at all testifies to the "art" involved in many immobilization methods. It is difficult to prevent the cellulose from becoming crosslinked by the activation step, and loss of 90% of the enzyme activity is quite usual. A more common version of this method, therefore, is first to synthesize dichloro-s-triazinyl ethanolamine, then mono-chloro-s-triazinylethanolamine-cellulose, and use this to couple the enzyme (5). Since the monochloro-compound is relatively unreactive, the process is much easier to control. This method has not been widely used, probably because it was covered by a fairly comprehensive patent up to the late 1970s. It has a number of inherent advantages over the cyanogen bromide, however. First the enzyme–polymer bond is quite stable. Second, cyanuric chloride is a trifunctional reagent; any suitable ligand may also be bound to the polymer along with the enzyme. Thus the polymer may be made polyanionic, polycationic, or even hydrophobic in character. Third, although it is not exactly harmless, cyanuric chloride is not as difficult to handle, nor are its reactions as potentially lethal as cyanogen bromide.

Glutaraldehyde, strictly a dialdehyde, will react with amino groups under very mild reaction conditions. In the presence of ammonium ions, it will undergo polymerization. It is thus a potentially useful bifunctional molecule for enzyme immobilization. It has been used in one of two different ways; either to crosslink enzyme molecules to each other or to "activate" amino group-containing polymers. When used as a crosslinking reagent, it has a tendency to react with the enzyme's active site, unless this is protected by the presence of a competitive inhibitor. Glutaraldehyde is used to good effect to crosslink already flocculated (dead) cells in the industrial-scale preparation of immobilized enzymes (e.g., xylose isomerase for high-fructose corn syrup production). The method given here is a very simple process for the formation of immobilized enzyme membranes. It can be adapted for any sheet or membranous material to which enzymes can be adsorbed.

2. Materials

2.1. Coupling via Cyanogen Bromide

1. Sepharose 4B.
2. 2M Phosphate buffer, pH 12.1.
3. 5M Phosphate buffer, pH 12.1.
4. Cyanogen bromide solution: 100 mg in 1 mL of 5M phosphate buffer, pH 12.1.
5. α-Chymotrypsin (EC 3.4.21.2): 100 mg dissolved in 4 mL of 0.3M $KHCO_3$.
6. Glycine.
7. 0.1M $NaHCO_3$.
8. $10^{-3}M$ HCl.
9. 0.5M NaCl.

2.2. Coupling via Tresyl Chloride

2.2.1. For Activation

1. Sepharose 4B: 10 mL of swollen gel (about 3.2 g dry weight).
2. Tresylchloride (2,2,2-trifluoro-ethyl sulfonate chloride). Caution: corrosive.
3. 30:70, 60:40, and 80:20 mixtures of acetone:water (100 mL of each).
4. Dried acetone (200 mL), dried over molecular sieve 30 g/L acetone.
5. Pyridine.
6. 30:70, 50:50, 70:30, and 85:15 mixtures of 1 mM HCl:acetone (100 mL of each).
7. 1 mM HCl (250 mL).

2.2.2. For Coupling

1. Coupling buffer: 0.2M Sodium phosphate plus 0.5M sodium chloride, pH 8.0 (500 mL).

   2.   Trypsin (EC 3.4.21.4): 100 mg dissolved in 10 mL of coupling buffer at 4°C.

   3.   0.1M Dithiothreitol (10 mL).

   4.   0.2M Sodium acetate buffer, pH 3, containing 0.5M sodium chloride (100 mL).

   5.   0.5M Sodium chloride (100 mL).

2.3. Coupling via Cyanuric Chloride

   1.   Ammonia solution (sp. gr, 0.88).

   2.   Cyanuric chloride.

   3.   Dioxane.

   4.   Toluene.

   5.   Acetone.

   6.   Cellulose powder.

   7. Na_2CO_3; 15% solution.

   8.   1M HCl.

   9.   HCl concentrated.

   10.   0.05M Phosphate buffer, pH 7.0.

   11.   0.05M Borate buffer, pH 9.0.

   12.   0.05M Borate buffer, pH 9.0, containing 1.0M NaCl.

   13.   Papain (EC 3.4.22.2).

2.4. Formation of Immobilized Enzyme Membrane with Glutaraldhyde

   1.   Cellulose acetate sheet, 30 µm thick.

   2.   Alkaline phosphatase (EC 3.1.3.1) solution : 10 mg/ mL in 0.02M phosphate buffer, pH 6.8.

   3.   0.1M Glycine solution adjusted to pH 9.0.

   4.   Glutaraldehyde solution: 25% solution diluted to 2.5% with 0.02M phosphate buffer, pH 6.8.

3. Methods

3.1. Coupling via Cyanogen Bromide

3.1.1. Activation of Sepharose 4B

1. Wash 10 g of Sepharose 4B with 2.0M phosphate buffer, pH 12.1, and filter dry on a Buchner funnel.
2. Slurry the Sepharose in 10 mL of cold 5M phosphate buffer, pH 12.1, and add 20 mL of cold distilled water.
3. To the Sepharose slurry, at 5°C, add the cyanogen bromide solution dropwise, such that the whole volume is added within 5 min.
4. Stir the mixture for a further 10 min.
5. Transfer the gel to a glass filter and wash with 100 mL of cold distilled water.

3.1.2. Coupling of α–Chymotrypsin to Cyanogen Bromide-Activated Sepharose 4B

1. Suspend the activated Sepharose in NaHCO$_3$ (50 mL) of and add to the solution of α-chymotrypsin.
2. Stir gently for 24 h at 4°C.
3. Add 1.5 g of glycine to the slurry and stir for a further 24 h.
4. Filter the α-chymotrypsin–Sepharose complex and wash in turn with 200 mL aliquots of water, 0.1M NaHCO$_3$, 10^{-3}M HCl, 0.5M NaCl, and finally distilled water.

3.2. Coupling via Tresylchloride

3.2.1. Activation of Sepharose 4B

1. Place 10 mL of swollen Sepharose 4B in a glass filter funnel.
2. Wash with 100 mL of water.
3. Wash with 100 mL each of 30:70, 60:40, then 80:20 acetone:water mixtures.
4. Wash twice with 100-mL aliquots of acetone.

5. Wash three times with 50-mL aliquots of *dry* acetone.

6. Place the gel in a small beaker containing 2 mL of *dry* acetone and 1 mL of pyridine and stir vigorously.

7. Add 500 mg of tresyl chloride (3 mmol), dropwise over 1 min.

8. Continue stirring for 10 min.

9. Transfer the gel to a glass filter and wash twice with 100-mL aliquots of acetone.

10. Wash with 100 mL each of 30:70, 50:50, 70:30, and 85:15 solutions of 1 mM HCl:acetone.

11. Wash twice with 100-mL aliquots of 1 mM HCl and store at 4°C in a minimum volume of 1 mM HCl.

3.2.2. Coupling of Trypsin to Tresylchloride-Activated Sepharose 4B

1. Wash the gel briefly with coupling buffer (10 mL) on a glass filter.

2. Add the washed gel to 10 mL of enzyme solution at 4°C.

3. Stir or shake gently for 18 h at 4°C (or 2 h at 25°C).

4. Filter off excess liquid.

5. Add the gel to 10 mL of 0.1M dithiothreitol.

6. Stir at 4°C for 2 h.

7. Wash three times with 100-mL aliquots of coupling buffer.

8. Wash successively with 100-mL aliquots of acetate buffer, sodium chloride solution, distilled water, and coupling buffer.

3.3. Coupling via Cyanuric Chloride

3.3.1. Activation of Cellulose

1. Slurry 184 g of cyanuric chloride in 1.2 L of dioxane:toluene solvent (5:1 v/v). Cool to 5°C.

2. Gas the slurry with a dried stream of gas produced by passing nitrogen through the ammonia solution. Continue the gassing until a thick suspension is formed.
3. Filter the reaction mixture and wash the precipitate with 0.5 L of dioxane. Discard the precipitate and retain both filtrate and washings.
4. Evaporate the filtrate and washings to dryness on a rotary evaporator.
5. Redissolve the 2-amino-4,6-dichloro-*s*-triazine in the minimum volume of 1:1 (v/v) acetone:water.
6. Evaporate the acetone under reduced pressure.
7. Redissolve the precipitate formed in boiling water.
8. Cool rapidly to room temperature.
9. Dissolve 1 g of the 2-amino-4,6-dichloro-*s*-triazine in 50 mL of 1:1 acetone:water at 50°C.
10. Add 10 g (dry weight) of cellulose and stir for 5 min.
11. Mix 12 mL of 1.0M HCl with 8 mL of the Na_2CO_3 solution, add to the cellulose slurry, and stir for a further 5 min.
12. Reduce the pH of the slurry to below pH 7.0 with conc. HCl.
13. Filter the reaction mixture and wash three times with 100-mL aliquots of the acetone:water solvent.
14. Wash the amino-chloro-*s*-triazyl-cellulose copiously with distilled water.
15. Wash with 100 mL of 0.05M phosphate buffer, pH 7.0, and store at 2°C.

3.3.2. Coupling of Papain to Cyanuric Chloride-Activated Cellulose

1. Suspend 5 g of the activated cellulose in 50 mL of borate buffer, pH 9.0.
2. Dissolve 0.5 g unactivated papain in 50 mL of borate buffer, pH 9.0.

3. Mix the papain solution and cellulose suspension and stir at 4°C for 24 h.

4. Filter the enzyme–cellulose complex and wash with 100 mL of borate buffer, pH 9.0, followed by 100 mL of borate buffer, pH 9.0, containing 1.0*M* NaCl and, finally, with 100 mL of borate buffer, pH 9.0.

3.4. Formation of Alkaline Phosphatase Membrane with Glutaraldehyde

1. Spread 5 mL of alkaline phosphatase solution over a 10-cm² piece of cellulose acetate in a Petri dish.

2. Leave at 4°C until desiccated.

3. Spread 5 mL of glutaraldehyde solution over the membrane and leave at 4°C until desiccated.

4. Soak the membrane in glycine solution for 10 h at 4°C.

5. Wash the membrane in distilled water until the washings show no absorbance at 280 nm.

6. Store wet at 4°C.

4. Notes

4.1. Cyanogen Bromide Method

1. *Cyanogen bromide is extremely hazardous; use of a fume cupboard is essential, and cyanide antidote should be at hand.*

2. It is essential during the reaction of the Sepharose with the cyanogen bromide that the pH does not fall below 11.5, hence the use of 5*M* buffer. It is prudent to monitor the pH with a meter and to have 10*M* sodium hydroxide available.

3. Any hydroxy-containing polymer can be activated by this method, and it can easily be adapted to the

activation of soluble polymers, for example, polyethylene glycol or dextran.

4.   Like most enzyme-immobilization methods, the
amount of enzyme coupled expressed as a fraction
of the amount added falls as the latter increases.
Thus, typically if 5 mg of enzyme is added to 1 g of
activated Sepharose, most of it will become bound,
whereas adding 20 mg may result in only 10 mg
being bound. For Sepharose 4B, the usual practical
limit to the binding of enzyme occurs at around 15–
20 mg/g. At these levels of substitution, however,
the specific activity of the enzyme declines markedly, probably as a result of mass transfer/diffu-
sion limitation or steric hindrance of the active sites
caused by the high local enzyme concentration (in
the region of 5 mg/mL). Thus there is an optimum
level of enzyme loading beyond which it may be
uneconomic to stray (*see*, however, ref. 7).

5.   In general, coupling occurs most efficiently between pH 8 and 10, although it can be carried out at
as low as pH 5 and is usually complete after 2 h at
25°C. For sensitive enzymes, coupling is best carried out at 4°C for at least 18 h.

6.   Where possible, the enzyme–Sepharose coupling
should be carried out at relatively high ionic
strength (approximately 0.5) in order to minimize
protein interactions.

7.   CNBr-activated polymers are not stable in aqueous
suspensions and neutral pH values; therefore they
should be used immediately or lyophilized and
stored cold.

4.2. Tresylchloride Method

1.   The important feature of this method is the ratios of
the various volumes or reagents.

2. The ratio of pyridine:tresylchloride should be kept at 2/1 (v/v).

3. Varying degrees of substitution may be obtained by altering the tresylchloride volume in the range of 0.05–0.5 mL.

4. With other types of support material, more acetone may have to be added in step 6 of the activation procedure in order to achieve adequate stirring.

5. Coupling may be carried out at pH values from 7.0 to 8.5.

6. The wash in acetate buffer (step 8, coupling procedure) may be omitted or the pH modified if the enzyme to be coupled is unstable in these conditions.

7. Similarly the concentration of sodium chloride used may be reduced if necessary, although at the expense of total desorption of physically bound enzyme.

8. Enzymes immobilized by this method generally show high stability and are best stored at 4°C, pH 7–8.

9. The coupling buffer can be modified (no amino groups) as necessary, depending on the enzyme to be coupled.

4.3. Cyanuric Chloride Method

1. Though not as hazardous as cyanogen bromide, cyanuric chloride should be handled with care. Certain individuals may be particularly sensitive, and even brief exposure may cause an irritant reaction.

2. This method has been used to bind enzymes to a variety of materials (filter paper, filter cloth, dialysis tubing) and invariably works, although fairly

large (e.g., 80%) losses of enzyme activity can occur as a result of coupling.

3. The use of this method for activation of soluble polymers is not recommended because of the difficulty of preventing cross-linking of the polymer.

4.4. Glutaraldehyde Method

1. This method either works well or not at all! A lot depends on the enzyme chosen and its specific activity.

2. Inclusion of a competitive inhibitor or substrate of the enzyme will help protect the active site.

3. Better yields (in terms of activity coupled to activity added) can often be achieved by adding bovine serum albumin to the enzyme solution at a ratio of up to 10:1, albumin:enzyme.

4. Despite, or perhaps because of, the impurities, technical-grade glutaraldehyde seems to work better than EM grade.

5. Sometimes drying down a particular enzyme solution causes it to become denatured. In such cases allowing the membrane to soak in a solution of the enzyme (at about 1–5 mg/mL) for a couple of days and then transferring this to the glutaraldehyde solution may solve the problem.

6. If for some reason a thick sheet of immobilized enzyme is required, the above method may be easily adapted thus. Cast a sheet of cellulose nitrate by dissolving cellulose nitrate in a 49:49:2 ether: ethanol:water solvent (at 4% concentration). Pour onto a flat glass surface to the desired thickness, allow the solvents to evaporate, and soak the sheet off the glass with water (the glass must be scrupulously clean). Keep the sheet moist at all times.

     Soak the sheet for a number of days (the thicker the sheet, the longer the time—allow about 5 d/1 mm) in a 0.1% solution of the desired enzyme with or without added albumin. Allow 100 mL of enzyme solution per cm³ of membrane, and proceed as above. If desired, the cellulose nitrate may subsequently be dissolved in methanol, leaving a pure protein sheet.

7. Glutaraldehyde may be used to activate amino group carrying polymers (e.g., amino ethyl cellulose or amino silane-treated control pore glass). The following guidelines may aid in method development. React the polymer with a 2.5% glutaraldehyde in phosphate buffer, pH 7–8, at a ratio of about 10 mL of glutaradehyde per g of polymer. The reaction is fairly rapid and should be complete in 1 h at 25°C. Wash off any unbound glutaraldehyde. Add enzyme solution at concentrations of the order of 1–10 mg/mL at 10 mL/g polymer. The enzyme should be in a neutral, nonamino group-containing buffer. React for 1–2 h at 25°C or overnight at 4°C. Quench any remaining aldehyde groups on the polymer with 0.1M glycine or ethanolamine, depending on the ionic nature required of the polymer. Wash appropriately with high ionic strength buffer.

Acknowledgment

The materials and methods used in this chapter for coupling cyanogen bromide, cyanuric chloride, and glutaraldehyde are copyrighted by John Wiley & Sons (1980), and are reproduced, with permission, from ref. 8.

References

1. Trevan, M.D., Boffey, S.A., Goulding, K.G., and Stanbury, P.F. (1987) Enzyme Applications, in *Biotechnology: The Biological Principles* Open University Press, Milton Keynes.

2. McManus, D.E. and Randerson, D.H. (1986) A covalently immobilized antibody: The influence of the bond chemistry and support polymer on the uptake, stability and immunoreactivity. *Chem. Engineer. J.* **32**, B19–B28.

3. Axen, R., Porath, J., and Ernback, S. (1967) Chemical coupling of peptides and proteins to polysaccharides by means of cyanogen halides. *Nature* **214**, 1302–1304.

4. Nilsson, K. and Mosbach, K. (1981) Immobilization of enzymes and affinity ligands to various hydroxyl group carrying supports using highly reactive sulphonyl chlorides. *Biochem. Biophys. Res. Comm.* **102**, 449–457.

5. Kay, G. and Crook, E.M. (1967) The coupling of enzymes to cyanuric chloride activated celluloses. *Nature* **216**, 514.

6. Kay, G. and Lilly, M.D. (1970) The coupling of enzymes to cellulose by 2-amino-4,6 dichloro-s-triazine. *Biochem. Biophys. Acta* **198**, 276–279.

7. Trevan, M.D. (1987) Diffusion limitation—friend or foe? *Trends Biotechnol.* **5**, 7–9.

8. Trevan, M.D. (1980) *Immobilized Enzyme: An Introduction and Their Applications in Biotechnology* Wiley, Chichester.

Chapter 38

Cell Immobilization

M. D. Trevan

1. Introduction

We will be concerned in this chapter with methods
that may be used to immobilize live rather than dead
cells. The use of immobilized viable cells is seen as an
alternative approach to extracellular metabolite production by traditional fermentation technology. There
are a number of reasons why this may be advantageous.
First, immobilized cells are easily utilized in continuous
flow reactors without, theoretically, the contamination
of the product stream by cellular material. Second, cell
densities in the reactor of at least one order of magnitude
greater than with traditional fermentation techniques
are possible. Third, immobilization has been reported
to enhance the longevity of cells, in particular plant cells,
which is therefore useful in extending the productive
life of slow-growing cells. Fourth, it has been suggested
that immobilization may enhance the release of secondary metabolites from cells, particularly when (as in

plant cells) these metabolites are normally stored intracellularly.

The European Federation of Biotechnology (1) has published guidelines on the classification of cellular biocatalysts, but these have not yet in practice been universally adopted; some confusion therefore remains in the literature. As we have seen (Chapters 35 and 37), nonviable cells (that is, effectively crude enzyme preparations) may be immobilized by many of the methods used for pure enzymes. With viable (resting) or growing (reproducing) cells, something more subtle is called for. When growing (i.e., reproducing) cells are to be immobilized, one particular problem is that caused by cell reproduction; that is, how to retain the daughter cells. Covalent immobilization would clearly be impractical, as would most absorptive methods. With the exception of anchorage-dependent cells (e.g., many animal cells) that may be immobilized onto polymer microcarriers using their natural adherent properties, the immobilization methods usually employed therefore fall into the general category of entrapment. Clearly the methods used must not debilitate the cells, at least not beyond the point of recovery/regeneration, and must therefore be relatively mild—methods based on polyacrylamide gel entrapment are liable not to work, and there are a number of other suitable entrapment methods.

In practice, however, cell "leakage" from immobilized cell preparations made by an entrapment method can still be a significant problem. For the most part, "leakage" seems to be related to cell growth, and thus stationary phase cells are much less prone to this problem (2), as well as being advantageous for their predicated superior secondary metabolite production.

A common observation of immobilized viable cells is that, after a period of time, the cells appear to be located only around the periphery of the immobilization matrix. This is probably a result of cell death in the center of the matrix coupled to cell growth at the periphery. For the alga *Chlorella emersonii* immobilized in calcium alginate beads and grown phototrophically, this has been shown to be caused by poor penetration of CO_2 into the alginate beads (3); the cells at the periphery of the beads effectively scavenge all the available CO_2.

Supplementing the medium with CO_2 in this study not only encouraged cell growth throughout the bead, but also resulted in a more stable bead structure, suggesting that the cells, in this case at least, form an integral part of the immobilized cell bead structure. It is likely that the limiting factor on cell growth and immobilization matrix colonization for any viable cell is poor mass transfer of necessary gases, not so much because the concentration is usually low, but because cellular demand is usually high, and thus significant concentration gradients of dissolved gas will be found in the immobilized cell particle. Recognizing this problem, Aldercrutz et al. (4) coimmobilized *Chlorella pyrenoidosa* with *Gluconobacter oxydans* in calcium alginate, on the grounds that the oxygen evolved by *C. pyrenoidosa* would stimulate the microorganism's growth and productivity. Interestingly, *C. emersonii* immobilized in calcium alginate and grown heterotrophically (i.e., illuminated and with a carbon source, glucose, in the medium) colonized the whole alginate matrix to a very high density (5). Clearly, for maximum productivity, maximum cell density in the immobilization matrix is required. If dissolved gas mass transfer cannot be improved in a given system, the problem can in part be overcome by producing smaller immobilized cell particles (approximately 0.2 mm in diameter) (*see* section 4).

Of the methods given here, neither is perfect, and many variants exist in the literature. Both methods involve mixing a cell suspension with a solution of the entrappping polymer and then causing the polymer to gel. In the case of κ-carrageenan, the solution-gel transition is effected by lowering the temperature, and in the case of alginate, by metal ion substitution. Thus the former gel is sensitive to disruption by increased temperature, and the latter is sensitive to disruption by the presence of metal-ion chelating agents (e. g., phosphate or citrate).

There is another form of immobilization by "entrapment" not given here, that is, "passive" entrapment. In this technique, cells are allowed to invade preformed polymer matrices (e.g., polyurethane foams). This immobilization method, however, results in a preparation that does not generally retain cells well, and it is difficult to prepare particles for use in packed-bed types of reactors.

Since the methods recorded here have almost universal applicability, no specific cell cultures are suggested. Hence the medium used in the methods must be suitable for the cells chosen. Experience shows that almost any microbial or plant cell can be immobilized successfully by these methods, with the possible exception of normally highly motile cells, which do not appear to favor being fixed.

2. Materials

All reagents should be autoclaved when necessary.

2.1. Calcium Alginate Entrapment

1. 5% Solution of sodium alginate in water (5 mL).
2. 0.1M Calcium chloride (500 mL).
3. Appropriate growth medium.

4. Cell suspension between 2 and 32% (w/v) (5 mL).
5. Laboratory detergent (nonionic).

2.2. κ–*Carrageenan Entrapment*

1. 5% Solution of κ-carrageenan (low-gelling temperature) in water (5 mL).
2. Ice-cold solution of 0.3*M* potassium chloride and 0.01*M* calcium chloride (500 mL).
3. Appropriate growth medium.
4. Cell suspension between 2 and 32% (w/v) (5 mL).
5. Laboratory detergent (nonionic).

3. Method

3.1. *Calcium Alginate Entrapment*

1. Mix the cell suspension and sodium alginate solutions thoroughly, but gently. Any air bubbles present may be removed by gentle centrifugation, but take care that the cells are not separated out.
2. Put the calcium chloride solution in a large beaker (at least 1 L) on a magnetic stirrer set to stir slowly, and add a few drops of detergent.
3. Load the alginate/cell mixture into a 20-mL sterile syringe or peristaltic pump.
4. Drop or pump the alginate/cell mixture into the calcium chloride solution from a height of about 10 cm. Gel beads will form almost instantly (*see* Fig. 1 and section 4).
5. Leave the beads stirring in the calcium chloride solution for at least 90 min to allow for complete substitution of the calcium into the gel.
6. Wash beads in appropriate medium.

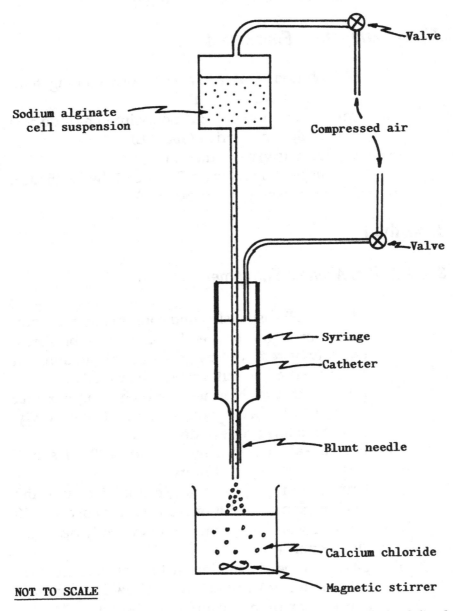

NOT TO SCALE

Fig. 1. Apparatus for production of small calcium alginate-immobilized cell beads.

3.2. κ-*Carrageenan Entrapment*

1. Once the κ-carrageenan solution has been made, it should be kept a few degrees above its gelling temperature.
2. Gently heat the cell suspension to the same temperature as the κ-carrageenan solution.
3. Mix the cell suspension and κ-carrageenan gently.
4. Place the potassium/calcium chloride solution in a large beaker with a few drops of detergent, and stir gently with a magnetic stirrer.
5. Load the κ-carregeenan/cell mixture into a 20-mL syringe or peristaltic pump and add dropwise to the potassium/calcium chloride solution from a height of about 10 cm above the liquid surface.
6. Leave the beads so formed to cure for 90 min.
7. Wash with appropriate medium.

4. Notes

4.1. *Calcium Alginate Entrapment*

1. Alginic acid is a polysaccharide found in many species of macro alga composed of D-mannuronic (M) and D-guluronic (G) acids, arranged in linear chains as M-G-M-G sequences, interrupted by M-M-M and G-G-G blocks. The precise composition, chain length, and so on, vary not only with source organism, but also with the part of the plant and time of year of harvesting. Since it is the D-guluronic acid that is primarily responsible for binding to the metal ion, alginic acids high in this component are recommended in order to obtain a gel with high strength. The most common alginic acid for immobilization is thus isolated from *Laminara hy-*

perborea and can be supplied by various companies (e.g., Kelco-AIL, London).

2. Gel strength is dependent on initial alginate concentration, but is more or less independent of metal cation concentration or type. Metal cation type markedly influences the resistance of alginate gels to disruption/dissolution by chelating agents, and changing the gelling cation may prove useful in certain circumstances. Thus barium alginate gels tend to be less easily disrupted than calcium alginate gels.

It is not just the binding affinity of the metal cation to the alginate that is important in this respect, however, but the ratio of the metal cation binding affinities to both the alginate and the chelator. Thus, although Cu^{2+} ions bind more strongly to alginate than Ca^{2+} ions do, calcium alginate resists disruption by phosphate better than copper alginate. Generally, barium alginate gels are the least susceptible to disruption by chelators, with calcium or aluminum alginate gels next best. The moral of all this is that stability of the beads must be tested in the media to be used.

One common approach to this problem with calcium alginate gels is to reduce the concentration of chelators to a minimum and then add a slight molar excess of calcium chloride to the medium. This appears to enhance the stability of the alginate cell beads when studied in batch culture. The effect may be caused more by removal of the chelator by the calcium, however, rather than by supplementation of the gel's calcium content. If the chelator is essential to the growth or viability of the immobilized cells (often the case with phosphate), its removal is obviously not desirable. With continuous-flow types of reactors, an additional problem is encountered, particularly in respect to phosphate chelation of calcium; that is, under the conditions often present, calcium phosphate is insoluble. Attempts to add calcium chloride to the medium reservoir will probably trap the phosphate there; without additional calcium chloride, even micromolar concentrations of phosphate will eventually chelate out most of the calcium from the alginate gel. Therefore, when phosphate or citrate are indispensable constituents of the medium, even at low concentration, it is probably better to evaluate the use of barium or aluminum alginate gels.

3. The best method for dissolving the sodium alginate prior to use is simply to autoclave a suspension of alginate in water (20 min at 15 psi). It should be noted, however, that autoclaving will probably shorten the polysaccharide chain lengths and thus lower the viscosity of the sodium alginate solution and the strength of the calcium alginate gel.

4. It is sometimes difficult to mix the cell suspension and sodium alginate solution without introducing bubbles into the mixture. Should this happen, these can be removed by gentle centrifugation.

5. The addition of detergent to the calcium chloride solution lowers the surface tension just sufficiently to allow the drops of sodium alginate/cell mixture to penetrate the surface. Without it, beads tend to float on the surface of the calcium chloride solution and become stuck together. The lessening of the impact on the falling drop may also help create a more spherical shape (particularly for small beads —see below).

6. Bead size is largely unaffected by the external diameter of the orifice from which the alginate/cell

mixture is dropped, because of the physical characteristics of the alginate in solution. Generally beads of between 2 and 4 mm diameter are formed. The mass transfer/diffusion limitation of gases and the effect of this on cell distribution within the gel-matrix has been discussed above (section 1), and for this or other reasons, smaller beads may be deemed desirable. A number of methods for producing small calcium alginate beads have been suggested, but perhaps the simplest and most reliable is the use of a stream of air flowing downward along the outside of the tubing from which the alginate/cell mixture is dropped (Fig 1). The drops are effectively blown off the tip of the aperture before they assume their usual size. By this means, beads of the order of 0.2 mm diameter are easily obtained. The apparatus for this purpose can be constructed easily from syringes, needles, and catheters of various sizes, and must be "tuned" by trial and error.

7. Metal ion alginate gels act as molecular sieves. The cut-off point will depend upon the precise composition of the gel, but in most cases it is around 100,000 dalton.

8. Although cell leakage from alginate gels is probably less problematic than from most other entrapment gels, it can still be significant (*see* section 1). One way around this, on a small scale, is to recoat preformed calcium alginate/cell beads with a thin layer of cell-less calcium alginate. This can be done by removing the formed calcium alginate beads from the calcium chloride solution and rinsing briefly in medium or distilled water. The beads are then placed carefully in a fresh solution of sodium alginate (of around 0.5–2%). Sufficient free calcium chloride will have been left in the gel beads to

diffuse out and precipitate the surrounding layer of sodium alginate. The thickness of this coating layer depends upon the initial calcium chloride concentration, rinsing efficiency, and concentration of, and time of exposure to, the fresh sodium alginate solution in a totally predicatable manner. Beads produced thus show zero cell leakage, even under conditions of active cell growth.

4.2. κ–*Carrageenan Entrapment*

1. κ-Carrageenan is extracted from the red algae (*Rhodophyceaea*). It is available from a number of suppliers. Since, however, a low-gelling-temperature κ-carrageenan is preferred in this application and this parameter varies with source and processing (see below), care must be exercised in its purchase. A suitable product is "Gelcarrin CIC," marketed by FMC Corporation, Marine Colloids Division, PO Box 70, Springfield, NY, 07081, USA.

    κ-Carrageenan is a polysaccharide that characteristically contains repeating units of the disaccharide D-galactose-4-sulfate β 1:4,3,6-anhydro-D-galactose linked together by 1:3 glycosidic bonds. The gelling mechanism of κ-carrageenan is one of formation of a network of (aggregated) double helices. Aggregation of the double helices is believed to be caused by the presence of calcium ions and results in an increase in gel strength. The nonaggregated helical structure, itself a gel, is formed on cooling the κ-carrageenan to below its characteristic gelling temperature. The gelling temperature is affected by the source and processing of the κ-carrageenan and is markedly elevated by the presence of potassium ions. The presence of calcium ions and the carrageenan concentration do not affect gelling temperature. Thus by following the method outlined in this text, κ-carrageenan/cell beads of fairly high melting temperature can be formed from a κ-carrageenan solution that will not gel spontaneously at room temperature. It should also be noted that even for a pure carrageenan solution, gelling and melting temperatures are not identical, the latter being about 20°C higher than the former. The gel structure will, however, eventually melt!

2. One problem that is often encountered is that caused by the presence of gelling cations, peptones, and so on, in the cell suspension. Mixing the cell suspension into the κ-carrageenan solution may therefore cause it to gel, unless it is kept well above its native gelling temperature.

3. Some workers (6) have proposed the use of polyethyleneimine or glutaraldehyde to increase the stability of the gel structure. Although this may have some utility with immobilized dead cells, however, these reagents are probably too toxic to be of much use with viable cells.

4. Although not subject to disruption by metal ion chelators as are alginate gels, κ-carrageenan gels are less strong and more porous and do break down in use.

5. The more porous open structure of the κ-carrageenan gel has two consequences: cell leakage can be a major problem; although higher plant and animal cells may be retained, algal and microbial cells are not well retained; the gel is rather squashy and spherical beads are difficult to form. Penetration of the gel structure by very high molecular weight compounds is, however, less of a problem than with alginate gels.

6. Compared to alginate, κ-carrageenan is expensive.
7. The easy dissolution of calcium alginate gels by the addition of phosphate or citrate may be a useful attribute if cell growth within the bead is to be measured directly (i.e., by counting cells) rather than indirectly (e.g., by pigment extraction). Purposeful dissolution of κ-carrageenan gels is more troublesome.

References

1. Lilly, M. D. (1986) Recommendations for nomenclature to describe the metabolic behaviour of immobilized cells. *Enzyme and Microbial Technol.* **8**, 315.
2. Dainty, A. L., Goulding, K. H., Robinson, P. K., Simpkins, I., and Trevan, M. D. (1986) Stability of alginate-immobilized algal cells. *Biotechnol. Bioengineer.* **28**, 210–216.
3. Robinson, P. K., Goulding, K. H., Mak, A. L., and Trevan, M. D. (1986) Factors affecting the growth characteristics of alginate entrapped *Chlorella*. *Enzyme Microbial Technol.* **8**, 729–733.
4. Aldercrutz, P., Olle, H., and Mattiason, B. (1982) Oxygen supply to immobilized cells. 2. Studies on a coimmobilized algae-bacteria preparation with *in situ* oxygen generation. *Enzyme Microbial Technol.* **4**, 395–400.
5. Robinson, P. K., Dainty, A. L., Goulding, K. H., Simpkins, I., and Trevan, M. D. (1985) Physiology of alginate-immobilized *Chlorella*. *Enzyme Microbial Technol.* **7**, 212–216.
6. Chibata, I. and Tosa, T. (1983) Continous production of L-malic acid by immobilized cells. *Trends Biotechnol.* **1**, 9–11.

Index